Advances in

Geosciences

Volume 11: Hydrological Science (HS)

ADVANCES IN GEOSCIENCES

Editor-in-Chief: Wing-Huen Ip *(National Central University, Taiwan)*

A 5-Volume Set

Volume 1: Solid Earth (SE)
ISBN-10 981-256-985-5

Volume 2: Solar Terrestrial (ST)
ISBN-10 981-256-984-7

Volume 3: Planetary Science (PS)
ISBN-10 981-256-983-9

Volume 4: Hydrological Science (HS)
ISBN-10 981-256-982-0

Volume 5: Oceans and Atmospheres (OA)
ISBN-10 981-256-981-2

A 4-Volume Set

Volume 6: Hydrological Science (HS)
ISBN-13 978-981-270-985-1
ISBN-10 981-270-985-1

Volume 7: Planetary Science (PS)
ISBN-13 978-981-270-986-8
ISBN-10 981-270-986-X

Volume 8: Solar Terrestrial (ST)
ISBN-13 978-981-270-987-5
ISBN-10 981-270-987-8

Volume 9: Solid Earth (SE), Ocean Science (OS) & Atmospheric Science (AS)
ISBN-13 978-981-270-988-2
ISBN-10 981-270-988-6

A 6-Volume Set

Volume 10: Atmospheric Science (AS)
ISBN-13 978-981-283-611-3
ISBN-10 981-283-611-X

Volume 11: Hydrological Science (HS)
ISBN-13 978-981-283-613-7
ISBN-10 981-283-613-6

Volume 12: Ocean Science (OS)
ISBN-13 978-981-283-615-1
ISBN-10 981-283-615-2

Volume 13: Solid Earth (SE)
ISBN-13 978-981-283-617-5
ISBN-10 981-283-617-9

Volume 14: Solar Terrestrial (ST)
ISBN-13 978-981-283-619-9
ISBN-10 981-283-619-5

Volume 15: Planetary Science (PS)
ISBN-13 978-981-283-621-2
ISBN-10 981-283-621-7

Advances in

Geosciences

Volume 11: Hydrological Science (HS)

Editor-in-Chief

Wing-Huen Ip

National Central University, Taiwan

Volume Editor-in-Chief

Namsik Park

Dong-A University, Korea

World Scientific

NEW JERSEY • LONDON • SINGAPORE • BEIJING • SHANGHAI • HONG KONG • TAIPEI • CHENNAI

Published by

World Scientific Publishing Co. Pte. Ltd.

5 Toh Tuck Link, Singapore 596224

USA office: 27 Warren Street, Suite 401-402, Hackensack, NJ 07601

UK office: 57 Shelton Street, Covent Garden, London WC2H 9HE

British Library Cataloguing-in-Publication Data
A catalogue record for this book is available from the British Library.

ADVANCES IN GEOSCIENCES
A 6-Volume Set
Volume 11: Hydrological Science (HS)

ISBN-13 978-981-283-610-6 (Set)
ISBN-10 981-283-610-1 (Set)
ISBN-13 978-981-283-613-7 (Vol. 11)
ISBN-10 981-283-613-6 (Vol. 11)

Typeset by Stallion Press
Email: enquiries@stallionpress.com

Printed by FuIsland Offset Printing (S) Pte Ltd, Singapore

EDITORS

Editor-in-Chief: Wing-Huen Ip

Volume 10: Atmospheric Science (AS)
Editor-in-Chief: Jai Ho Oh
Editor: Gyan Prakash Singh

Volume 11: Hydrological Science (HS)
Editor-in-Chief: Namsik Park
Editors: Joong Hoon Kim
Eiichi Nakakita
C. G. Cui
Taha Ouarda

Volume 12: Ocean Science (OS)
Editor-in-Chief: Jianping Gan
Editors: Minhan Dan
Vadlamani Murty

Volume 13: Solid Earth (SE)
Editor-in-Chief: Kenji Satake

Volume 14: Solar Terrestrial (ST)
Editor-in-Chief: Marc Duldig
Editors: P. K. Manoharan
Andrew W. Yau
Q.-G. Zong

Volume 15: Planetary Science (PS)
Editor-in-Chief: Anil Bhardwaj
Editors: Yasumasa Kasaba
Paul Hartogh
C. Y. Robert Wu
Kinoshita Daisuke
Takashi Ito

CONTENTS

Advances in Geosciences
Vol. 11: Hydrological Science (2007)
Eds. Namsik Park *et al.*

GA APPLICATION TO DETERMINE OPTIMAL PUMPING POLICY IN HETEROGENEOUS UNCONFINED AQUIFER

B. APARNA* and A. K. RASTOGI†
Department of Civil Engineering
Indian Institute of Technology Bombay, India-400076
**baparna@iitb.ac.in*
†akr@civil.iitb.ac.in

In this paper, an irregular heterogeneous unconfined aquifer is simulated by finite element method. Groundwater head distribution is obtained at various nodes of the flow domain involving a set of boundary conditions, source, and sink terms. Finite element model is coupled with genetic algorithm techniques to minimize the cumulative drawdown in the 11 wells subject to constraints on aquifer heads and pumping rates. The drawdowns and the corresponding optimal pumping rates in each well are determined by real-coded GA, and the results are compared with binary-coded genetic algorithm. The values of the GA parameters were chosen based on the sensitivity analysis of crossover, mutation, and number of generations. The results showed that real-coded GA yielded marginally lower values of cumulative drawdowns in comparison to binary-coded GA. The results also suggest near-global optimality of the pumping rates, which are further examined by statistical reliability analysis.

1. Introduction

Groundwater is an important source of water supply in many parts of the world today. Its use in agriculture, industries, municipalities, and rural homes continues to increase. The ever-increasing demand on groundwater has resulted in many environmental imbalances. Excessive extraction rates have caused undesirable effects of critical lowering of water table, increased pumping costs, land subsidence, salination, desertification, changes in hydraulic pressure, and underground flow directions. When several pumping wells are involved particularly in a heterogeneous anisotropic aquifer, a careful decision is required by water managers to withdraw water from these wells. Optimal assessment of groundwater withdrawal rates and the corresponding drawdowns are of utmost importance for adequate management of groundwater resources.

In the present work, a simulation model based on Galerkin's finite element approach has been developed that predicts the head in the flow domain. Aquifer head distribution and its sensitivity to pumping and recharge rates are also discussed. Real-coded genetic algorithm is coupled to the flow simulation model to assess the optimal drawdowns for this combinatorial optimization problem involving 11 pumping wells in a heterogeneous unconfined aquifer. Composite scaled sensitivity of parameters is evaluated to assess the reliability of estimated results.

2. Problem Description

The unconfined aquifer flow domain is divided into nine zones based on hydraulic conductivity and specific yield values. The flow region is discretized into 90 linear triangular elements involving 56 nodes (Fig. 1). The head at constant boundary is 100 m and the aquifer consists of 11 pumping wells (Fig. 2). Initially, the pumping rate is assumed to be 1500 m^3/d at each well. An average annual recharge of 0.002 m/d is considered applicable in the region.

The zones of the flow domain are classified based upon the hydraulic conductivity and specific yield values as given in Table 1. Coefficient of anisotropy is equal to 2.

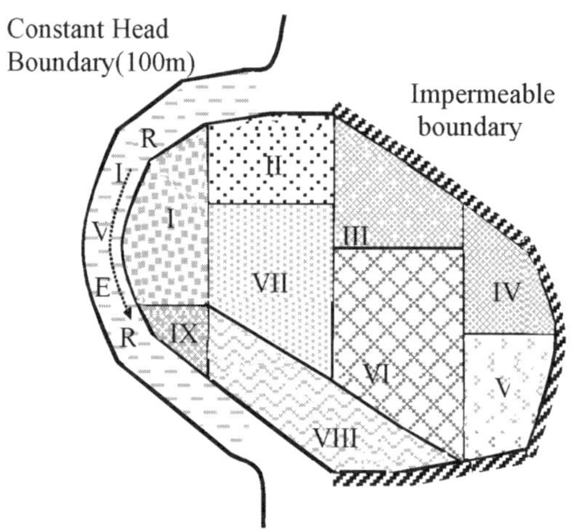

Fig. 1. Flow domain of heterogeneous isotropic unconfined aquifer.

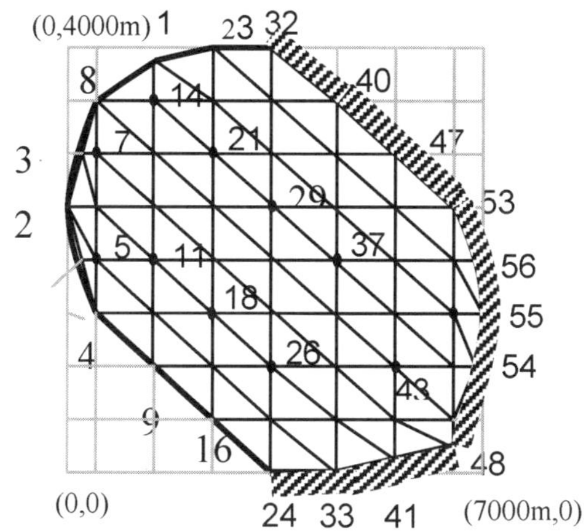

Fig. 2. Finite element discretization. No. of nodes: 56; No. of elements: 90.

Table 1. Aquifer properties in different zones.

Zone	I	II	III	IV	V	VI	VII	VIII	IX
K_x (m/d)	200	50	100	200	150	50	120	150	150
K_y (m/d)	100	25	50	100	75	25	60	75	75
S_y	0.25	0.05	0.12	0.25	0.18	0.05	0.15	0.18	0.18

The governing equation describing the flow in a two-dimensional heterogeneous anisotropic unconfined aquifer is given as[1]

$$\frac{\partial}{\partial x}\left[K_x h \frac{\partial h}{\partial x}\right] + \frac{\partial}{\partial y}\left[K_y h \frac{\partial h}{\partial y}\right] = S_y \frac{\partial h}{\partial t} + Q_w \delta(x - x_i)(y - y_i) - q \tag{1}$$

subject to the following initial and boundary conditions:

$$\begin{aligned} h(x,y,0) &= h_0(x,y) && x,y \in \Omega, \\ h(x,y,t) &= h_1(x,y,t) && x,y \in \partial\Omega_1, \\ Kh\frac{\partial h}{\partial n} &= q(x,y,t) && x,y \in \partial\Omega_2, \end{aligned}$$

where $h(x,y,t)$ is the piezometric head (m), K_x and K_y are hydraulic conductivities in x- and y-directions (m/d), S_y is specific yield, x and y are space variables (m), Q_w is the source or sink term ($-Q_w$ = source, Q_w = sink in $\text{m}^3/\text{d}/\text{m}^2$), t is time in days, Ω is the flow region, $\partial\Omega$ is the boundary region ($\partial\Omega_1 \cup \partial\Omega_2 = \partial\Omega$), $\frac{\partial}{\partial n}$ is the normal derivative, $h_0(x,y)$ is the initial head in the flow domain (m), $h_1(x, y, t)$ is the known head value of the boundary head (m), $q(x, y, t)$ is the known inflow rate (m^3/d/m), and δ is the Dirac delta function.

After Galerkin's finite element formulation, the above partial differential equation is transformed to

$$[G]\{h_I^{t+\Delta t}\} + [P]\{h_I^t\} = \{F\}, \tag{2}$$

where $[G]$ is the conductance matrix containing hydraulic conductivity terms; $[P]$ is the storage matrix containing storativity terms; Δt is the time step size; $\{F\}$ is the net flux vector; $h_I^{t+\Delta t}$ is the unknown head vector; and h_I^t is the known head vector at time t.

The linear simultaneous equations derived above are solved to obtain the head distribution at nodal points using Gauss–Seidel iterative method for the given initial and boundary conditions, recharge, pumping, hydraulic conductivity, and the specific yield values. Iterations are terminated when the difference in heads between two successive iterations is below a set of tolerance level ε, which is presently considered as 0.0001.

3. Groundwater Head Distribution

The groundwater levels may not vary uniformly throughout the flow domain due to pumping by various wells from a heterogeneous anisotropic unconfined aquifer. These are influenced by the pumping conditions, boundary conditions, properties of the porous media (hydraulic conductivity, specific yield, saturated aquifer thickness), aquifer recharge, and also by the distribution of wells in the aquifer domain. Flow pattern in the aquifer and the groundwater head distributions obtained by FEM simulation at a section in the flow domain is shown in Fig. 3. Drawdown in well at nodes 7 and 21 is different due to their location. Adequacy of the model is established by examining its sensitivity to pumping and recharge rates by satisfying a daily mass balance criterion. Head variation near the river (node 5) is shown in Fig. 4 which approaches nearly a steady state due to river recharge. The optimization problem is aimed at minimizing the collective drawdown from the wells, and is represented as

$$\text{Min } F = \sum_{t=1}^{t\max} \sum_{i=1}^{n} H_I(i) - H_F(i,t) \tag{3}$$

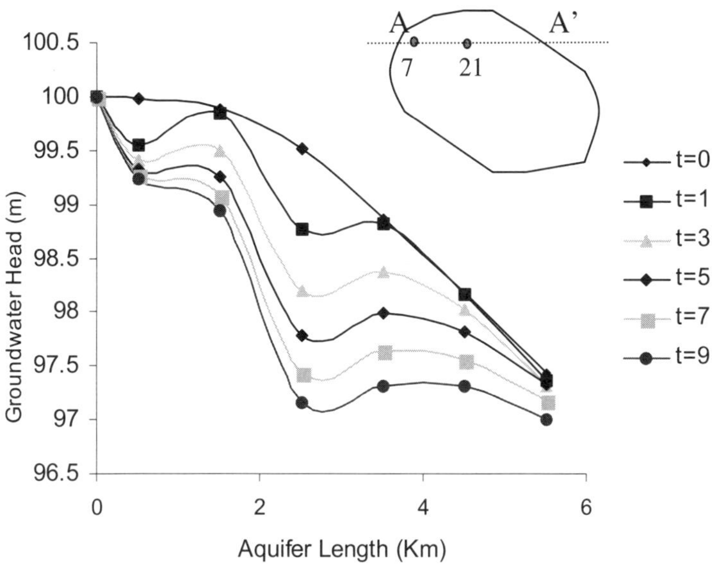

Fig. 3. Groundwater head distribution along section A-A'.

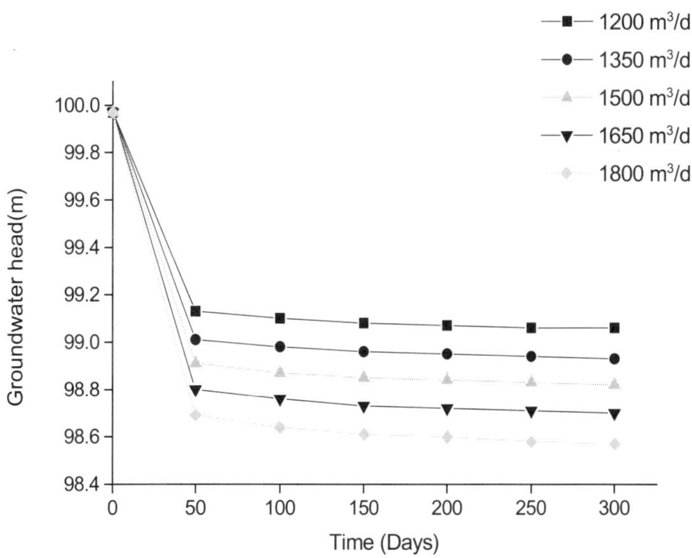

Fig. 4. Head variation in well at node 5 for different pumping rates.

subject to the following set of constraints:

$$Q_L \leq \sum Q \leq Q_U,$$
$$h_L \leq h(i,t) \leq h_U,$$

and

$$q_L \leq q_i \leq q_U.$$

The objective function F represents the cumulative drawdown due to continuous pumping from the 11 wells; t is the time period and $t_{\max}$ is the maximum time for which the pumping is done (1 year presently); n represents the number of wells; Q is the total pumping rate during operational period t; $H_I(i)$ represents the initial head in well i while $H_F(i,t)$ is the final head in well i at the end of the operational period; Q_L and Q_U represent the lower and upper bounds of total pumping rate which are 5000 m^3/d and 10,000 m^3/d, whereas h_L and h_U are the lower and upper bounds of the groundwater heads in the wells which are 85 m and 100 m, respectively depending upon the availability of the pump; q_i is the pumping rate in well i; and q_L and q_U are the lower and upper bounds on individual pumping rates of each well.

4. Genetic Algorithm

Genetic algorithm (GA) is a heuristic method to find approximate solutions to combinatorial optimization problems. Genetic algorithms are a particular class of evolutionary algorithms that use techniques inspired by evolutionary biology such as inheritance, mutation, natural selection, and recombination (or crossover). Important steps involved in GA methodology can be summarized as follows:

- Generation of initial population of strings;
- Selection of coding scheme;
- Running the flow simulation finite element model for all the sets (strings) of variables;
- Evaluation of each string in a population (computing fitness or objective function);
- Performing the computation with genetic operators;
- Selection of the best strings and formulation of mating pool;
- Crossover of the selected strings;
- Mutation of the strings;
- Terminating condition.

In this study, real-coded GA is applied, and later on the results are compared with binary-coded GAs which are discussed in the following sections.

4.1. *Real-coded GA*

The above optimization problem is solved by real-coded GA in which the variables are used directly without adopting any coding scheme. In order to estimate the cumulative drawdown, an initial population size which linearly varies with the number of variables ($N = 10n$, where n is the number of variables) is chosen. For this case, the number of variables is 11, i.e. the pumping rates at each of the 11 wells. GA starts with an initial population of a set of individual pumping strategies, which are randomly generated and meet the constraint on the total withdrawal rate. For each particular solution, the finite element model is run to evaluate the groundwater heads. Fitness of each individual is evaluated subsequently based upon the cumulative drawdown function. Individuals are selected by tournament selection based on their fitness values. The winners of the tournament selection are further processed by crossover and mutation operations (directly applied on real parameter values) to produce new individuals, but keeping the population size fixed. One generation refers to one complete cycle of selection, crossover, and mutation. If the optimization criteria are not met, the creation of a new generation starts. This cycle of generations is performed until the optimization criterion is reached.

4.1.1. *Simulated binary crossover for real-coded GA*

The crossover that has been applied for the present case of real-coded GA is a simulated binary crossover (SBX) operator, which is particularly suitable here, because the spread of children solutions around parent solutions can be controlled using a distribution index η_c. A large value of η_c allows only near-parent solutions to be created, whereas a small value of η_c allows distant solutions to be created. For that reason, η_c value of 8 is considered for the present problem. Another aspect of this crossover operator is that it is adaptive, allowing any solution to be created in the beginning to have a more focused search when the population is converging. The procedure for calculating the children solutions (c_1 and c_2) from parent solutions (p_1 and p_2) is as follows[2]: A uniform random number (u) between 0 and 1 is generated, and the spread factor β is computed as

$$\begin{aligned} \beta &= (2u)^{\frac{1}{n_c+1}} && \text{if } u \leq 0.5 \\ &= \left(\frac{1}{2(1-u)}\right)^{\frac{1}{n_c+1}}, && \text{otherwise,} \end{aligned} \tag{4}$$

where η_c is a distribution index of SBX and a non-negative real number. The children solutions are subsequently calculated as follows:

$$\begin{aligned} c_1 &= 0.5[(1+\beta)p_1 + (1-\beta)p_2] \\ c_2 &= 0.5[(1-\beta)p_1 + (1+\beta)p_2] \end{aligned} \tag{5}$$

These two children solutions are symmetric about the parent solutions. A value of the distribution index (η_c) allows that the children solutions are closer to the parent solutions. A smaller value of η_c makes a more uniform distribution in the range, and if η_c equals to 0, then it makes a nonuniform distribution in the range.

4.1.2. *Polynomial mutation for real-coded GA*

A polynomial probability distribution is used to create a solution c in the vicinity of a parent solution p under the mutation operator. The following procedure is used for a parameter $p \in [p_l, p_u]$.

1. Create a random number u between 0 and 1.
2. Calculate the parameter δ_q as follows:

$$\begin{aligned} \delta_q &= [2u + (1-2u)(1-\delta)^{n_m+1}]^{\frac{1}{n_m+1}} - 1 && \text{if } u \leq 0.5 \\ &= 1 - [2(1-u) + 2(u-0.5)(1-\delta)^{n_m+1}]^{\frac{1}{n_m+1}} && \text{otherwise,} \end{aligned} \tag{6}$$

   where $\delta = \min[(p - p_u), (p - p_l)]/((p_u - p_l)$.
3. Then, the mutated child c can be calculated as follows:

$$c = p + \delta_q(p_u - p_l). \tag{7}$$

From the above equations, the normalized perturbance $= ((c-p)/(p_u - p_l))$ of the mutated solutions in both positive and negative sides separately. We observe that this value is $O(\frac{1}{n_m})$. In order to get a mutation effect of 1% perturbance in solutions, we set $n_m = 100 + t$, and the probability of mutation is changed as follows:

$$p_m = \frac{1}{n} + \frac{t}{t_{\max}}\left(1 - \frac{1}{n}\right), \tag{8}$$

where t and $t_{\max}$ are the current generation number and the maximum number of generations allowed, respectively. Thus, in the initial generation, we mutate on an average, one variable $p_m = \frac{1}{n}$ with an expected 1% perturbance and as generations proceed, we mutate more variables with lesser expected perturbance. In the real-coded mutation, the amount of perturbation in a variable can be controlled by fixing the parameter η_m (distribution index for mutation). An η_m value of 30 is taken after some careful initial experimentation.

4.2. *Binary-coded GA*

For the binary coding of GA, the string length for each parameter is taken as 7. The string length is reasonable in keeping view of the accuracy of the solution. The selection operator considered is the same as that taken for real-coded GA. Single point crossover and bitwise mutation operators are applied to evaluate the cumulative drawdown.

5. Sensitivity of GA

The success and performance of GA is dependent on several factors like number of generations, probabilities of crossover, mutation rate, and population size. An acceptable balance between exploration and exploitation can also be accomplished through appropriate selection of these factors. Figure 5 shows the convergence of GA process while optimizing the population size. It is evident from the figure that the optimum value of population for which the objective function is minimum is 110. Similar tuning of GA parameters was carried out for a number of generations (50), crossover, and mutation probabilities, to select the most applicable values for the present problem. Goldberg[3] suggested that good performance might be achieved from GA using high crossover probability and low mutation probability. Sensitivity to crossover probability is carried out using a population size of 110 and mutation probability of 0.1. The results show (Fig. 6) that the objective function is minimum for a crossover probability of 0.8. For the determination of sensitivity to mutation probability, a

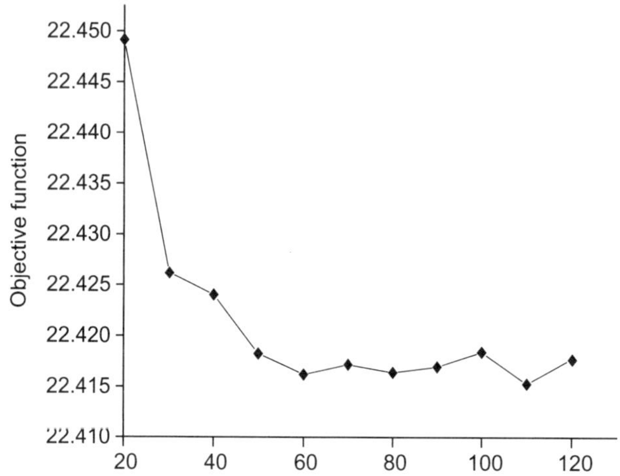

Fig. 5. Effect of population size on objective function.

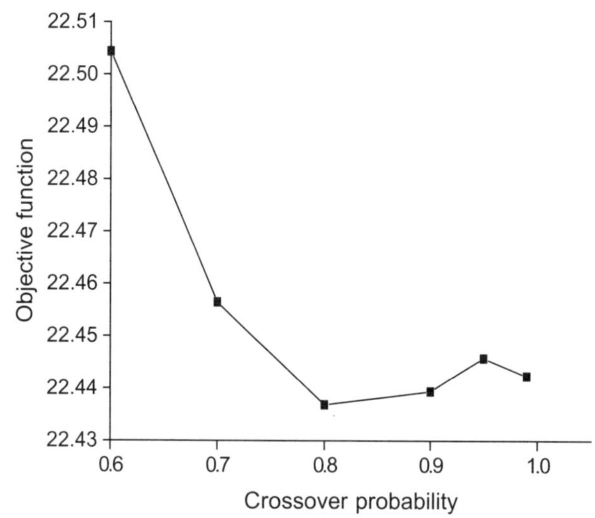

Fig. 6. Effect of crossover probability on objective function.

crossover probability of 0.8 is used. With the initial population taken as 110, the objective function attains a minimum at mutation probability 0.1, and for this value it was found that the solution converged faster in comparison to any other probability value.

6. Results and Conclusions

Individual and cumulative drawdowns in 11 wells by real and binary GAs are shown in Tables 2 and 3, respectively. Optimal pumping rates for the drawdown are summarized in Table 4 for binary GA. However, real GA results are very similar and not reproduced here for brevity. Cumulative optimal drawdown in 11 wells by real and binary GAs is shown in Fig. 7, where a close match can be observed. For the reliability analysis, the composite scale sensitivity parameter[4] was worked out which suggested that the drawdown near the river wells was more reliable compared to the wells near the impervious boundary of the flow region. FEM aquifer analysis

Table 2. Optimal drawdown (m) in 11 wells by real-coded GA.

Well No.	$T = 100$ days	$T = 200$ days	$T = 365$ days
5	0.484767	0.497149	0.504237
7	0.423515	0.433589	0.439417
11	1.029083	1.095222	1.13293
14	0.572364	0.59372	0.60597
18	1.545848	1.725679	1.82821
21	2.310775	2.554207	2.692876
26	2.256996	2.696994	2.948904
29	3.087477	3.617759	3.921039
37	4.146849	5.09629	5.64315
43	3.368417	4.761373	5.56794
51	3.211266	4.727302	5.606377
Cumulative	22.43736	27.79928	30.89105

Table 3. Optimal drawdown (m) in 11 wells by binary-coded GA.

Well No.	$T = 100$ days	$T = 200$ days	$T = 365$ days
5	0.516797	0.529983	0.537549
7	0.456912	0.467675	0.473912
11	1.079607	1.149942	1.190144
14	0.585693	0.60844	0.621516
18	1.648056	1.839187	1.948465
21	2.368883	2.627478	2.775208
26	2.346852	2.814479	3.083002
29	3.174618	3.738085	4.061333
37	4.277928	5.28749	5.87092
43	3.591697	5.073979	5.935438
51	3.436777	5.050179	5.989259
Cumulative	23.48382	29.18692	32.48675

Table 4. Optimal pumping rates (m^3/d) in pumping wells by binary-coded GA.

Well No.	$T = 100$ days	$T = 200$ days	$T = 365$ days
5	704.7232	703.5173	666.3831
7	721.44	715.8291	683.5269
11	683.2956	673.7941	619.8667
14	720.1143	713.0156	670.7007
18	706.3097	689.6951	619.2156
21	719.2965	701.2842	629.7749
26	675.6772	656.5659	575.1869
29	692.1406	669.9077	586.8978
37	692.9689	662.5927	571.4718
43	609.4508	593.2166	506.4367
51	588.4892	578.2648	493.6202
Cumulative	7513.906	7357.683	6623.081

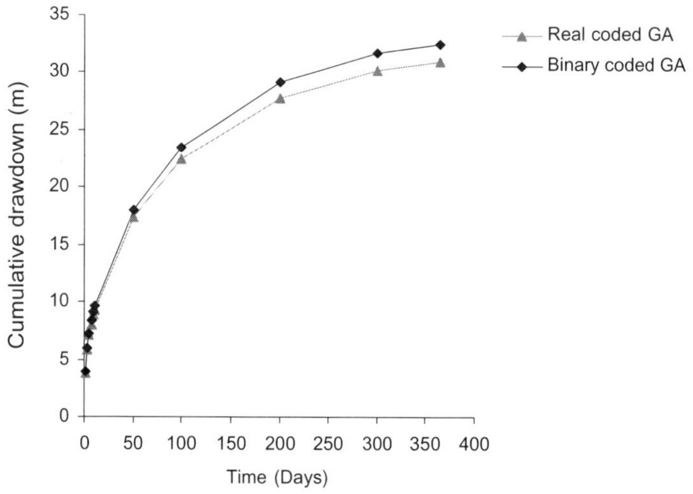

Fig. 7. Cumulative drawdown at various times by real and binary GA.

suggested that there is a rapid decline in groundwater heads for a period of almost 50 days. After 50 days, the groundwater levels tend to recover in some wells located near the river boundary due to a dynamic balance developed between the groundwater withdrawn with the aquifer boundary recharge. However, near the impervious boundary it took nearly 300 days to achieve constant drawdown conditions in the wells. This indicates that the water levels, in general, would recover at a faster rate if the wells are appropriately located near the river due to the contribution of the boundary flux to the aquifer, but at a slower rate elsewhere, depending on the distance from the boundary. Drawdown increase near impermeable boundary is attributed to the fact that the cone of depression cannot expand across the boundary due to the absence of lateral flow conditions. Coupling of real-coded GA with the simulation model assessed optimal cumulative drawdown for the 11 wells as 30.9 m with the accompanying total withdrawals of 6298.18 m^3/d. However the optimal results varied only marginally when compared with the binary-coded GA which estimated 5% higher total drawdown followed by a matching increase in cumulative pumping. Both the results satisfied the demands and constraints on the pumping and aquifer head throughout the year. Therefore, the present study concludes that the nonconventional techniques of real- and binary-coded GA can give meaningful results for groundwater withdrawal in heterogeneous anisotropic aquifers from practical consideration.

References

1. J. Bear, *Hydraulics of Groundwater* (McGraw-Hill, New York, 1979).
2. K. Deb and M. Goyal, A combined genetic adaptive search GeneAS engineering design, *Comp. Sci. Informatics* **26**(4) (1996) 30–45.
3. D. E. Goldberg, *Genetic Algorithms in Search, Optimization and Machine Learning* (Addison-Wesley, Reading, MA, 1989).
4. M. C. Hill, R. L. Cooley and D. W Pollock, A controlled experiment in groundwater flow model calibration, *Groundwater* **36**(3) (1998) 520–535.
5. M. P. Samuel and M. K. Jha, Estimation of aquifer parameters from pumping test data by genetic algorithm optimization technique, *J. Irrigation Drain. Eng.* **129**(5) (2003) 348–35.

Advances in Geosciences
Vol. 11: Hydrological Science (2007)
Eds. Namsik Park *et al.*

STUDY OF THE FRESH GROUNDWATER RESOURCES, NEIL ISLAND, INDIA

V. K. SAXENA
National Geophysical Research Institute
Hyderabad 500 007, India
vks_9020010@yahoo.co.in

India has a large number of islands and among these, Andman Nicobar, which is located in the South East of India, formed by Bay of Bengal, has 352 islands within an area of 8250 km^2. Most of these islands, and various other coastal areas such as Krishna, Godavari, and Kaveri delta are facing sea water intrusion problems and this in turn affect not only surface/groundwater quality but also contaminate soil structure, which further affect agriculture production and tourism. Agriculture and tourism are the main sources for the survival of local people. Because of such water problems, most of the inhabitants face lots of problems, which include health hazards, as a result of which, they are forced to migrate to some other places. Almost no hydrochemical work has been done so far, in most of these islands. The present preliminary work is an attempt to identify the problem in Neil Island and present the data. A systematic hydrochemical study was carried out to assess the chemical quality of groundwater in Neil Island. This study indicates the following:

- Water is nearly neutral to mildly alkaline in pH (7.2–7.9), and its TDS varies from 374 to 1270 mg/l.
- Water is classified into Na–Ca–Cl–HCO_3 and mixed type.
- 86% of samples have shown hardness ($CaCO_3$) >250 mg/l, and are classified as hard water.
- About 30% of area of Neil is covered with limestone, and its probable interaction with groundwater may cause the hardness. The Ca vs TDS; HCO_3 vs TDS, and SO_4 vs EC have good correlation.
- At present, the influence of seawater intrusion on groundwater is almost negligible, but if excessive withdrawal of groundwater is continued, the aquifers will be contaminated. Presently, groundwater is suitable for various types of agricultural production.
- Major cations, anions, and trace elements are within the permissible limit of portable water. Na and Cl are the most dominant ions; good correlation exists between Na vs Σ cations and Cl vs Σ anions.

1. Introduction

The Neil Island is located in the Andaman Nicobar group of islands in the Bay of Bengal, India. From the groundwater studies' point of view, not much work has been done or reported so far. More or less, this area is known as virgin in hydrochemical studies point of view. In 2003, CSIR started a groundwater network program and the studies are being carried out in various parts of India, in particular in hard rock areas and more emphasis has been given to rural areas and islands. The National Geophysical Research Institute, Hyderabad, has started such work in the Neil Island. This island is located about 35 km NE of Port Blair (Capital of Andaman–Nicobar). The area is known for rich agricultural production, where a large number of fruits and vegetables have been cultivated and supplied to nearby islands, which also include Port Blair. In October/November 2003, a hydrochemical study was carried out for the assessment of fresh groundwater resources and the suitability for irrigation purpose. About 159 well inventions were done with 24 km^2. The purpose of this work is to assess the fresh groundwater potential zones and identify the hydrochemical processes in this island. Groundwater sampling stations are shown in Fig. 1.

2. About the Study Area

Andaman–Nicobar Islands are summits of a submerged mountain range lying on the greater tectonic structure zones that extend from eastern

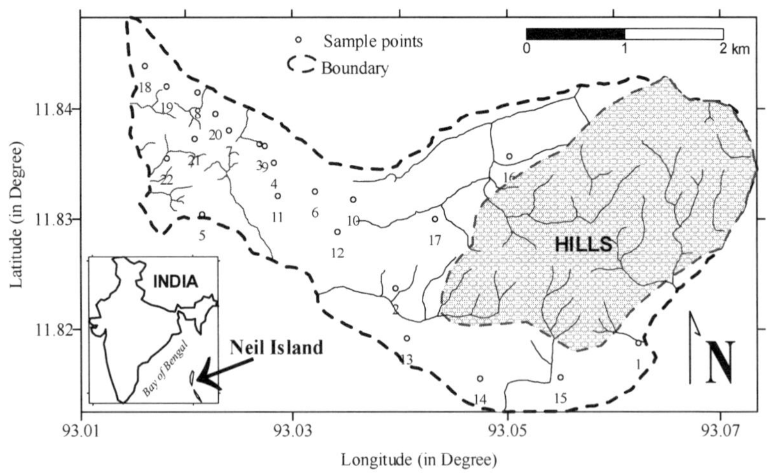

Fig. 1. Location of the sampling stations, Neil Island.

Himalayan along Myanmar through the Arakars and finally to Sumatra and lesser Sundas. The Archipelago consists of 352 islands and a total area of 8249 km^2 of which the Andaman group covers 6408 km^2 and the Nicobar 1841 km^2. Geologically, the Andaman–Nicobar Islands can be considered as two concentric areas: (1) The outer/western sedimentary areas include the major island, and (2) inner (eastern) may include the only volcano, which is known as Narcondem and Barren. It is noted that this is the only active volcano in India. Neil Island is located about 35 km NE of Port Blair, which is the capital of Andaman–Nicobar. Port Blair is known for cellular Jail and has become popular with the name of Kalapani (Black Water). The prisoners were kept in this Jail during 1910–1944. Neil Island is spread within 24 km^2 with the population of 7000. The elevation of this island varies from sea level to 750 m. The depth of water table varies from 1.5 to 6.5 m and the total depth of the well varies from 2.0 to 8.0. The average rainfall is around 3200 mm. This island is mainly covered with quaternary formations. Outcrop of limestone is easily visible in the eastern side of the island and covers almost one-third of the total area. The total thickness of the limestone is approximately 45.0 m.[4,5] The Holocene sediments are widespread over a large area which includes alluvium, raised beaches, terraces, wave-cut platforms, coral rags, calcareous tuff and shell limestone, and beach rock.[6] The middle and western parts of the island are comparatively moderately hard, fine- to medium-grained, and foraminiferal algal biocalcarenite. Because of its beautiful greenness and beaches, Neil Island has become a tourist spot and has a fairly good floating population. Neil is connected by ship and helicopter from Port Blair. A large number of spots have been developed during the past few years. Groundwater is the only source for agriculture and drinking needs.

3. Collection of Water Samples and Analytical Techniques

A total of 22 groundwater samples were collected from open wells, bore wells, and hand pumps. Insite measurement was done for EC, temperature, and NO_3. The samples were also collected (treated with nitric acid for pH adjustment) for trace elements (Fe, Sr and B) in 100 ml polythene bottles. The chemical analysis was done for TDS, TH, pH, and all major and minor elements as well as trace elements.[1,2] The results of the chemical analysis for major elements are shown in Table 1 and for trace elements in Table 2.

Table 1. Chemical composition of groundwater samples, Neil Island.

Sample No.	EC	TDS	pH	TH	Na	K	Ca	Mg	Cl	HCO_3	SO_4	NO_3
1	1868	1270	7.8	608	155	20	193	30	295	415	142	25
2	741	519	7.4	280	62	12	70	10	85	210	62	10
3	538	382	7.4	198	48	10	66	10	58	132	48	10
4	1245	859	7.7	275	120	15	120	22	255	155	160	15
5	656	466	7.3	266	42	15	64	12	67	192	55	10
6	668	468	7.3	255	42	15	65	12	70	190	54	20
7	780	585	7.3	307	65	12	75	12	102	132	58	20
8	573	418	7.4	225	55	8	65	10	62	160	58	25
9	1050	777	7.4	438	82	12	128	12	175	310	40	10
10	700	525	7.2	297	60	10	57	10	95	225	60	10
11	1399	1063	7.8	530	120	13	170	12	155	360	122	20
12	946	681	7.5	373	78	10	118	16	138	255	58	20
13	773	564	7.5	295	65	10	75	12	92	220	60	15
14	752	560	7.5	294	65	10	72	10	92	222	62	15
15	1412	1070	7.8	450	132	15	125	20	158	325	128	15
16	696	487	7.3	252	65	10	57	14	65	195	48	15
17	757	562	7.3	340	65	10	70	14	95	220	65	15
18	1578	1152	7.9	530	152	22	155	18	205	375	143	25
19	839	629	7.7	375	72	10	80	12	102	292	48	15
20	535	374	7.6	195	47	8	65	12	260	133	49	10
21	1235	914	7.9	427	120	14	152	14	225	275	92	15
22	1587	1012	7.9	455	128	10	62	20	245	293	110	10

Notes: Concentrations are in mg/l, EC in μS/cm, TH: Total hardness as $CaCO_3$, pH at 25°C.

Table 1 shows that the TDS of groundwater varies from 374 to 1270 mg/l with an average value of 697 mg/l. Of all the samples, 27% have shown less than 500 mg/l, which is considered to be of very fresh water type[7]; 50% samples are with the TDS range of 501 to 1000 mg/l, which is the desirable limit of the groundwater quality[3]; 23% samples have shown more than 1001 mg/l of TDS, which is known to be the higher side of permissible limits; 86% samples show $>$ 250 mg/l as $CaCO_3$ hardness. In view of high rate of hardness, enrichments of Ca, Mg, and HCO_3 are visible in spite of low TDS. It seems more likely to have a host rock, which may have more calcium carbonate. The geology of this area shows the presence of huge amount of limestone, which covered almost one-third of the total area. Thus, the majority of these groundwaters may be interacted with limestone. With such interactions, Ca and HCO_3 have been dissociated and disintegrated and due to such dissolution make the groundwater enriched in hardness. HCO_3 vs TDS, and Ca or Mg vs TDS have been plotted (Figs. 2a and 2b), which showed a correlationship $<$ 900 mg/l. The pH of water is

Table 2. Trace elements in the Neil Island.

Sample No.	F	B	Sr	Fe
1	1.2	1.4	3.503	0.5
2	1.1	0.5	0.81	0.3
3	1.1	0.5	0.721	0.2
4	1.1	0.8	2.897	0.6
5	1.2	0.6	0.823	0.3
6	1.1	0.6	0.828	0.3
7	1	1.2	0.721	0.3
8	1	1	0.73	0.4
9	1.1	0.8	0.721	0.4
10	1.2	1	0.7	0.4
11	1.1	1.1	0.982	0.4
12	1	0.5	0.804	0.2
13	1	0.5	0.808	0.1
14	0.5	0.6	0.792	0.1
15	1	0.6	0.921	0.2
16	0.5	0.5	0.568	0.2
17	0.5	0.5	0.591	0.2
18	1.2	1.4	3.721	0.5
19	0.5	0.6	1.601	0.2
20	0.5	0.5	0.625	0.2
21	1	1	0.985	0.2
22	1	1.1	0.925	0.3

Note: Concentrations are in mg/l.

nearly neutral to mildly alkaline (7.2–7.9). Mg vs TDS and Na vs TDS have been plotted and shown in Figures 2(c) and 2(d), respectively, which were almost similar as indicated by TDS vs HCO_3 and Mg. The source for the bicarbonates is more likely from limestone–groundwater interaction. A good linear relation has been seen in Ca vs TDS. Similarly, Na vs TDS indicates the seawater intrusion in the fresh water aquifers and simultaneous dilution.

Table 2 indicates that the composition of fluoride varies from 0.5 to 1.2 mg/l, which is within the permissible limits of F in potable water. Similarly, B, Fe, and Sr are also within the permissible limits.[1,7]

EC contours have been made and shown in Fig. 3.

Figure 3 shows that the EC value observed is more in the middle part and in particular South East of the island. It may be due to over-exploitation of groundwater and also due to the interaction of groundwater with limestone. Furthermore, in the western side of the island, TDS is comparatively more in the highly cultivated areas, where a large number of

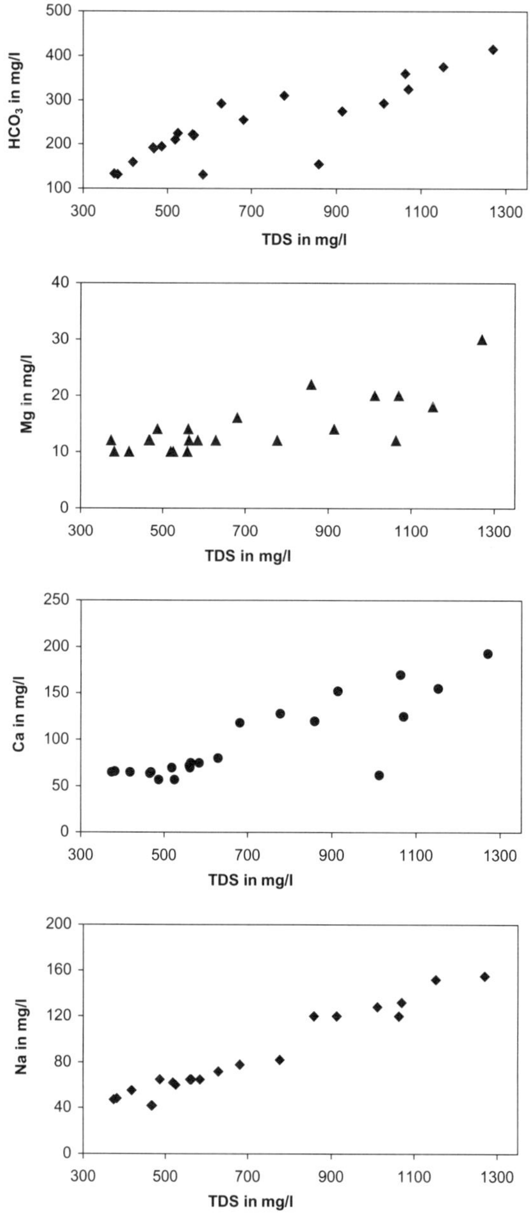

Fig. 2. Cross plot of HCO_3 and TDS. (b) Cross plot of Mg and TDS. (c) Cross plot of Ca and TDS. (d) Cross plot of Na and TDS.

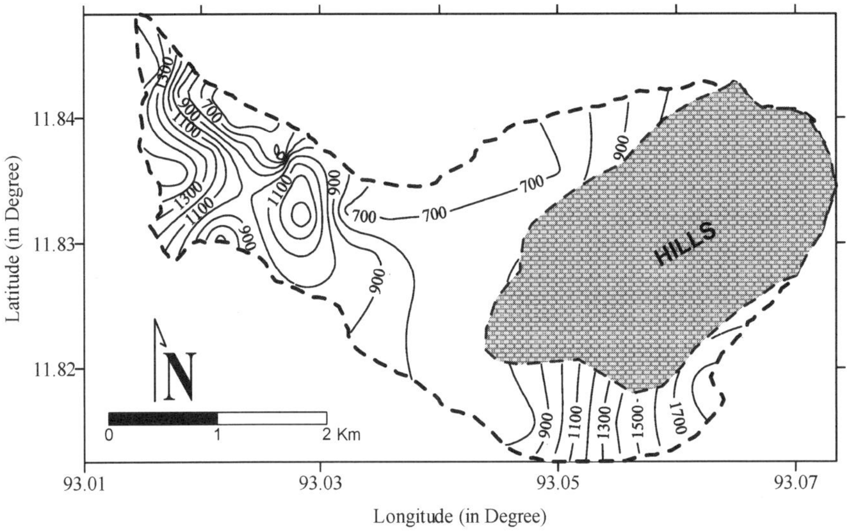

Fig. 3. EC contour map of Neil Island.

borewells have been drilled in the last few years. Hence, the possibility of seawater intrusion/ingression may increase if overexploitation is continued. Furthermore, the concentrations of F, B, Sr, and Fe are shown in Table 2, which indicate that these are within the permissible limits of potable water.

4. Conclusions

- The water is classified into Na–Ca–Cl–HCO_3 and mixed type.
- F, B, Sr, Fe are within the permissible limit of potable water.
- Water–limestone interaction may be the cause for the hardness of groundwater.
- Though none of the samples has reported the brackish/saline type, the trend of enrichment of ionic concentrations and their mixed character may alert the possible seawater intrusion. Therefore, care should be taken to control the excessive withdrawal of groundwater.

Acknowledgment

Thanks are due to Dr V. P. Dimri, Director, National Geophysical Research Institute, Hyderabad for his kind support to carry out this work and for his permission to publish this paper.

References

1. American Public Health Association (APHA), *Standard Methods for the Examination of Water and Waste*, 16th edn. (American, Public Health Association, Washington, DC, 1985), pp. 1–100.
2. E. Brown, M. W. Skougstad and M. J. Fishman, *Methods for Collection and Analysis of Water Samples for Dissolved Minerals and Gasses* (US Govt. Printing Office, Washington, 1974), pp. 1–160.
3. A. Mericado, Quality of groundwater, in *Groundwater Assessment Development and Management*, ed. K. R. Karnath (Tata McGraw Hill, New Delhi, 1985), pp. 217–275.
4. C. Rajshekhar and P. P. Reddy, Quaternary stratigraphy of Andaman–Nicobar Islands, Bay of Bengal, *J. Geol. Soc. India* **62** (2003) 485–493.
5. C. Rajshekhar, Studies in late Cenozoic smaller benthic foraminifera of Ritchie's Archipelago, Andaman Sea, Unpublished PhD thesis, BHU, Varanasi (1979).
6. M. S. Srinivasan, The Neogen-Quaternary boundary in the marine sequences of Andaman–Nicobar Islands, Northern Indian Ocean, is *Field Conf. IGCP-41* (Neogene/Quaternary Boundary, India, 1979), pp. 1–20.
7. World Health Organization (WHO), *Guidelines of Drinking Quality* (World Health Organization, Washington, DC, 1984), pp. 333–335.

Advances in Geosciences
Vol. 11: Hydrological Science (2007)
Eds. Namsik Park *et al.*

MULTIGEOLOGICAL STUDIES FOR ILLUSTRATION OF THE WATER–ROCK INTERACTIONS IN AL JAWW PLAIN, EAST OF AL AIN AREA, UNITED ARAB EMIRATES (UAE)

AHMED A. MURAD*, AYMAN EL-SAIY and HIND S. AL-NUAIMI
Department of Geology, College of Science, UAE University
P. O. Box 17551, Al-Ain, United Arab Emirates
**ahmed.murad@uaeu.ac.ae*

Detailed petrographic and mineral investigations of quaternary sediments accompanied with chemical analyses of groundwater were carried out in four aquifer wells located in Al Jaww plain, east of Al Ain area. The megascopic and microscopic examinations were conducted for 49 rock-cutting samples at different depths collected from these wells to investigate their detailed textural and compositional characteristics in order to evaluate the water–rock interactions as one of the processes that release elements to the groundwater. X-ray diffraction analyses were also conducted for most of the bulk samples to determine the detailed mineral composition. Petrographically, there are four facies types recorded in the studied aquifer sediments: ophiolite clasts conglomerate, ophiolite brecciated conglomerate, calclithite and lithic carbonate facies. Mineralogically, the aquifer sediments consist of serpentine, calcite (major components), dolomite and plagioclase (subordinate), with minor contents of quartz, pyroxene, and amphibole. Groundwater analyses for the samples showed that the Ca^{2+} concentrations were increasing toward the west and this may be ascribed to the dissolution of carbonate contents of Neoautochthonous units (Asmari, Dammam, and Rus formations) and diagentically formed as carbonate cements. This is supported by the presence of calcite as revealed from the mineral compositions of the rock-cutting samples. In addition, the Mg^{2+} concentrations for the groundwater samples were increasing toward the east and this might be related to the formation of serpentine and dolomite which are released to the groundwater due to the weathering of Allochthonous units (Sumeini Group, Hawasina Assemblage, Haybi Complex, and Semail Ophiolite) and diagentically formed. The vertical and lateral variations in their mineral composition percentages were also assessed to illustrate the distribution of mineral constituents percentages throughout the Quaternary aquifer.

1. Introduction

The arid countries are facing a shortage in availability of water resources for different uses. United Arab Emirates (UAE) is among those countries. The availability of sufficient quantity of water of high quality is important for future development in Al Ain area. The quaternary aquifer is the most promising and economic source for the groundwater supply in Al Ain area. The investigated area is Al Jaww Plain which is located in Al Ain area. Al Jaww Plain lies in the eastern part of Al Ain city and receives a considerable share of Abu Dhabi's fresh water resources. The geology and geomorphology are affecting the quality of groundwater that is stored in main aquifers in the area. The interaction between the groundwater and sediments of the aquifer lead to release cations and anions to the groundwater and even to the surface water and due to that, the quality of water has been changed. The present study is aimed to determine the petrographic, minerologic, and hydrochemical characteristics of the subsurface Quaternary aquifer system in north of Al Jaww Plain. The obtained results are used to highlight the interaction between groundwater and aquifer materials.

2. Geology and Hydrogeology of AI Jaww Plain

Al Jaww Plain is an especially large (15 km) wide and prominent piedmont situated east and southeast of the city of Al Ain between the Oman Mountains and Jabal Hafit. It consists of gently inclined gravelly materials transported by wadis dissecting the northern Oman Mountains. The plain is transversed by numerous wadis such as Wadi Shik, Al Ain, and Muraykhat. Three alluvial fans were identified within the plain; namely: the Zarub fan in the north, the Moundassah fan in the middle, and the Ajran fan in the south.[1]

Al Jaww Plain is mostly covered with the Quaternary deposits. Hunting Geology and Geophysics[2] recognized five sediment types (Fig. 1) which are alluvial deposits, desert plain deposits, mixed deposits, sabkha deposits, and Aeolian sand. All the rocks of Al Jaww Plain in Al Ain area are sedimentary including massive, fossiliferous Tertiary limestone of Lower Eocene, limestone with intercalations of Pelagic blue grey marls, gypsum, and gypsiferous marls of Miocene age overlain by marls and clays with conglomerates of the Fars formation. The latter forms the base of the Quaternary of recent deposits which include the aquiferous zones of the study area.

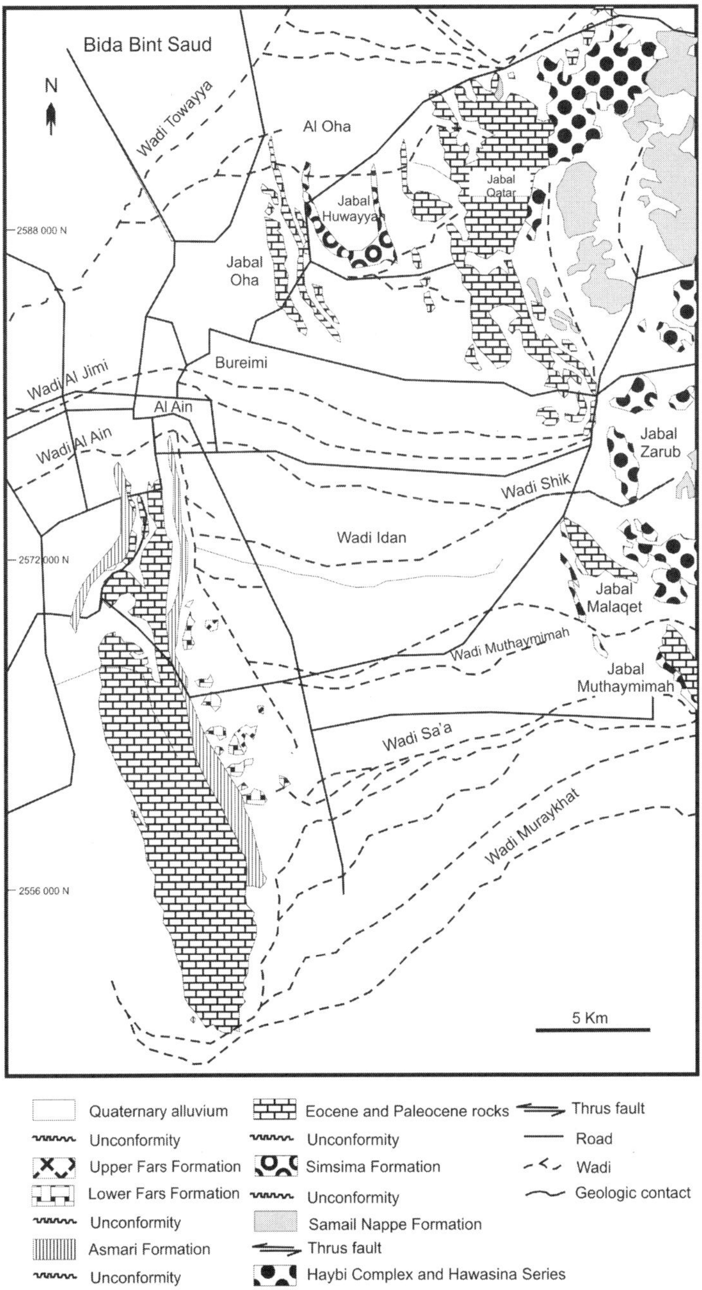

Fig. 1. Geologic map of Al Jaww Plain (modified from Hunting Geology and Geophysics, Ltd., 1979).[7]

All of the major wadis crossing Al Jaww Plain originate from the Oman Mountains range which forms the catchment and feeding area of the water resources in the studied area. These catchments areas are made up of barren high mountains and hills rising up to elevations of 1500 m and is relatively impervious except for fault zones. Al Jaww Plain is a large spread of an area made up of gravel and sand outwashes from the Oman Mountains and deposited in the main wadies including Al Ain, Shik and Hamad which traverse the plain. The main source for the recharge to the Quaternary aquifer at Al Jaww Plain is the rainfall on the Oman Mountains. The Quaternary aquifer may also be recharged by the infiltration of precipitation in the interdune areas and gravel plain and from Jabal Hafit where the precipitation percolates rapidly through the permeable limestone rocks of Jabal Hafit.[3]

The main existing aquifers in the UAE are fractured ophiolite rocks in the east, gravel aquifers flanking the eastern mountain ranges on the east and west, and sand dune aquifers in the south and west.[4] The most important aquifer in the study area is the Quaternary aquifer. Quaternary-age deposits consist of near-surface and surfacial sediments of mixed alluvial, aeolian, and locally sabkha (Evaporites) origins. Quaternary alluvium represents the principle water bearing lithostratigraphic unit relative to other units. Quaternary alluvium is composed of a sequence of about 60 m of sand and gravel with interbeds of silt and clay. Most of the coarse clastic units contain a clay-rich matrix that is usually calcareous.

Jabal Hafit is composed of 1500 m thick limestone and marl interbeds with gypsum and dolomite and evaporate formations of Lower Eocene to Miocene age. Limestone of the Middle Eocene of Dammam Formation constitutes an aquifer in Jabal Hafit. The aquifer is characterized by extensive dolomitization and is affected by numerous faults and fractures.[5]

3. Methodology

A total of 49 samples representing the various lithologies of the Quaternary aquifer sediments from wells no. 18, 17, 466, and 302 (Fig. 2) were provided from National Drilling Company (NDC), Al Ain area. These samples were investigated megascopically and their thin sections were subjected to microscopic examination to investigate their detailed textural and compositional characteristics. X-ray diffraction analysis was also conducted on 34 bulk samples using Philips X-ray diffractometer (Model PW/1840) with a Ni-filter, Cu-Kα radiation ($\lambda = 1.542$Å) at 40 kV, 30 mA potential,

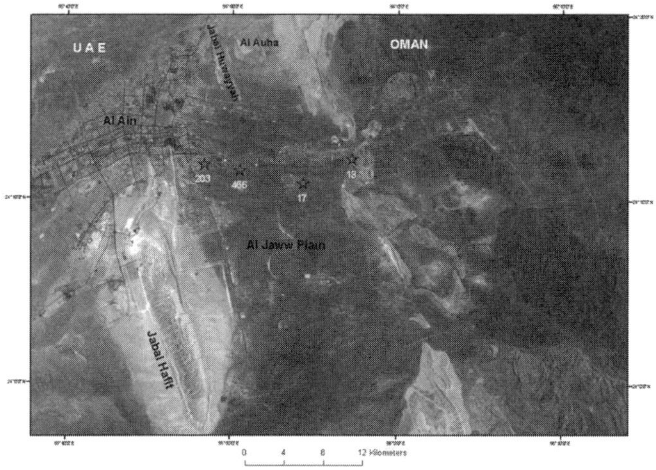

Fig. 2. Map showing the locations of studied wells.

and scanning speed of 0.02° per second. Mineral identification was achieved by comparing the obtained X-ray data with those published by the American Society of Testing and Materials (ASTM). The water chemistry of some from the study area was obtained from Ref. 6.

4. Results and Discussion

The results of X-ray diffraction analysis of the Quaternary aquifer sediments and the vertical variations of mineralogy of bulk aquifer (Figs. 3 and 4) revealed remarkable variations in their mineral composition throughout each of the studied wells. Sediments of well No. 18 consist of serpentine and calcite with minor concentrations of dolomite and plagioclase and traces of quartz, pyroxene, and amphibole in some stratigraphic levels. The percentages of serpentine, dolomite, and pyroxene increase while those of calcite, quartz, and plagioclase decrease upward in the aquifer. In well No. 17, the aquifer sediments are dominated by serpentine and calcite and contain smaller proportions of dolomite, plagioclase and traces of pyroxene, quartz, amphibole, clay minerals and talc in some stratigraphic levels. The proportions of serpentine and dolomite increase while those of calcite decrease upward in the aquifer but with increasing percentages of calcite in the upper part. The proportion of plagioclase oscillates throughout the aquifer. In well No. 466, the sediments are essentially made up of

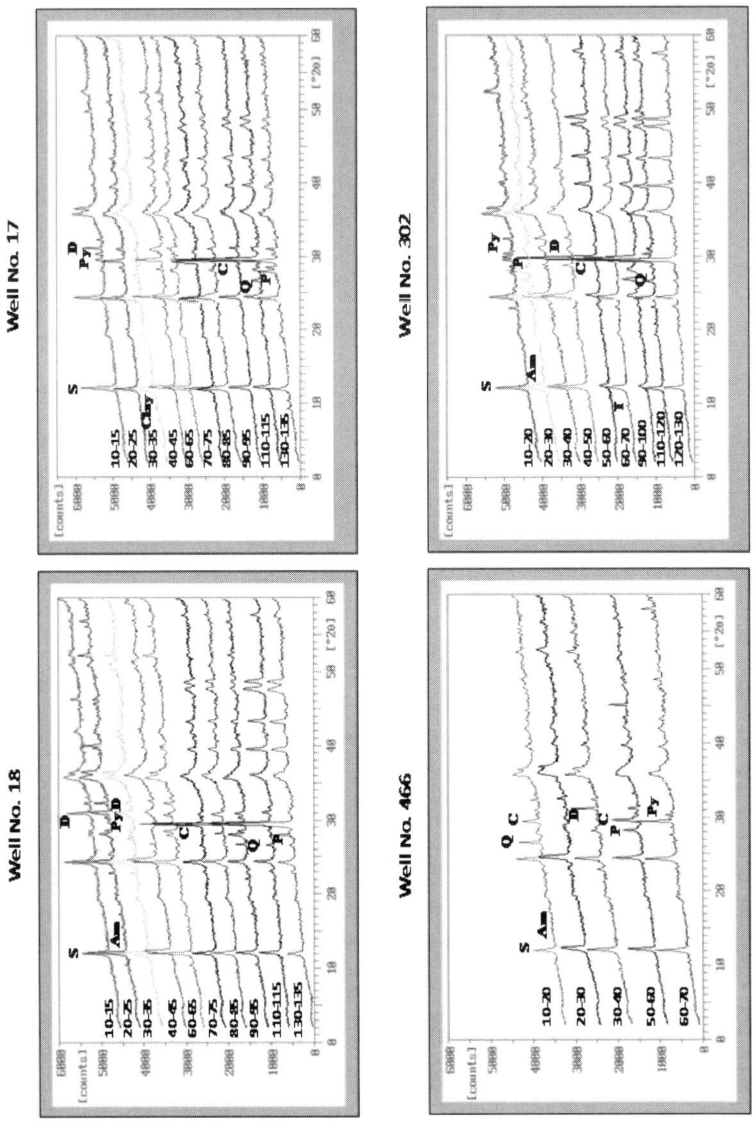

Fig. 3. X-ray diffraction patterns of some selected bulk samples of the studied wells. (C = Calcite; Clay = minerals; D = Dolomite; P = Plagioclase; Py = Pyroxene; Q = Quartz; S = Serpentine; Am = Amphibole; T = Tale).

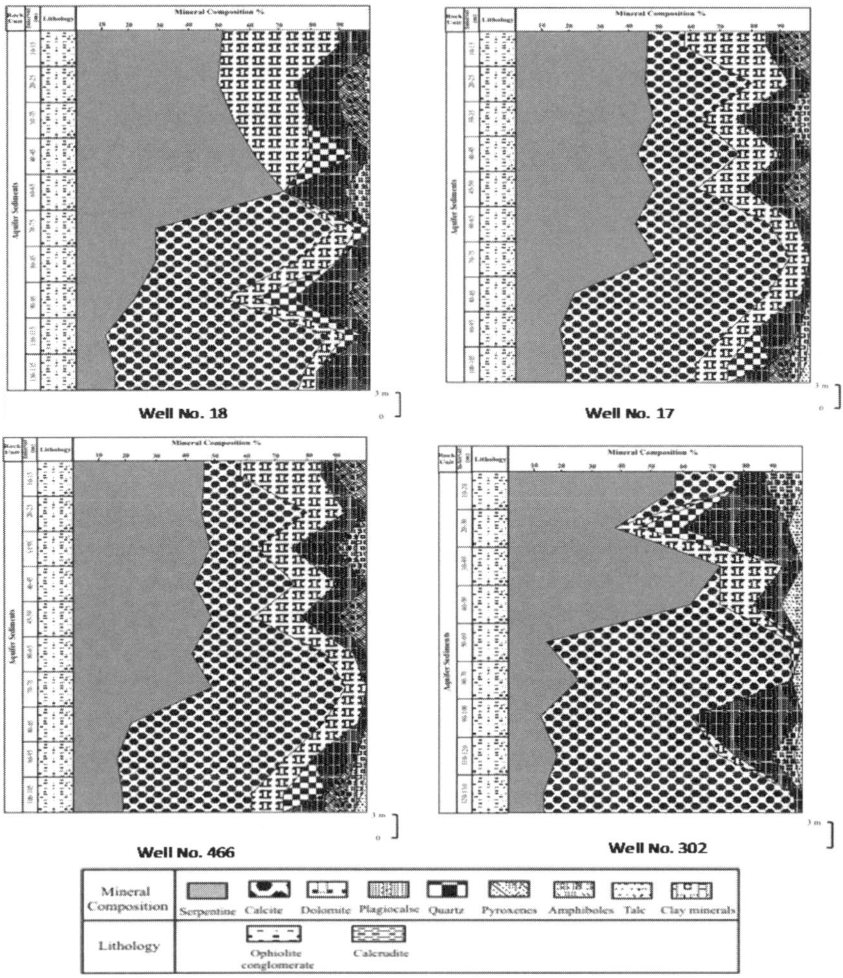

Fig. 4. Vertical variation in mineral composition of the aquifer sediments of Al Jaww Plain.

serpentine followed by calcite and minor concentrations of dolomite and plagioclase with traces of quartz, pyroxene, and amphibole. The proportions of serpentine oscillate throughout the succession but increase markedly at its top to the expense of those of calcite. Dolomite, plagioclase, and pyroxene increase in the middle part of the sequence while quartz is recorded at its top. On the other hand, the sediments of well No. 302 consist of serpentine and calcite with minor concentrations of plagioclase, dolomite, quartz, and traces of pyroxene, talc and amphibole. The concentrations of serpentine, dolomite, and plagioclase increase while those of calcite, quartz decrease upward in the aquifer sequence. The remaining minerals were detected in some stratigraphic levels. The lateral variation in the average mineral composition percentages of the aquifer sediments shown in Fig. 5 shows that the major minerals represented by serpentine decreases towards the west except in well No. 466 it increases, while calcite increases

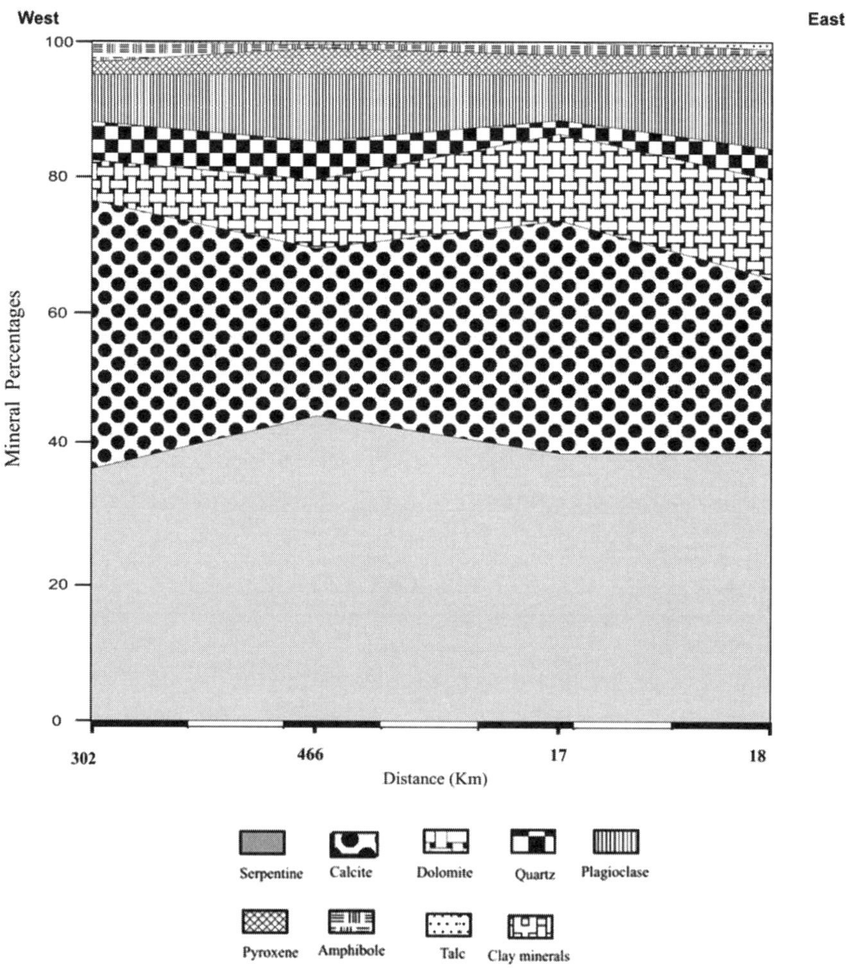

Fig. 5. Lateral variations in the mineralogy of the bulk aquifer sediments of Al Jaww Plain.

particularly in well No. 302 where it closed of Jebel Hafit. Dolomite and plagioclase existing in subordinate show increasing eastward where it closed from some ophiolites. With respect to the minor minerals, quartz increases westward while the pyroxene increases in the center. Amphibole occurs nearly the same in all wells while the clay mineral occurs as traces only in well No. 17. Traces of talc were recorded only in the western and central part of the studied wells.

Textural and compositional characteristics of the sediments of Al Jaww aquifer have essential effect on the quality of water. Fourty thin sections were prepared from the selected samples collected at different depth intervals in four wells. These wells are directed in the line nearly ENE–SWS direction in Al Jaww Plain. The various rock types constituting the aquifer system in the studied wells have been examined microscopically. The results of this examination revealed presence of the following four facies types: the Ophiolite clasts conglomerate, ophiolite brecciated conglomerate facies, calclithite facies, and lithic carbonate facies. The Ophiolite clasts conglomerate (Figs. 6A and 6B) was recorded in the upper part of the aquifer particularly eastward of the study area (well Nos. 18 and 17). The framework of the sediments consists of silty to coarse pebbly, poorly sorted clasts which are subangular to subrounded and have low spericity. These clasts are essentially ophiolitic consisting of serpentine, olivine, pyroxenes, plagioclase, and amphiboles with minor of cherty made up of microquartz and megaquartz. Some clasts are rimmed with isopachous calcite crystals. Some clasts are cemented by intergranular meso-to-coarse sparry calcite cements.

Two phases of cementation were recognized in few cases. The earlier is isopachous calcite cement which is followed by intergranular calcite cement. The ophiolitic brecciated conglomerate facies (Figs. 6C and 6D) were recognized in the upper part of the western aquifer (well No. 302). The sediments composed of medium-to-coarse pebble sized, poorly sorted, mainly ophiolitic clasts composed mainly of serpentine, altered olivine, and pyroxenes. Stockwork texture was detected due to filling of the network of altered ophiolite by subhedral to euhedral calcite crystals. The calclithite facies (Figs. 6E and 6F) was detected in the lower part of the aquifer particularly westward (well No. 302). The facies is mainly made up of sand-sized, moderately sorted mainly angular to subangular consisting of altered serpentine, chert, and subordinate quartz with minor components of pebble-sized clasts ophiolites, micro-quartz, polyquartz, and micrite limestone. Scattered silt-sized, angular grains of iron oxides are recorded. Some clasts

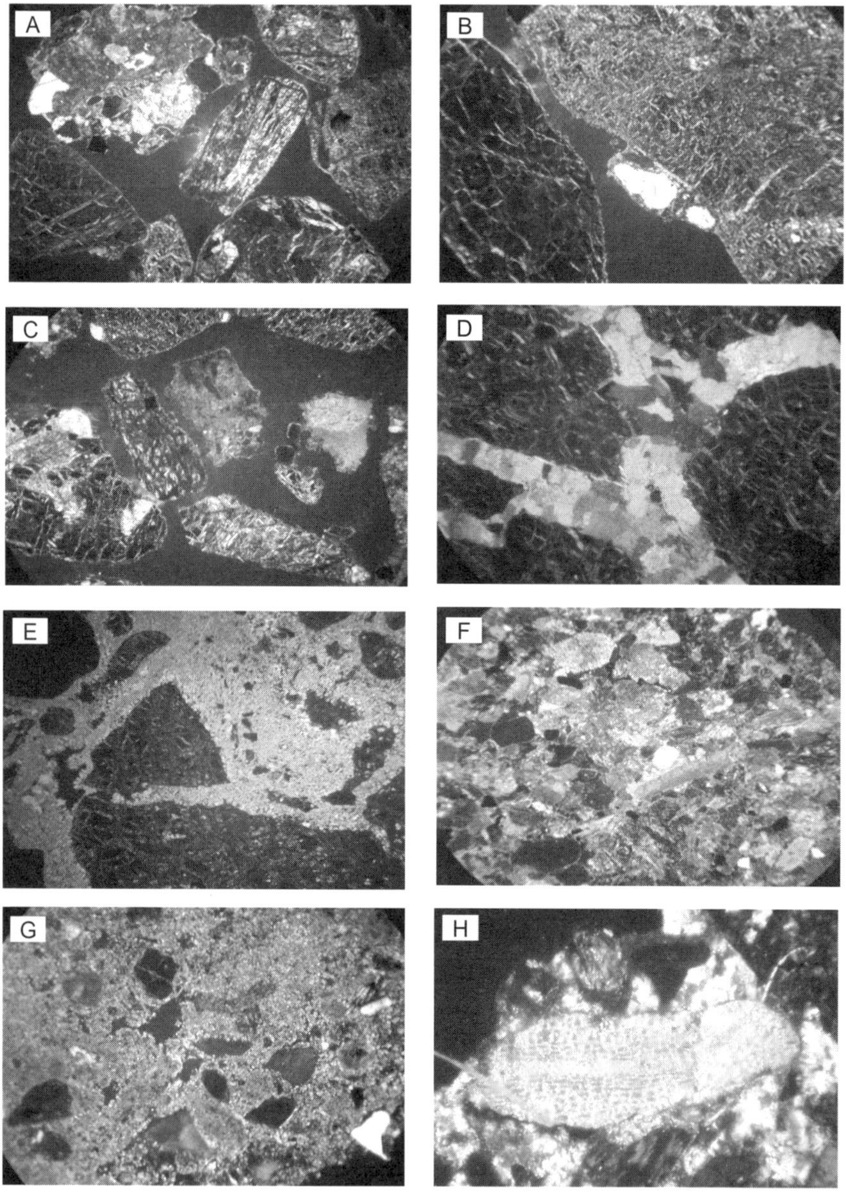

Fig. 6. Sections of facies of the aquifer.

are cemented by meso-to-coarse calcite crystals and rhombohedral dolomite crystals. The lithic carbonate facies (Figs. 6G and 6H) was recognized in several hoizon especially in westward aquifer (well No. 302). It consists of ophiolite grains which are made up of fine sand-sized, well-sorted, subangular to subrounded ophiolite, quartz, chert, and micrite limestone grains scattered in groundmass composed of partially dolomitized calcite or micritic matrix.

Groundwater samples were analyzed for chemical analysis of major cations namely: Ca^{2+}, Mg^{2+}, K^{+}, and Na^{+} (Fig. 7)[6] in order to determine the interactions between the groundwater and sediments of the aquifer in Al Jaww Plain. The results clearly indicated that the proportions of analyzed major cations increase from the east to the west of the Quaternary aquifer of Al Jaww Plain toward Jabel Hafit (Fig. 8). The increasing percentage of calcium (Ca) and magnesium (Mg) westwards is due to the fact that the positions of the wells are close to Jabal Hafit which mainly consists of calcite and dolomite ($CaMg(CO_3)$). On the other hand, the increase of potassium (K) and sodium (Na) is due to the weathering of feldspars and pyroxenes which are rich in Oman Mountains, east of the study area.

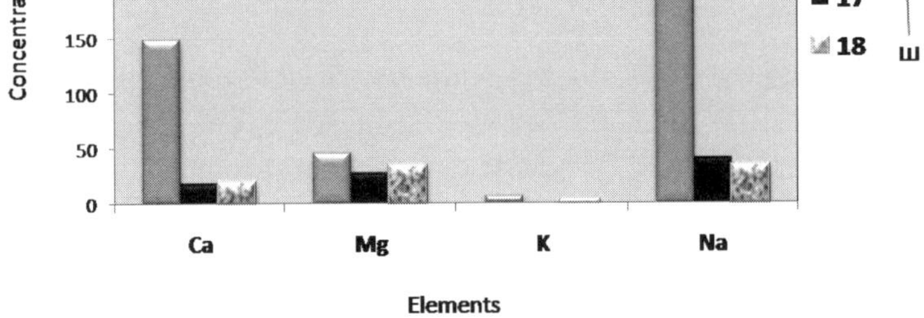

Fig. 7. The distribution of major cations along the Quaternary aquifer of Al Jaww Plain (ppm).

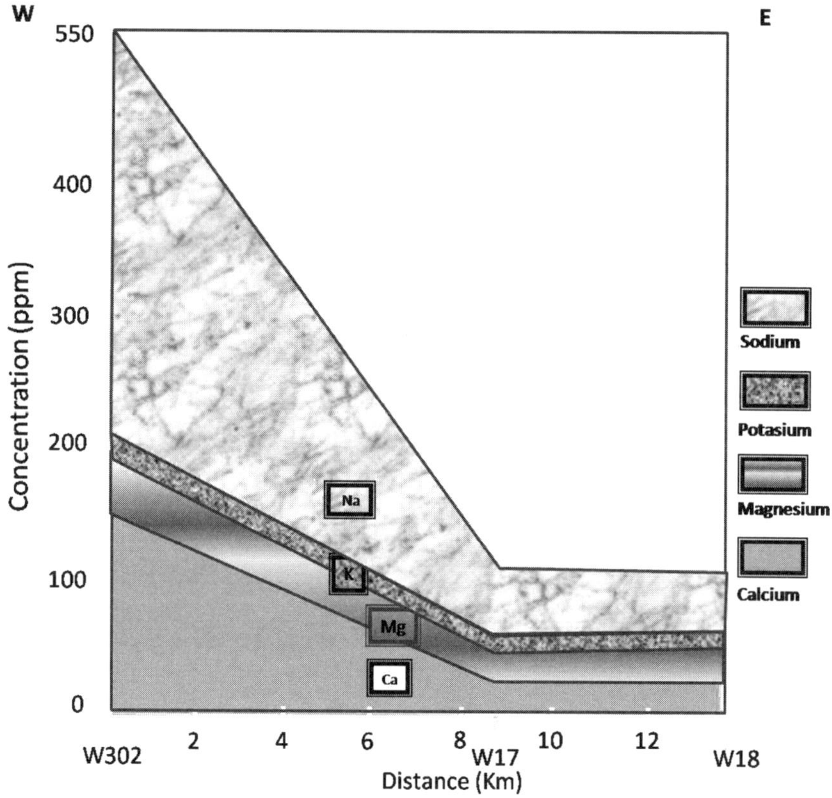

Fig. 8. Lateral variation of some major cations for the Quaternary aquifer of Jaww Plain (ppm).

5. Conclusions

Different geological studies were conducted to determine the interactions between the water and the aquifer sediments. The results of X-ray diffraction clearly showed the variations of mineral compositions of aquifer materials, and this conclusion is supported by the vertical distribution of the mineralogy of the aquifer. Four different facies were determined using the microscopic examinations and these facies are ophiolitic clasts conglomerate, ophiolite brecciated conglomerate, calclithite facies, and lithic carbonate facies. The percentages of major cations of groundwater increased from the east to the west of the Quaternary aquifer of Al Jaww Plain toward Jabal Hafit. It is clear that the lithology of Jabal Hafit affected the water quality.

References

1. M. H. Al-Shamsei, Drainage basins and flash flood hazards in Al Ain area, United Arab Emirates, M.Sc. Thesis, United Arab Emirates University (1993), 151p.
2. Hunting Geology and Geophysics Ltd., Report on a mineral survey UAE, Al Ain Area, Ministry of Petroleum and Mineral Resources, Abu Dhabi, Vol. 9 (1979), 1–22p.
3. A. S. Alsharhan, Z. A. Rizk, A. E. M. Narin, D. W. Bakhit and S. A. AlHajari, *Hydrogeology of An Arid Region: The Arabian Gulf and Adjoining Areas* (Elsevier Publishing Company, New York, 2001), 331p.
4. Z. S. Rizk, A. S. Alsharhan and S. Shino, Evaluation of groundwater resources of United Arab Emirates. *Proceedings of 3rd Gulf Water Conference*, Muscat, Sultanate of Oman, Vol. 1 (1997), pp. 95–122.
5. G. L. Whittle and A. S. Alsharhan, Dolomitization and certification of the Early Eocene Rus Formation in Abu Dhabi, UAE, *Sediment. Geol.* (91) (1994) 273–285.
6. H. S. Al-Nuaimi, Hydrogeological and geophysical studies on Al Jaww Plain, Al Ain Area, United Arab Emirates, M.Sc. Thesis, United Arab Emirates University (2003), 176p.
7. M. Warrak, Synchronous deformation of neoautochthonous sediments of the Northern Oman Mountains, *5th Conf. S.P.E.*, Bahrain (1987), pp. 129–136.

Advances in Geosciences
Vol. 11: Hydrological Science (2007)
Eds. Namsik Park *et al.*

A MODIFIED COMBINED MODEL OF A GEYSER INDUCED BY INFLOW OF GAS: CONSIDERING EVAPORATION EFFECT OF GAS DISSOLVED IN HOT SPRING WATER

HIROYUKI KAGAMI
Department of Preschool Education
Nagoya College, 48 Takeji, Sakae-cho
Toyoake, Aichi, 470-1193 Japan
kagami@nagoyacollege.ac.jp

We have proposed a mathematical model (a static model), a dynamical model, and a modified dynamical model of a geyser induced by inflow of gas (a periodic bubbling spring) based on observation and model experiments of the geyser and have also proposed a combined model combining the above two models. But, evaporation effect of gas dissolved in hot spring water during spouting has not been added to the above models yet. In this paper, we derive a modified combined model to which evaporation effect of gas dissolved in hot spring water during spouting is added. Then we discuss the effects of increase of total volume of hot spring water and evaporated gas packed in the spouting pipe through numerical simulation of the modified combined model.

1. Introduction

Geysers are classified into two types, depending upon the inducer. One is a geyser induced by boiling and the other is a geyser induced by inflow of gas (a periodic bubbling spring). The latter is the target of this study. There are many geysers induced by boiling, and some theories about its mechanism have been proposed.[1] On the other hand, there are a few geysers induced by inflow of gas (a periodic bubbling spring), and a few theories about its mechanism have been proposed.[2] Iwasaki[2] did model experiments of a geyser induced by inflow of gas and showed that injection of higher pressure gas spouted water from a spouting spout intermittently. Then he calculated spouting time and pause time using the gas supply rate as a parameter based on the mathematical model of gas balance. But this model had not discussed spouting dynamics of a geyser induced by inflow of gas.

We have proposed a mathematical model (a static model),[3,4] a dynamical model,[4,5] and a modified dynamical model of a geyser (a periodic bubbling spring),[4,6,7] based on the observation of Hirogawara Geyser (Yamagata, Japan)[8] and model experiments of the geyser,[9] and have also proposed a combined model combining the above two models.[4] And, numerical simulations of the modified dynamical model or the combined model have reappeared the dynamics of spouting of geysers (periodic bubbling springs). Then, through comparison between the results of numerical simulation and those of observation, we could estimate values of some underground parameters of geyser, that is, volume of the underground space, depth of spouting hole, and so on.[4,7] Moreover, we have verified the above models through geological exploration, analysis of hot spring water, and radioactive prospecting.[10]

In this study, we add the evaporation effect of gas dissolved in hot spring water during spouting to the combined model so as to recreate more practical spouting of a geyser induced by inflow of gas (a periodic bubbling spring). When hot spring water goes up from the underground deep region during spouting mode, vapor pressure of gas dissolved in it saturates before it reaches a spouting spout because of a drop in pressure. As a result, the dissolved gas evaporates one after another in various places where its vapor pressure saturates and volume of hot spring water and the evaporated gas packed in a spouting hole increases greatly. These phenomena will affect spouting dynamics of a geyser (a periodic bubbling spring) very much. In this study, we take the effects of increase of volume of hot spring water and the evaporated gas packed in a spouting hole into account.

2. Model

In this chapter, at first we introduce the main fruits of the former dynamical model, the mathematical model (a static model), and the combined model of a geyser induced by the inflow of gas (a periodic bubbling spring). Then, we introduce the derivation of the modified combined model to which the above-mentioned effects are added. Then we will introduce some results of numerical simulation of the modified combined model and minutely discuss the effects of the above-mentioned increase of volume of hot spring water and the evaporated gas.

2.1. *The former model*

Detailed derivation of the former dynamical model, the mathematical model (a static model), and the combined model of a geyser induced by inflow of gas (a periodic bubbling spring) was already stated in Refs. 4 and 7. So we show only the main fruits of the above models here.

At first, in case friction between walls of a spouting pipe and water packed in it is not taken into account, an evolution equation of temporal variations of the height of the top of a water pole packed in the spouting pipe of a periodic bubbling spring is written as

$$(n_0 + \beta t)(V_0 + Sx)\rho H \frac{d^3x}{dt^3} + (n_0 + \beta t)pS\frac{dx}{dt} = (V_0 + Sx)p\beta, \qquad (1)$$

where n_0 represents the molar number of gas in an underground space just before the water pole begins to move up to the upper entrance of the spouting pipe, β is constant concerning gas supply rate, V_0 represents volume of gas packed in an underground cave, S represents an area of a cross section of the spouting pipe, H represents length (height) of a lump of water packed in the pipe, p represents pressure of gas packed in the underground cave, x is regarded as a position of the lower interface between water and gas of the water pole, and an upper direction of a vertical line is regarded as a plus direction of x-axis.

In case friction between walls of a spouting pipe and water packed in it is taken into account, Eq. (1) is changed to the following Eq. (2):

$$\begin{aligned}&(n_0 + \beta t)(V_0 + Sx)\rho H \frac{d^3x}{dt^3}\\ &\quad + \frac{8\pi\eta H}{S}(n_0 + \beta t)(V_0 + Sx)\frac{d^2x}{dt^2} + (n_0 + \beta t)pS\frac{dx}{dt}\\ &= (V_0 + Sx)p\beta, \end{aligned} \qquad (2)$$

where η represents viscosity coefficient. As a result, the second term of Eq. (2) is added, and it represents the effects of viscosity.

Using the mathematical model (a static model), a spouting period τ and so on are represented as functions of various parameters. For example, a spouting period T can be written as

$$\tau = \frac{V_0}{\alpha\beta} + \frac{f_k S}{\rho g \alpha\beta}(f_k + P_0 + \rho g H), \qquad (3)$$

where $\alpha = RT$, R represents gas constant, T represents temperature, f_k represents pressure due to surface tension on an interface between water packed in the spouting pipe and gas in the underground cave, P_0 represents pressure of the atmosphere, and g represents gravity acceleration.

And the combined model is the one combining the former dynamical model and the mathematical model (a static model). The combined model enables more reliable estimation of underground parameters of a geyser induced by inflow of gas.

2.2. *A modified model*

In this section, we explain the introduction of evaporation effect of gas dissolved in hot spring water during spouting of the combined model. When hot spring water goes up from the underground deep region during spouting mode, vapor pressure of gas dissolved in it saturates before it reaches a spouting spout because of the drop in pressure as mentioned above. Or in the first place, hot spring water of a geyser induced by inflow of gas (periodic bubbling spring) may be a saturated solution of gas in the underground deep region. We can give the following two examples as evidences of the above conjectures.

1. Underground caves (spaces) are filled by gas. This is one of the important elements in all models of geyser induced by inflow of gas.
2. Spouting hot spring water includes bubbles of gas. A snapshot of the beginning of spouting at Kibedani Geyser (Shimane, Japan) is shown in Fig. 1. This snapshot enables us to understand the above fact.

Fig. 1. A snapshot of the beginning of spouting at Kibedani Geyser (Shimane, Japan). We can see that the spouting hot spring water includes many bubbles of gas.

As a result, the dissolved gas evaporates one after another in various places where its vapor pressure saturates, and the volume of the hot spring water and the evaporated gas packed in a spouting hole increases greatly. So, we estimate the effects of increase of volume of hot spring water and the evaporated gas packed in a spouting hole.

At first, it is assumed that the dissolved gas is CO_2, $T = 40°C$, and the volume of dissolvable CO_2 in $1\,cm^3$ of water under any pressure is $0.53\,cm^3$ for simplicity. Therefore, when a saturated solution of CO_2 at a depth of h goes up x in height, CO_2 of

$$\frac{\rho g(h-x)+P_0}{\rho gh+P_0}\times 0.53\,\mathrm{cm}^3 \qquad (4)$$

in volume is extracted per $1\,cm^3$ of water. Integrated volume of the above extracted CO_2 concerning all height increases the volume of the lump of water and gas packed in the spouting pipe.

In general, volume of extracted gas is written as

$$\frac{\rho g(h-x)+P_0}{\rho gh+P_0}\times a, \qquad (5)$$

where a represents the volume of dissolvable gas. These effects are added to the above-combined model.

3. Results and Discussion

In this chapter, we show some results of numerical simulation of the above-modified combined model and discuss them.

At first, temporal variation of the top of a water pole depended on a, and is shown in Fig. 2. We can see the slope of the graph of the top of the water pole's temporal variation under 0 m (the surface of the earth) to be steeper than the one over 0 m (the surface of the earth), which is characteristic for real spouting of a periodic bubbling spring. The larger the a is, the steeper the slope of the graph under 0 m is. For larger a, the larger total volume of dissolved gas is shown in Eq. (5).

Further, temporal variation of the top of water pole also changes depending upon the depth in which the gas solution saturates. This situation is shown in Fig. 3. We can see that the deeper the saturated depth is, the steeper the slope of the graph under 0 m is. For the deeper saturated depth, the total volume of dissolved gas is larger.

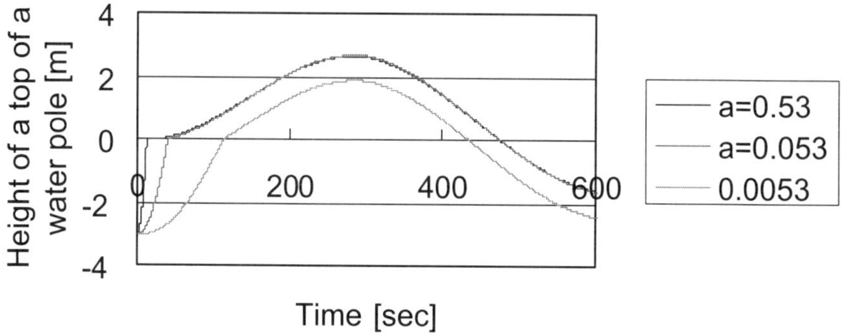

Fig. 2. Temporal variation of the top of a water pole dependent on a (in case gas solution is saturated over 100 m in depth).

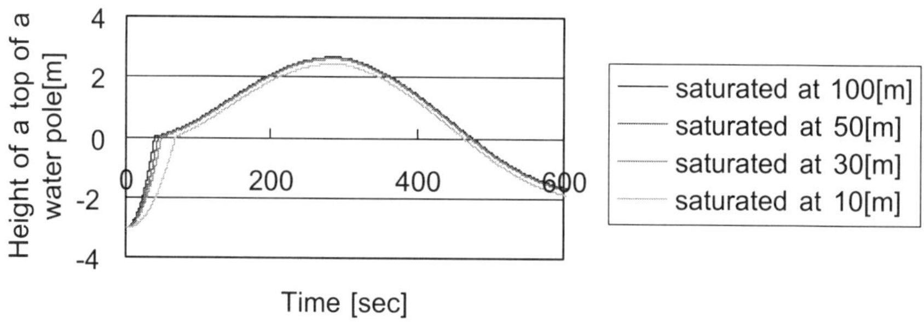

Fig. 3. Temporal variation of the top of the water pole dependent on the depth in which gas solution saturates.

4. Conclusions

We derived the modified combined model to which evaporation effect of gas dissolved in hot spring water during spouting was added. Then, we introduced some results of the numerical simulation of the modified combined model and discussed the effects of the increase of total volume of hot spring water and the evaporated gas packed in the spouting pipe.

After this, quantitative comparison with observational data will be needed.

References

1. K. Honda and T. Terada, *Publ. Earthq. Inv. Com.* **22B** (1906).
2. I. Iwasaki, Bulletin of Tokyo Institute of Technology, No. 46 (1962).

3. H. Kagami, *Abstracts of the 53rd Annual Meeting of the Balneological Society of Japan* (2000), p. 27.
4. H. Kagami, *Advances in Geosciences*, Hydrological Science (HS), Vol. 4 (2006), p. 191.
5. H. Kagami, *Abstracts of the 55th Annual Meeting of the Balneological Society of Japan* (2002), p. 33.
6. H. Kagami, *The 2003 IUGG General Assembly*, HW04/09P/C31-004 (2003).
7. H. Kagami, *Proc. 38th Conf. Sciete Internationale des Techniques Hydrothermales and The 56th Annual Meeting of the Balneological Society of Japan* (2003), pp. 55–60.
8. E. Ishii, *et al.*, *Abstracts of the 52nd Annual Meeting of the Balneological Society of Japan*, (1999), p. 28.
9. M. Katase, *et al.*, Abstracts of a meeting for presenting research papers of Kanto Gakuin University College of Engineering (1999), pp. 99–100.
10. H. Kagami, *Advances in Geosciences*, Hydrological Science (HS), Vol. 6 (2007), p. 203.

Advances in Geosciences
Vol. 11: Hydrological Science (2007)
Eds. Namsik Park *et al.*

EMERGING CONCEPTS IN HYDROLOGY FOR TROPICAL PACIFIC REGIMES

JAMES P. TERRY
Department of Geography, National University of Singapore
1 Arts Link, Singapore 117570
geojpt@nus.edu.sg

Work over the last decade in Fiji, New Caledonia, Samoa and Solomon Islands has begun to reveal some emerging concepts in Pacific tropical island hydrology. Tropical cyclones (TCs) are a major cause of floods and several features of cyclone behavior, such as slow and unusually sinuous tracks, increase both the amount of precipitation received and the consequent size of the flood produced. Strong orographic influences of rugged volcanic terrain mean that "cyclone sides" of islands may be identified in relation to the common directions of TC approach, but these cyclone sides are not the same as the windward sides of islands under normal climatic conditions. Large and/or numerous landslides during storms may temporarily dam headwater channels, but then subsequently fail, giving unpredictable storm hydrographs with multiple peaks, where the timing of the floods is not clearly associated with the temporal patterns in rainfall.

1. Introduction and Aim

When compared to other regions of the world with humid tropical regimes, the hydrology of rivers in the Pacific islands remains poorly studied. This is because many rivers in the islands are not properly gauged on a continuous basis, owing to the small budgets and limited resources of national hydrological services. Information on river flow behavior, responses to large and/or intense precipitation, and characteristics of storm hydrographs is therefore generally piecemeal in nature and lacking in detail. Even for those rivers where good long-term flow records do exist without significant gaps in the data sets, there is often insufficient personnel in small island countries to analyze and then interpret the hydrological information collected.

For these reasons, research led by the University of the South Pacific has examined some of the available records for a number of rivers in Fiji, New Caledonia, Samoa, and Solomon Islands, to determine the principal controls on fluvial behavior, especially in relation to high-flow events. This work forms part of a broader program of investigation on the climate, hydrology, and fluvial geomorphology of selected river basins across Oceania.

Previous investigation has already established an important principle — that tropical cyclones (TCs), elsewhere known as typhoons or hurricanes, are a primary cause of either the largest or most frequent overbank flood events in the steep and mountainous rivers of Pacific volcanic islands[1] (Fig. 1). Because of the widely-scattered distribution of islands in the South Pacific and their generally small size, not all cyclones make landfall on an island.[2] Nonetheless, many still pass close enough to island archipelagoes to deliver intense precipitation on land, generating enormous river discharges that easily cause widespread inundation and damage to towns, infrastructure, and farms across lowland areas. Cyclone-generated floods are therefore a major problem for both rural and urban communities occupying island floodplains.[3]

For example, in the Tontouta River on the island of Grande Terre in New Caledonia, 65% of the largest historical flows were caused by cyclones

Fig. 1. Flood in the lower Lungga River on Guadalcanal in Solomon Islands, near the capital Honiara, generated by TC Namu in May 1986. Photo courtesy of the Solomon Islands Department of Mines. Severe inundation of the Guadalcanal Plains by flooding of most of the rivers draining northward from the southern highlands led to widespread destruction and population resettlement after this event.

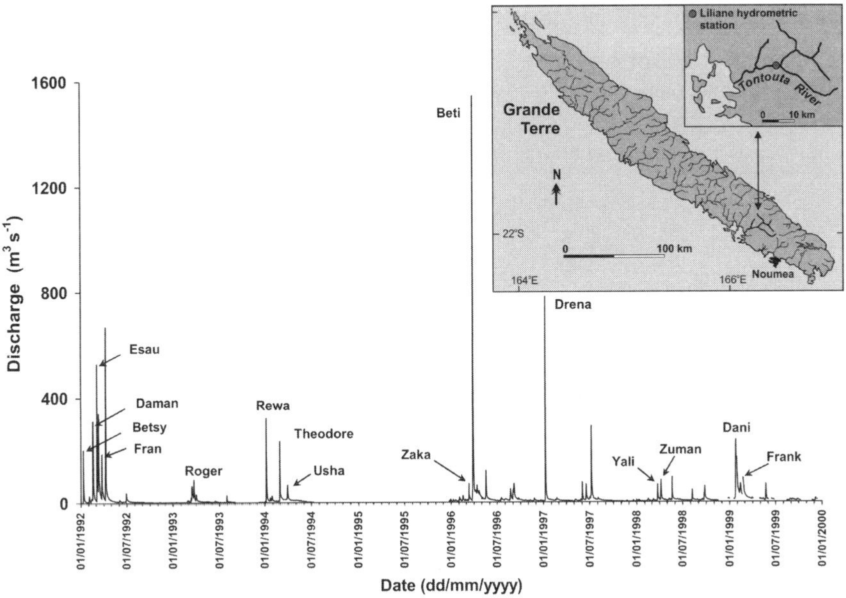

Fig. 2. Flow peaks, based on mean daily flows, produced by named tropical cyclones and unnamed tropical storms from 1992 to 1999 in the Tontouta River on Grande Terre island, New Caledonia. Bankfull discharge at the hydrometric station is approximately $600\,m^3\,s^{-1}$. Instantaneous flows (not shown) far exceeded the daily means. (*Source*: Observatoire de la Ressource en Eau, New Caledonia.)

and 75% of cyclone-induced floods were overbank events[4] (Fig. 2). The remaining 35% of floods were generated by other types of (non-rotational) tropical storms. In contrast, for the major Rewa River on Viti Levu island in Fiji, noncyclone floods are more common than the cyclone floods, but the latter are bigger events because of the high intensity of tropical cyclonic precipitation.[1]

2. Data Sources

On average, there are 9 or 10 tropical cyclones annually across the South Pacific ocean, which falls under the responsibility of the Fiji Meteorological Service (FMS) for recording and archiving cyclone activity.[5] The FMS operates the Regional Specialized Meteorological Centre Nadi-Tropical Cyclone Centre (RSMC-Nadi TCC) and has the task of monitoring all tropical cyclones originating or moving into its region (0° to 25°S; 160°E to 120°W). The Climate Services Division of FMS maintains the regional archive, utilized for this project, of historical tropical cyclone reports and associated meteorological information (weather charts, satellite imagery, track data, wind speeds, barometric pressures, etc.). Rainfall records held at the national meteorological offices of other island countries were also accessed when needed. River stage heights and discharge data for cyclone events (where these exist), measured at established hydrometric stations, were collected from national hydrological service providers — usually Hydrology Divisions of Public Works Departments, or equivalent government sections.

3. Observations and Results

3.1. *Cyclone speed*

Analysis of historically important tropical cyclone events revealed that several features of their speed and track shape play an important role in the amount of rainfall received on an island and the consequent size of river floods. The often slow approach of cyclones toward islands (usually from the north or north west) during the early part of their lifespan, especially while still in the intensification phase, tends to deliver high rainfall and therefore generates large stream discharges. This is in contrast to many fast-moving storms in their mature phase, which may also give intense precipitation over catchments, but for shorter durations, causing less extreme flood peaks. This is a general observation, and clearly there are many possible rainfall and resulting flood scenarios for individual cyclones, but the example of TC June which affected the Fiji Islands in May 1997 (Fig. 3), illustrates the idea.

TC June was a weak "midget" cyclone that never really formed a visible eye due to an insufficiently favorable environment.[6] TC June displayed erratic track behavior, making sudden changes in both direction and speed. In particular, the storm moved slowly on a southeast course toward Fiji during 4 May (Fig. 3), and then decelerated northwest of the Yasawa Islands group on 5 May 1997, before turning southward. This gave an unexpected distribution of high-intensity rainfalls across parts of Fiji, and many people were taken by surprise when severe localized stream flooding occurred. Several other cyclones that affected Fiji earlier in the same year were more intense systems than TC June, and yet produced smaller stream responses and less flooding (Fig. 4), because they progressed more rapidly through the Fiji Islands archipelago.

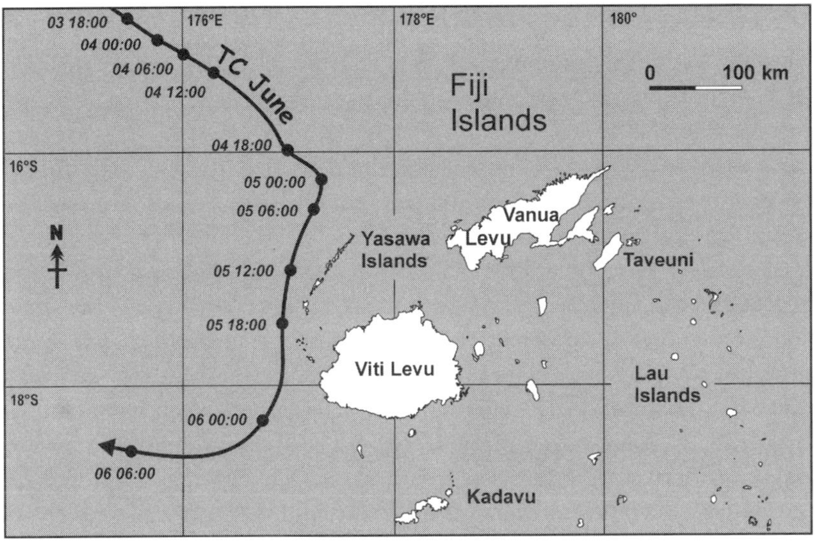

Fig. 3. TC June was a slow-moving cyclone that rested almost stationary over north western Fiji (Yasawa group) on 5 May 1997. The slow advance meant that in spite of the weak and poorly-developed structure of this cyclone, large hydrological responses were produced in streams on several islands.

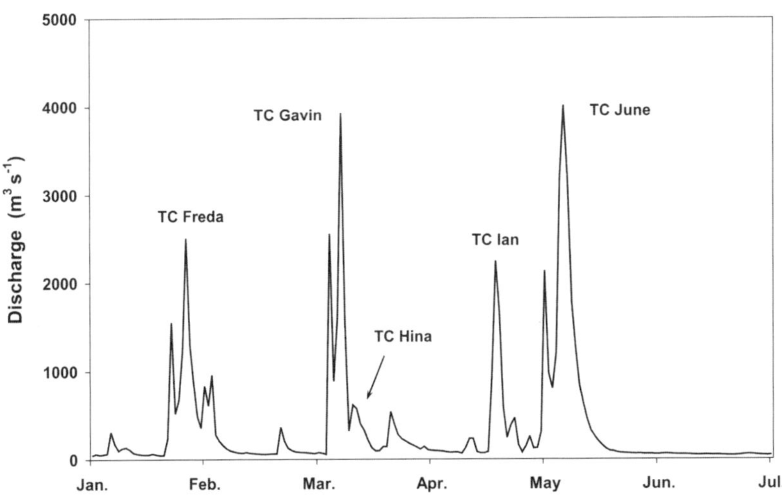

Fig. 4. Early to mid-1997 hydrograph for the major Rewa River, measured at Navolau hydrometric station, on Viti Levu island in Fiji. TC June was a poorly-structured cyclone, but still generated a big flood discharge because of its slow speed through Fiji waters, compared to e.g. TC Gavin, which was a much more intense but rapidly-moving cyclone.

3.2. *Cyclone track sinuosity*

Unusually looping or sinuous tracks are another feature of cyclone behavior that can have consequences on island hydrological responses. A looping track may "hold" a storm in the vicinity of a group of islands for many hours longer than if a straighter path were followed, therefore prolonging the period of both rainfall delivery and elevated river discharges. For example, TC Val was a cyclone that completed a small, tight loop over the Samoan island chain in December 1991. The system developed southeast of Tuvalu on 4 December, but as TC Val approached Savai'i island in Samoa on 5 December, the storm began making a clockwise turn and passed across the center of the island. The major extent of the damage was attributed to TC Val reaching peak intensity during the period when it looped over Samoa. Heavy, continuous rain (with synchronous high winds and enormous seas) devastated both Samoa and American Samoa with the loss of 16 lives.[7]

Similarly, TC Rewa with its rare double looping track (Fig. 5), delivered a total of 173 mm of rain between 31 December 1993 and 6 January 1994 (Table 1) at Liliane hydrometeorological station on Grande Terre island in New Caledonia. This was not such an extreme event compared to a later pair of cyclones that struck New Caledonia in 1996, but the extended

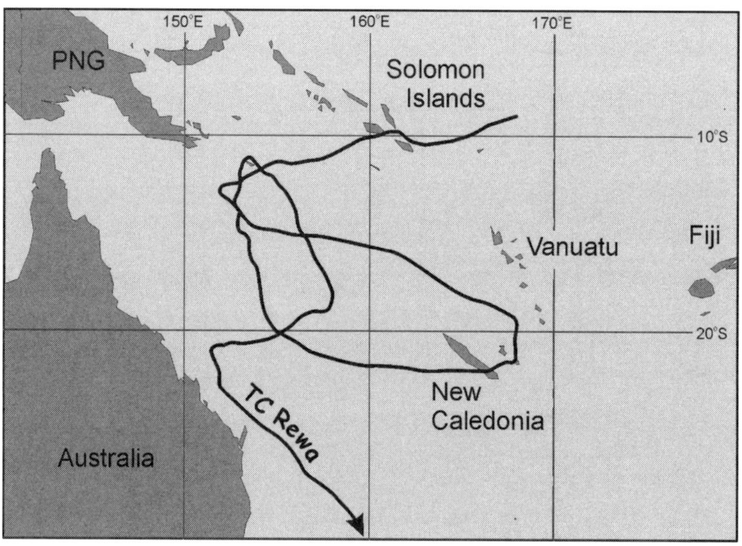

Fig. 5. Many rivers on Grande Terre island in New Caledonia experienced high flows in early January 1994 as TC Rewa traversed the island, but then backtracked across the north of the territory, eventually describing a rare double loop in its overall track shape.

Table 1. Rainfall–discharge relationship in the Tontouta River basin on Grande Terre island, New Caledonia, during TC Rewa which followed a rare double-looping path in late 1993 to early 1994.

Date	Rainfall (mm)	Discharge ($m^3 s^{-1}$)
28/12/1993	0	1.67
29/12/1993	2	1.48
30/12/1993	3	1.65
31/12/1993	11.5	2.97
01/01/1994	9	2.30
02/01/1994	7	3.72
03/01/1994	1.5	5.02
04/01/1994	0	3.85
05/01/1994	84.5	4.29
06/01/1994	59.5	321.69
07/01/1994	1	148.43
08/01/1994	0	63.59
09/01/1994	0.5	35.64
10/01/1994	0	25.12
11/01/1994	0	18.98
12/01/1994	0	15.16
13/01/1994	0	12.49
14/01/1994	0	10.74

Source: Observatoire de la Resource en Eau, New Caledonia.

temporal pattern in precipitation during TC Rewa gave big discharges in the Tontouta River for over a week (Table 1). Cyclone track direction in relation to river-basin orientation is also important. Where storms travel parallel to the long axis of large and elongated river basins, then more tributaries receive high rainfall. This gives a greater chance of big floods being generated, compared to where the storm tracks run perpendicular to the principal drainage networks and thus deliver rainfall to only some of the tributaries.[1]

3.3. *Orographic influences*

For the high islands of the western South Pacific, the windward sides face the prevailing moist south east Trade Winds, whereas north west sides are drier, lying in the lee of volcanic ranges. Yet, the strong orographic influences of volcanic terrain on the spatial distribution of big rainfalls during tropical cyclones means that "cyclone sides" of islands may be identified in relation to the common direction of storm approach.

Fig. 6. Rivers on the relatively dry, leeward, northern sides of high volcanic islands in the South Pacific (in this case Viti Levu island in Fiji) are also on the "cyclone side" since they face the direction from which most cyclones arrive. These rivers are therefore likely to show more extreme changes in flow from normal (low) baseflow conditions during intense cyclonic precipitation.

Since most cyclones develop and advance from the north of the main island archipelagoes, the "cyclone side" of islands is therefore not the same as the windward side under normal climatic conditions. The implication for hydrology is that rivers in the northern dry zones of islands (Fig. 6) are generally more vulnerable to flood impacts, because they tend to experience greater fluctuations in discharge (above normal) during cyclones, compared to wet zone rivers which have relatively higher wet-season baseflows.

3.4. *Landslide dams*

Several individual cyclone reports indicate that hillslope failures (especially landslides and debris flows) in steep sections of upper catchments (Fig. 7), caused by regolith saturation, have a major control on stormflood characteristics. Landslides sometimes have the effect of temporarily damming headwater channels, but they subsequently fail. This produces unpredictable storm hydrographs, sometimes with multiple flow peaks, where the timing of the peaks is not clearly related to temporal patterns in rainfall. In such circumstances, rainfall–runoff relationships cannot usefully

Fig. 7. Numerous small landslides in the upper basin of the Labasa River on Vanua Levu island in Fiji, caused by TC Ami in January 2003, led to temporary damming of headwater channels before subsequently failing, giving unpredictable hydrological behavior downstream in the main river. Widespread flooding of the Labasa River lowlands drowned 17 people.

be applied. Instead, understanding rainfall thresholds associated with soil saturation and slope failure, and the length of time before landslide dam rupture, is much more important for explaining post-event (and perhaps forecasting) cyclone-hydrograph shapes. In addition, landslide sediments that are deposited directly in the rivers cause channel aggradation that exacerbates the size of overbank floods.

A good illustration is the effect of TC Namu in May 1986, which resulted in the worst catastrophe to afflict Solomon Islands in recent decades. The cyclone traversed the island of Guadalcanal, where the climate station at Gold Ridge (290 m elevation) recorded 874 mm of precipitation over the duration of the storm. TC Namu inflicted terrible destruction because of the numerous landslides. In the river channels of upper watersheds, the so-called debris floods (a process combining debris flows and severe floods in valley bottoms) were responsible for spectacular amounts of sedimentation. The enormous quantities of debris choked valley bottoms, causing swollen rivers on the Guadalcanal Plains to burst their banks, and led to unprecedented river flooding and channel aggradation.[8] Huge water (and sediment) discharges smashed bridges on the

Lungga, Ngalimbiu, and Mbalisuna rivers. At least 100 lives were lost, and evacuation of thousands of people led eventually to permanent relocation by about 5% of the Guadalcanal Plains population.

4. Summary

It is well known that intense rainfalls associated with tropical cyclones have a major impact on the hydrology of Pacific islands, often causing the largest annual flood events. Not so clearly understood, however, are some of the other characteristics of cyclones that influence island river behavior and flood magnitude. This study, focusing on selected TC events in the South Pacific, has established that the following cyclone features are also important, and should therefore be taken into consideration when assessing, post-event, fluvial responses to severe storms:

(1) Speed of cyclone approach — slow moving or stationary storms can produce large floods, even for storms that are poorly structured or classified as "weak" systems.
(2) Track sinuosity — sinuous or looping tracks that keep a cyclone in the vicinity of an island group for an extended period can prolong the duration of high river discharges.
(3) Track orientation — cyclones approaching parallel to the long axis of elongated river basins are more likely to deliver large rainfalls to more of the main tributaries (and therefore generate big floods) than cyclones moving perpendicular to drainage network orientation.
(4) The north and west sides of islands in the South Pacific are identified as the "cyclone sides," and rivers here are especially vulnerable to cyclone-induced floods compared to rivers on the normally wet, windward sides of islands.

Based on these findings, it is recommended that further research needs to be directed at investigating (and modeling) the relationships between river flow dynamics and cyclone speed, track orientation, and sinuosity in affected regions. Eventually this will help to improve our flood-forecasting capabilities on tropical Pacific islands.

Acknowledgment

Financial support from the University Research Committee (project no. 6D104-1431) is gratefully acknowledged.

References

1. R. Kostaschuk, J. Terry and R. Raj, The impact of tropical cyclones on river floods in Fiji, *Hydrol. Sci. J.* **46** (2001) 435–450.
2. R. Krishna, *Tropical Cyclones* (Fiji Meteorological Service, Publ. No. 4, 1984).
3. J. P. Terry, S. McGree and R. Raj, The exceptional floods on Vanua Levu Island, Fiji, during Tropical Cyclone Ami in January 2003, *J. Natural Disaster Sci.* **26** (2004) 27–36.
4. R. A. Kostaschuk, J. P. Terry and G. Wotling, Tropical storms and associated flood risk on Grande Terre, New Caledonia, in *Technical Documents in Hydrology*, eds. S. Sethaputra and K. Promma, UNESCO International Hydrological Programme, No. 6 (2006), pp. 207–210.
5. FMS, *List of Tropical Cyclones in the South West Pacific 1969/70 — Present*, Fiji Meteorological Service, Information sheet No. 121, Climate Services Division, Nadi Airport, Fiji (2003).
6. FMS, Preliminary Report on Tropical Cyclone June, 3–5 May 1997, Fiji Meteorological Service, Climate Services Division, Nadi Airport, Fiji (1997).
7. FMS, Tropical Cyclone Val: 4–13 December 1991, Fiji Meteorological Service. Tropical Cyclone Report 92/1, Climate Services Division, Nadi Airport, Fiji (1992).
8. N. A. Trustrum, I. E. Whitehouse and P. M. Blaschke, Flood and landslide hazard, northern Guadalcanal, Solomon Islands, Unpublished report for United Nations Technical Cooperation for Development, New York, *Department of Scientific and Industrial Research, New Zealand*, 6/89 SOI/87/001.43 (1989).

Advances in Geosciences
Vol. 11: Hydrological Science (2007)
Eds. Namsik Park *et al.*

SCALING PARAMETER ESTIMATION FOR RAINFALL EVENTS IN BANGKOK USING EM–SRE ALGORITHM*

VIRAT CHATDARONG
Water Resources Engineering Department, Faculty of Engineering
Chulalongkorn University, 254 Phyathai Road, Patumvan
Bangkok 10330, Thailand
cvirat@gmail.com

HANSA VATHANANUKIJ
Water Resources Engineering Department, Faculty of Engineering
Kasetsart University, So Phahon Yothin Road
Chatuchak, Bangkok 10900, Thailand
hansa.v@ku.ac.th

PORNPIMOL PIPATTHAWORNSUK
Data and Information Services, Department of Drainage and Sewage
Mitmaitee Road, DinDang, Bangkok 10400, Thailand
p_pipat47@yahoo.com

This paper illustrates the use of EM–SRE algorithm to study the fractal behavior of rainfall event in Bangkok, Thailand. The EM–SRE algorithm for rainfall simulation consists of three components. First, the scale-recursive estimation (SRE) framework is employed to propagate information among many scales given noisy or incomplete observations. Second, the Multiplicative Cascade (MC) rainfall model is used to describe the spatial rainfall structure. Finally, the expectation-maximization (EM) algorithm is integrated for automatic estimation of scaling parameters. In this research, EM–SRE algorithm is employed to evaluate all scaling parameters and study spatial rainfall characteristics in Bangkok from 40 storm events during the 2005 rainy season. The result confirms that the scaling parameters estimated from the EM–SRE algorithm converge quickly after a few iterations. In addition, plots of the total variance at each scale versus the corresponding relative scale size exhibit relatively linear relationships as described by the fractal behavior. Conclusively, the EM–SRE algorithm is an effective yet simple algorithm to simulate spatial rainfall event and describe its spatial structure. The merits of this paper lead to further applications for estimating rainfall in Thailand including spatial rainfall downscaling or rainfall data assimilation from multiple sources of rainfall measurements such as radar and rain-gauges.

*This work was partially supported by the Thai Research Fund under the SSWM project. It was presented in AOGS 4th Annual Meeting (AS07-HS20-A0021) on July 30–August 4, 2007 at the Queen Sirikit National Convention Center, Bangkok, Thailand.

1. Introduction

Rainfall is a complex environmental variable that is difficult to describe either deterministically or statistically. It is controlled by turbulent and chaotic physical processes, varies over a wide range of spatial and temporal scales, and is intermittent. In order to understand rainfall behaviors, many types of instruments are employed to detect and collect rainfall information. However, these instruments vary greatly in their characteristic scales, coverage, and accuracy. For example, rain gauges can provide relatively accurate amount of rainfall at a fixed location while radar gives more coverage but relatively lower accuracy. Therefore, the best estimate of rainfall does not come from one source but from combining multiple sources of measurements to make maximum use of all available information.

The process of combining measurements using a physical or statistic model is normally referred to as data assimilation. Nevertheless, data fusion from various sources may not be easy or straightforward especially for a complex phenomenon such as spatial rainfall. It requires an appropriate assimilation algorithm, rainfall model, and parameter estimation technique that can account for multi-scale natures of the rainfall phenomena and differences in measurement characteristics.

Over the past decades, researchers have proposed many data assimilation techniques to combine measurements from different characteristics, time/spatial scale, and uncertainty. Of many schemes available, the scale-recursive estimation (SRE) has been proven to be one of the most efficient algorithms for modeling various stochastic phenomena given noisy and incomplete observations from multiple scales.[2] To use the SRE framework, a multi-scale stochastic model is required to describe scale-to-scale relationships in a recursive additive form. For precipitation processes, several spatial models were explored in the past. These studies include multi-fractal characterizations,[8] multiplicative cascade model,[11] clustered point processes,[4,9] and wavelet models.[7] However, the model that best suits the SRE framework is the multiplicative cascade (MC) model. The MC rainfall model embedded in the SRE framework has been used in spatial downscaling, model verification,[10] and measurement data assimilation purposes.[3] In addition, it is possible to adapt the expectation-maximization (EM) algorithm in the SRE form to straightforwardly estimate unknown scaling parameters.[4,6] The combined algorithm, named the EM–SRE algorithm, is efficient and simple.

The paper is organized as follows. In the next section, we introduce the SRE representation and outlines of the framework. Section 3 presents the multiplicative rainfall model in the form that fits the SRE algorithm. We also point out some special properties and limitations of this rainfall representation. Next in Sec. 4, we discuss the EM algorithm for estimating the SRE scaling parameters. Then the EM–SRE algorithm is tested in Sec. 5 using synthetic data with known scaling parameters. In Sec. 6, a series of spatial rainfall images from radar station in Bangkok is examined to illustrate the ability of EM–SRE algorithm and observe the fractal behavior of rainfall in Bangkok. Finally, Sec. 7 concludes the paper with suggestions for possible applications and future researches.

2. A Scale-Recursive Representation and Estimation Algorithm

The scale-recursive structure is generally represented by an inverse tree in 1D space or a pyramidal-like grid in 2D space as illustrated in Fig. 1(a) and 1(b), respectively. The top node on the tree, called the "root node," embodies total area of interest, while the bottom nodes, each referred to as a "leaf node," represent random variables at the finest scale where observations are presented.

Let $m(s)$ be a level of an interested node s on an M-level tree, whose root node has $m(s) = 0$ and leaf nodes have $m(s) = M$. From node s at level $m(s)$, we denote its parent at level $m(s) - 1$ by $s\gamma$, and its q children at level $m(s) + 1$ by $s\alpha_i$; $i = 1, \ldots, q$. In addition, let $x(s)$ and $z(s)$ denote

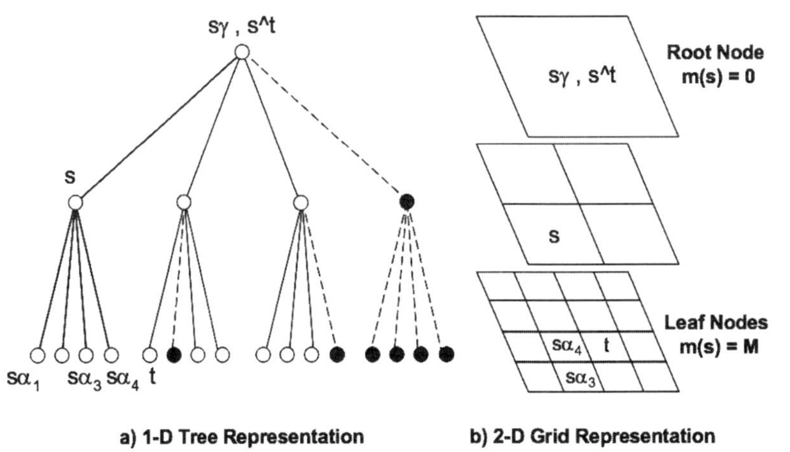

Fig. 1. A scale-recursive representation using (a) M-level inverse-tree and, (b) M-level 2D grid. The black nodes represent pruned nodes that will be excluded from the calculation.

a state vector and a measurement vector at node s, respectively. The scale-recursive dynamic model propagates state information from node $s\gamma$ to s, and the measurement equation linearly relates an observation to a state at node s via:

$$x(s) = F(s)x(s\gamma) + w(s), \tag{1}$$

$$z(s) = H(s)x(s) + v(s), \tag{2}$$

where a process noise, $w(s)$, and a measurement noise, $v(s)$, are independent, zero mean white noise processes with covariance matrices $Q(s)$ and $R(s)$, respectively. The term $F(s)x(s\gamma)$ represents a coarse-to-fine scale prediction with $w(s)$ representing higher resolution details. From Eq. (1), a covariance matrix at node s and a cross covariance matrix between node s and t are in the form:

$$P_x(s) = F(s)P_x(s\gamma)F^T(s) + Q(s), \tag{3}$$

$$C_{xx}(s,t) = \varphi(s, s\wedge t)P_x(s\wedge t)\varphi^T(t, s\wedge t), \tag{4}$$

where node $s\wedge t$ is the finest common predecessor of both node s and node t (see Fig. 1). The function ϕ is given by

$$\phi(s_1, s_2) = \begin{cases} I & s_1 = s_2, \\ F(s_1)\phi(s_1\gamma, s_2) & m(s_1) > m(s_2). \end{cases} \tag{5}$$

Thus, the scale-recursive structure and parameters $F(s)$ and $Q(s)$ govern the correlation among all nodes and implicitly represent the full covariance matrix with a full state consisting of state vectors from all nodes on the tree. Given the scale-recursive model equations (1) and (2), the SRE algorithm optimally estimates a state on each node based on all observations. The SRE algorithm consists of an upward fine-to-coarse filtering sweep followed by a downward coarse-to-fine smoothing sweep. These sweeps are similar to the traditional forward Kalman filtering and backward Rauch–Tung–Striebel (RTS) smoothing, respectively. Reader can refer to Ref. 2 for further details.

It is important to emphasize that if we combine all optimal state estimates calculated from the SRE framework to represent an "update" state vector, it will be exactly equal to the one obtained by performing one step Kalman filtering algorithm using a full covariance matrix explicitly. Estimates from both methods are the best linear least-square estimator and are optimal if all variables are jointly Gaussian. However, because the SRE algorithm never explicitly calculates the full covariance matrix, it is much more efficient and practical for solving large problems. Finally, the

SRE framework can be applied on a pruned tree when some nodes are excluded from the calculation as shown by the black nodes in Fig. 1(a). This pruning technique is beneficial for modeling rainfall with spatial intermittency characteristics.

3. Scale-recursive Description of Rainfall

In order to employ the SRE framework, a multi-scale stochastic model is required to describe scale-to-scale relationships of an interested variable in a recursive additive form according to Eqs. (1) and (2). For precipitation processes, several spatial models were explored in the past, but a group of model that best suits the SRE framework is the multiplicative cascade (MC) model.[5,9] The MC model assumes that rainfall at node s, denoted by capital $X(s)$, is log-normal distributed. It evolves from one scale to the next coarser scale, $X(s\gamma)$, by multiplying with a cascade weight, $W(s)$:

$$X(s) = X(s\gamma) \cdot W(s), \tag{6}$$

where W's are mutually independent random variables. Assume $W(s)$ and $X(s)$ are lognormal distributed, e.g.

$$X(s) = \exp\{x(s) - \sigma_x^2(s)/2\}, \tag{7}$$

$$W(s) = \exp\{w(s) - \sigma_w^2(s)/2\}. \tag{8}$$

Similarly, let the rainfall observation, $Z(s)$, be related to the state $X(s)$, by a log-normal measurement noise, $V(s)$, via

$$Z(s) = X(s) \cdot V(s), \tag{9}$$

Taking the log of Eqs. (6) and (9) gives the additive form noted by Eqs. (1) and (2) with $F(s)$ and $H(s)$ equal to 1.0 at all nodes. In other words, if we denote $x(s)$ and $z(s)$ as a log-rainfall state and a log-rainfall measurement at node s, respectively, the scale-recursive dynamic model and the measurement equation for logarithmic variables of a multiplicative rainfall model are given by

$$x(s) = x(s\gamma) + w(s), \tag{10}$$

$$z(s) = x(s) + v(s), \tag{11}$$

where process noise $w(s)$ and measurement noise $v(s)$ are mutually independent white noises with zero-means and variances $\sigma_w^2(s)$ and $\sigma_v^2(s)$, respectively. Reader can refer to Ref. 3 for more details.

Some properties of the log-rainfall dynamic system on the SRE framework are emphasized. First, the log-rainfall state and log-rainfall measurement are scalar variables representing logarithmic value of rainfall intensity at a node on the grid representation. Therefore, all covariance matrices are scalar and their inverse in the Kalman filter algorithm are easy and cheap to calculate. Second, all parameters are generally assumed to be homogeneous across scales but may vary from scale to scale, e.g. $\sigma_w^2(s) = \sigma_w^2(m(s))$ for all s on level $m(s)$. Let us define scalar parameters $P_x(s) = \sigma_x^2(m(s))$, $Q(s) = \sigma_w^2(m(s))$, and $R(s) = \sigma_v^2(m(s))$ to be the log-rainfall state variance, the process noise variance, and the measurement error variance at any node on the scale $m(s)$, respectively. Hence, on the SRE algorithm there will be $M+k+1$ unknown parameters on any M-level tree when k represents the number of levels that have observations. These unknown parameters are one-$P_x(0)$ at the root node, $M - Q(s)$'s at all transition level $m(s) = 1, \ldots M$, and $k - R(s)$'s at all observation levels $m(s) \in k$.

Finally, since all $F(s)$'s in Eqs. (3)–(5) are 1.0, a cross-covariance between any two nodes is simply a variance at their common parent node. With these constant $F(s)$'s and the scale homogeneous assumption, the scale-recursive representation usually generates a blocky covariance matrix and may be inconsistent with the physical property of rainfall. To illustrate this issue, we again refer to Fig. 1. With the scale recursive representative, the cross-covariance between node $s\alpha_3$ and $s\alpha_4$ is equal to the variance at node s, while the cross-covariance between node $s\alpha_4$ and t is equal to the variance at the root node. It shows that both cross-covariances are not the same even though their physical distances from $s\alpha_3$ to $s\alpha_4$ and from $s\alpha_4$ to t are. As a result, blocky estimated states are usually obtained from the scale-recursive algorithm. However, this error is usually considered minor and may be minimized by using a tall tree-structure with the smallest branch number possible.

4. The Expectation-Maximization Algorithm

Recall that when the MC rainfall model is applied to the SRE framework to estimate the log-rainfall state under a scale-homogeneity assumption, there are $M + k + 1$ unknown parameters that need to be estimated. The accuracy of these unknown parameters determines the quality of rainfall estimates from the SRE algorithm. Although there are a number of parameter estimation techniques, the EM seems to fit the SRE framework best. The EM algorithm derived for the scale-recursive estimation algorithm is effective and straightforward for estimating all unknown parameters. The EM algorithm gives the maximum likelihood estimates by iteratively maximizing the overall expected log-likelihood function of the observation data. The EM algorithm is known to converge but possibly to a local optimum; however, in many speech processing applications, the EM estimation usually converges to an optimum solution within typically a few iterations.[6]

The iteration in the EM algorithm consists of two steps: (1) the expectation or E-step, where conditional expectations of sufficient statistics are computed using the SRE framework; and (2) the maximization or M-step, where those statistics are used to re-estimate the new model parameters. These two steps are repeated until all unknown parameters converge as illustrated in Fig. 2. Details of the EM algorithm for the

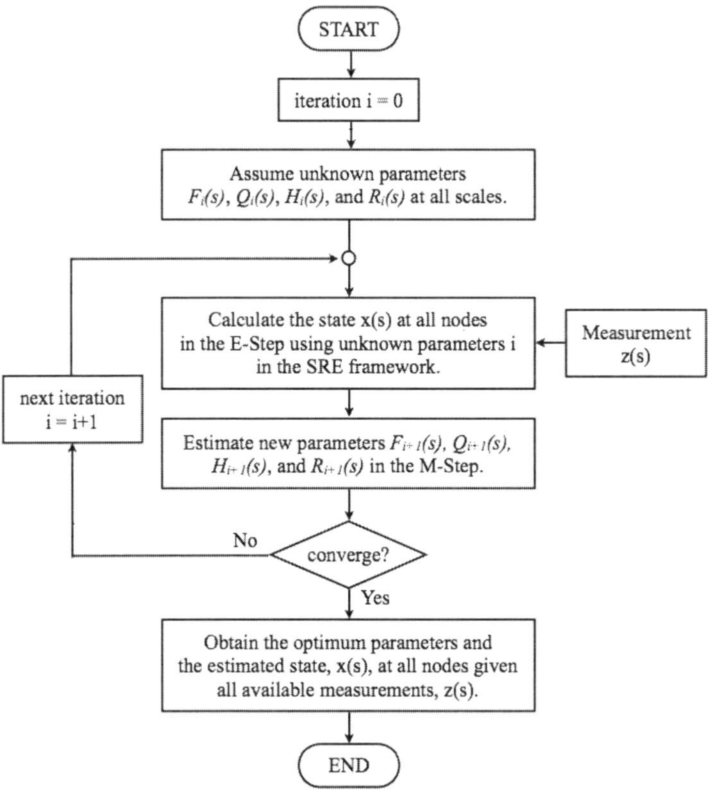

Fig. 2. A schematic diagram for the EM–SRE algorithm to iteratively estimate states $x(s)$ at all nodes and unknown parameters $F(s)$, $Q(s)$, $H(s)$, and $R(s)$ at all scales given all available measurement $z(s)$.

SRE framework and formula for calculating unknown parameters (e.g. $F(s)$, $Q(s)$, $H(s)$, and $R(s)$ in Eqs. (1) and (2)) are nicely summarized in Ref. 6.

For the log-rainfall problem, the parameters to be estimated are a variance at the root node $P_x(0)$, M-process noise variances $Q(s)$ at all transition scales, and k-measurement noise variances $R(s)$ at all observation scales. An independent study by Gupta *et al.* (2004) shows that if all measurement noise variances $R(s)$ are given and the process noise $Q(s)$ is a constant over all scales, this unknown process noise can be estimated from the EM–SRE algorithm. In fact, the EM–SRE algorithm can be used even though $Q(s)$'s at all scales are different because of the spatial correlation pattern of the observation data. Note that the accuracy of the estimated parameters depends on the number of nodes that share the same parameter; hence, parameters of the upper nodes (coarse scales) may be inaccurate because the numbers of nodes are not sufficient.

5. Synthetic Experiment on the EM–SRE Algorithm

The synthetic experiment is performed to test the EM–SRE algorithm before applying it to the real rainfall problem. In the experiment, 100 replicates of synthetic noisy observations are generated from an 8-level tree with branching number of 2×2. The parameters $F(s)$ and $H(s)$ are fixed at 1.0, while the scaling parameters $Q(s)$ and $R(s)$ are arbitrarlly chosen to be different for different scales. A value of 0.1 is used to initialize the EM algorithm. We test the EM–SRE algorithm twice: first given all measurements $z(s)$ at all scales, and second given the measurements $z(s)$ at only the finest scale. In both tests, unknown parameters converge quickly to the true values. Figure 3(a) shows the process noise variances $Q(s)$ and Fig. 3(b) shows the measurement error variances $R(s)$. Note that the estimated parameter at the finest scale is omitted from the plot because it cannot be uniquely identified unless the measurement noise variance $R(s)$ at the finest scale is given.[1]

The error bars in Fig. 3 represent one standard deviation of estimated parameters from their mean using the Monte Carlo technique. These results imply that errors in estimating "true" parameters decrease with increasing scale level. This is because the parameter estimation accuracy depends on the number of nodes that share the same parameter; therefore, parameters at finer scale with many nodes presented are more accurate than those at the coarser scale with only a few nodes. More interestingly,

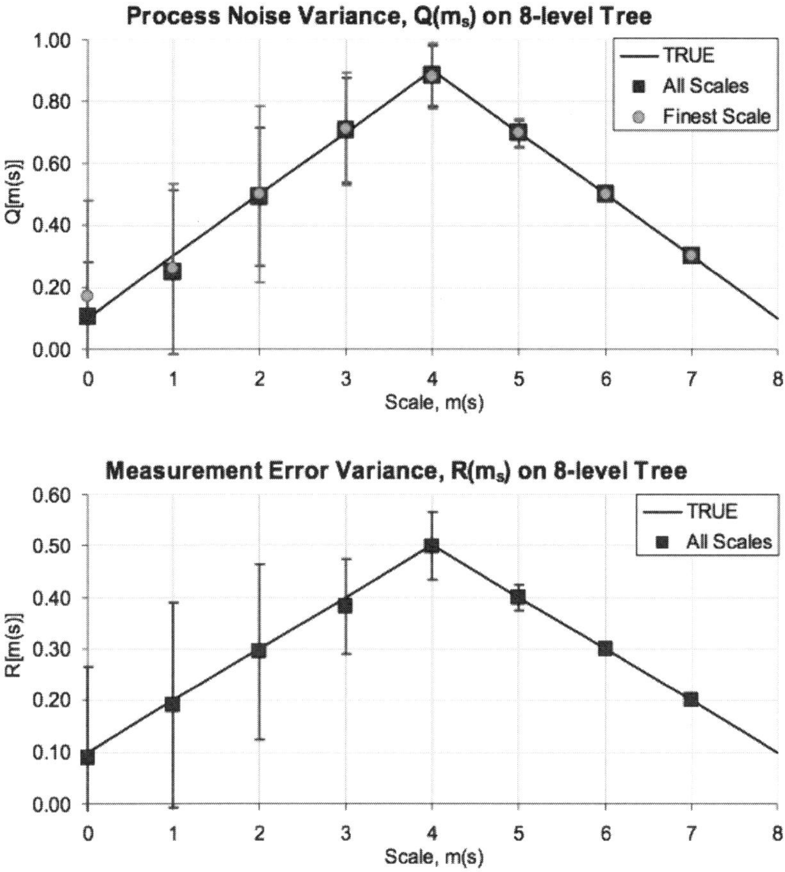

Fig. 3. Estimated parameter $Q(\text{ms})$ and $R(\text{ms})$ from the EM algorithm using 8-level tree. "True", "All Scales", and "Finest Scale" denote true parameter, estimated parameter from observations at all scales, and from observation only at the finest scale, respectively. Estimated parameters at the finest scale are omitted from the plot.

the difference in estimation errors when observations are given at all scales and when they are given only at the finest scale are insignificant. This may be because observations at the finest scale are complete and very accurate; thus, observations at other scales rarely provide additional information. However, when the observations at the finest scale contain many missing data or are inaccurate, having measurements at coarser scales should noticeably improve the accuracy of the algorithm.

6. EM–SRE Algorithm on Rainfall Event in Bangkok

The purpose of this section is to demonstrate that the EM–SRE algorithm is practical to study the scaling characteristics of rainfall events in Bangkok using only one level of real observation data. Although the algorithm is capable of merging observations from multiple scales, this point will not be included in this study. The EM–SRE algorithm is employed to evaluate all scaling parameters (e.g. variance matrices $Q(s)$'s at all scales) of the MC rainfall model for rainfall events over Bangkok and its Metropolitans. Approximately 40 storm events are selected during the 2005 rainy season. The raw data at 1-km radial resolution is provided by the Department of Drainage and Sewage (DDS), Bangkok, in the form of radar reflectivity. The radar reflectivity is converted to rain rate by the conventional Z–R relationship with the constant "a" and "b" equal to 200 and 1.6, respectively. All rainfall images are then tested for the log-normal assumption using the goodness of fit test with type I error of 5%. Because the MC model cannot deal with rainfall intermittency, any nodes where the converted rain rate is less than 0.1 mm/h is pruned out and excluded from the calculation. Most of the spatial rainfall measurements fit log-normal distributions well after dry (no rain) nodes have been removed as shown in Fig. 4. The results suggest that the log-normal assumption in the MC model is acceptable for most rainfall events in Bangkok.

Next, the SRE framework is applied to the converted rain rate images. We construct the 8-level pyramidal-like structure similar to those in Fig. 1(b) with the branch number of 2×2. The finest scale is chosen to be

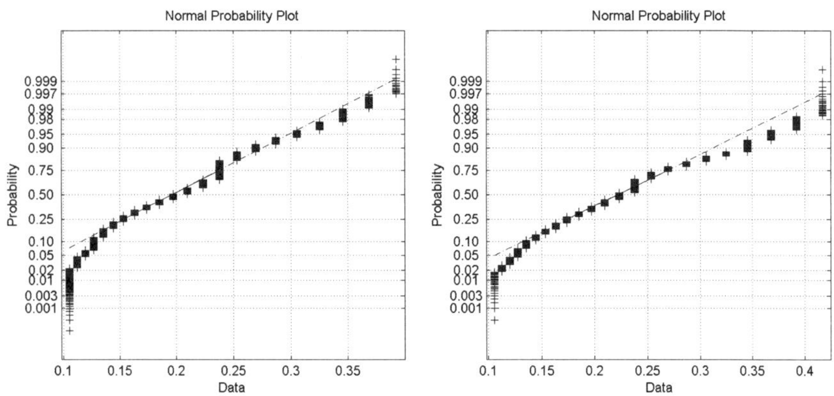

Fig. 4. Samples of normal plots of rain-rate from Prasri Chalearn radar station.

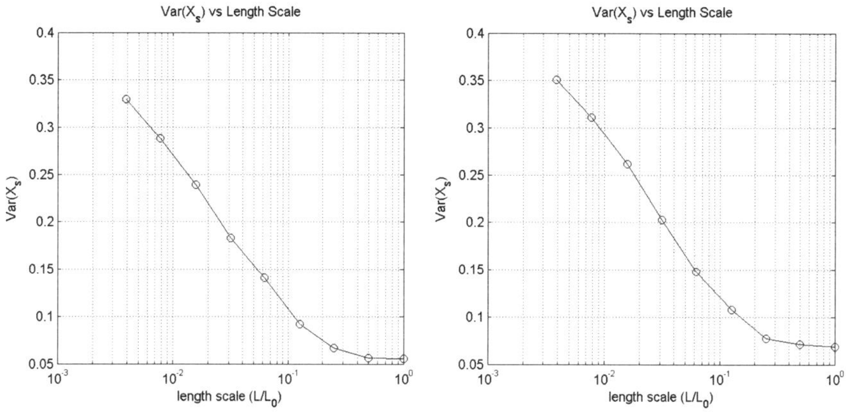

Fig. 5. Plots of the total variance from some sample storm events versus the log of length scale. The plots exhibit almost linear trends at the finer scales for most of the event.

consistent with the radar images at the resolution of $1 \times 1\,\text{km}^2$; hence, the coverage area is equal to $28 \times 28 = 256 \times 256\,\text{km}^2$. The structure is imposed on each storm event over the rainy region. Then the converted radar data is fed to the EM–SRE algorithm in order to produce the scaling parameter $Q(s)$ at each scale. Note that the pruning technique is applied and any cell with zero rainfall intensity is neglected during the calculation. The value of 0.1 is used as an initial condition. In most cases, only a few iterations are required for the parameters to converge.

In studying multi-fractal behavior of rainfall, it is more conventional to observe the total variance, $\text{Var}(x_s)$, at each scale. This total variance of the log rain-rate process is given by the sum of the scaling parameters, $Q(s)$, from the root node to the scale of interest. Figure 5 shows samples of total variance estimated from the EM–SRE algorithm plot versus the length scale defined as a ratio between the cell widths on the level of interest to that on the root node. The results reveal the total variances are almost linearly dependent on the length scale if we neglect the coarser scales where the estimated parameters are inaccurate due to insufficient number of nodes. Thus, it suggests that rainfall events in Bangkok can be described by the multi-fractal approach. Despite similar linear trends, the scaling parameters may vary greatly among storm events with scaling parameters ranging between 0.2 and 0.7. The results agree well with the experiments done by Chatdarong[1] who studies the multi-fractal behavior of the convective storm in the central great plain of the United States.

Nevertheless, more research is needed to identify the causes of variations and to study the sensitivity of these scaling parameters to many factors such as a storm type or an invert-tree structure.

7. Conclusion

This paper illustrates the use of EM–SRE algorithm to study the scaling characteristic of rainfall event in Bangkok. The EM–SRE algorithm is efficient yet simple. And when combined with the MC rainfall model, the EM–SRE algorithm can be used to estimate all scaling parameters and estimates rainfall state given noisy or incomplete measurements at a single or multiple scales. In the study, the EM–SRE converges after a few iterations and gives unknown scaling parameters that are consistent with multi-fractal property of rainfall. Although more research is needed to better understand the scaling property of rainfall, the EM–SRE algorithm and the MC rainfall model can be applied to many rainfall applications. These applications include the spatial rainfall downscaling and rainfall data assimilation. The later application is very useful for merging multiple sources of rainfall measurements and providing the most optimal and accurate estimate of spatial rainfall at any time scale in real time.

Acknowledgments

There are a number of people who have contributed significantly to this research. The authors would like to thank the Department of Drainage and Sewage (DDS), Bangkok, for providing radar data without any cost, the Water Resources System Research Unit and Department of Water Resources Engineering, Chulalongkorn University for providing computers and supporting the experimental and publication cost, Prof. Jun Matsumoto and MAHASRI group for organizing the session in AOGS 2007 and support on the presentation. This research was partially funded by the Karnchana Pisek Somphoat Fund, Chulalongkorn University.

References

1. V. Chatdarong, Multi-sensor rainfall data assimilation using ensemble approaches, in *Civil and Environmental Engineering* (Massachusetts Institute of Technology, Cambridge, MA, 2006) p. 204.

2. K. C. Chou, A. S. Willsky and A. Benveniste, Multiscale recursive estimation data fusion and regularization, *IEEE Trans. Automat. Contr.* **37**(3) (1994) 464–478.
3. I. P. Gorenburg, D. Mclaughlin and D. Entekhabi, Scale-recursive assimilation of precipitation data, *Adv. Water Resour.* **24** (2001) 941–953.
4. R. Gupta, V. Venugopal and E. Foufoula-Georgiou, A methodology for merging multisensor precipitation estimates based on expectation-maximization and scale-recursive estimation, *J. Geophys. Res.* **111**(D02102) (2006) 1–14.
5. V. K. Gupta and E. C. Waymire, A statistical analysis of mesoscale rainfall as a random cascade, *J. Appl. Meteorol.* **32** (1993) 251–267.
6. A. Kannan, M. Ostendorf, W. C. Karl, D. A. Castanon and R. K. Fish, ML parameter estimation of a multi-scale stochastic process using the EM algorithm, *IEEE Trans. Sig. Proc.* **48**(6) (2000) 1836–1840.
7. P. Kumar and E. Foufoula Georgiou, A multi-component decomposition of spatial rainfall fields: 1. Segregation of large and small scale features using wavelet transforms, *Water Resour. Res.* **29**(8) (1993) 2515–2532.
8. D. Marsan, D. Schertzer and S. Lovejoy, Casual space-time multifractal processes: Predictability and forecasting of rainfall fields, *J. Geophys. Res.* **101**(D21) (1996) 26333–26346.
9. T. M. Over and V. K. Gupta, Statistical analysis of mesoscale rainfall: Dependence of a random cascade generator on large scale forcing, *J. Appl. Meteorol.* **33**(12) (1994) 1526–1542.
10. B. Tustison, E. Foufoula-Georgiou and D. Harris, Scale-recursive estimation for multisensor quantitative precipitation forecast verification: A preliminary assessment, *J. Geophy. Res.* **108**(D8) (2003) 8377–8390.
11. E. Waymire, V. K. Gupta and I. Rodriguez-Iturbe, A spectral theory for rainfall intensity at the meso-beta scale, *Water. Resour. Res.* **20** (1984) 1453–1465.

Advances in Geosciences
Vol. 11: Hydrological Science (2007)
Eds. Namsik Park *et al.*

STOCHASTIC GENERATION OF MONTHLY RAINFALL DATA

R. SRIKANTHAN
Water Division, Bureau of Meteorology, GPO Box 1289
Melbourne 3001, Australia
r.srikanthan@bom.gov.au

Monthly rainfall data is generally needed in the simulation of water resources systems, and in the estimation of water yield from large catchments. Monthly streamflow data generation models are usually applied to generate monthly rainfall data, but this presents problems for most regions, which have significant months of no rainfall. In an earlier study, Srikanthan *et al.* (*J. Hydrol. Eng., ASCE* **11**(3) (2006) 222–229) recommended the modified method of fragments to disaggregate the annual rainfall data generated by a first-order autoregressive model. The main drawback of this approach is the occurrence of similar patterns when only a short length of historic data is available. Porter and Pink (*Hydrol. Water Res. Symp.* (1991) 187–191) used synthetic fragments from a Thomas–Fiering monthly model to overcome this drawback. As an alternative, a new two-part monthly model is nested in an annual model to generate monthly rainfall data which preserves both the monthly and annual characteristics. This nested model was applied to generate rainfall data from seven rainfall stations located in eastern and southern parts of Australia, and the results showed that the model performed satisfactorily.

1. Introduction

A number of models have been developed to generate monthly streamflows for use in design and operation of water resources systems. These models can be broadly classified into two categories, namely, monthly flow models and disaggregation processes. A monthly flow model, as its name implies, generates monthly flows using an autoregressive model.[1] In its normal application, it does not preserve the annual characteristics. This inadequacy was overcome by Harms and Campbell[2] by an adjustment procedure. However, for highly variable rivers, the adjustment procedure greatly distorted the monthly characteristics. Srikanthan[3] modified the adjustment procedure to overcome this deficiency. Two different methods are available to disaggregate the annual flows: one uses the distribution of historical monthly flows within a year[3,4] whereas the other uses a set of coefficients, usually in the form of two matrices[5] obtained from the historical monthly flows. Srikanthan and McMahon[6] compared the above methods and their variations and recommended the method of fragments to disaggregate the annual rainfall data generated by a first-order autoregressive model. The main drawbacks of this approach are the inability to preserve the monthly correlation between the first month of a year and last month of the previous year and the occurrence of similar patterns when only a short length of historic data is available. Maheepala and Perera[7] proposed a modification to the selection of fragments that preserves the year-end monthly correlation to improve on the first shortcoming. Porter and Pink[8] used synthetic fragments from a Thomas–Fiering monthly model to overcome the second drawback. For sites with a considerable number of zero rainfall months, there will be problems with the application of the Thomas–Fiering monthly model to generate the synthetic fragments. Recently, Sharma and O'Neil[9] developed a nonparametric approach to model the interannual dependence in monthly streamflows while Bayazit and Aksoy[10] proposed a method based on wavelets. The nonparametric method was compared with the modified method of fragments and found to give similar results.[11] The wavelet model tended to give the same historical mean value for all the replicates when the replicate length is the same as the historical record. This is because all the wavelets are selected once in each of the replicates generated. This is a major drawback with the wavelet model.

In this paper, a simple monthly data generation model, which can handle zero rainfall months, is developed and nested in an annual model. Consequently, the developed model is able to preserve the monthly as well as the annual characteristics of the historical rainfall data. In addition, it overcomes the problem of the same within a year pattern being repeated from short historical records in the method of fragments.[11] The model is applied to generate rainfall data from seven rainfall stations located in various parts of Australia. The results from this model are compared with the historical values.

2. Rainfall Data

Seven rainfall stations were selected to represent the rainfall variability of the Australian continent. The locations of the selected rainfall stations are shown in Fig. 1 while the details are shown in Table 1. The number of months of zero rainfall varies from 0% to 90%.

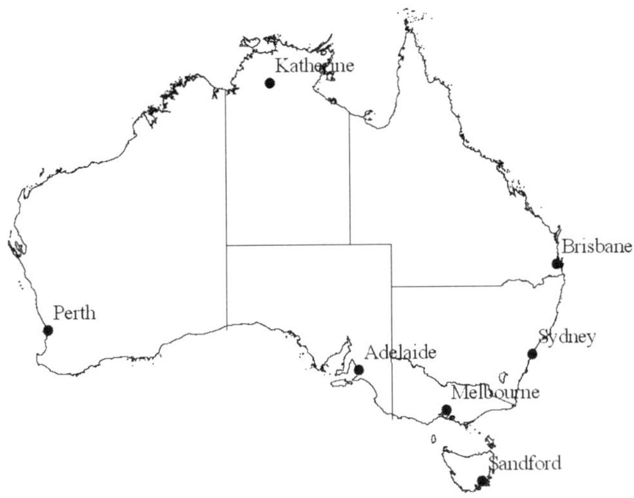

Fig. 1. Location of the rainfall stations used in the study.

Table 1. Details of the rainfall stations selected.

Number	Name	Latitude (degree)	Latitude (degree)	Length of record (years)	Annual mean (mm)	Range of no rainfall months (%)
23,000	Adelaide	−34.925	138.587	140	530	0–5
40,214	Brisbane	−27.478	153.031	134	11,548	0–4
14,902	Katherine	−14.459	132.257	112	975	0–90
86,071	Melbourne	−37.807	144.970	144	663	0–1
9034	Perth	−31.950	115.867	116	868	0–11
94,061	Sandford	−42.928	147.517	112	576	0–1
66,062	Sydney	−33.861	151.205	141	1225	0–1

3. Nested Monthly Model

In the method of fragments an annual rainfall model is first used and then the generated annual rainfall data are disaggregated into monthly rainfall using the historical patterns. The model developed in this paper generates monthly rainfall first and then the generated monthly rainfalls are adjusted by using an annual model. As noted before there are difficulties in generating monthly rainfall using the Thomas–Fiering monthly model due to the large number of no rainfall months. This problem is overcome by using a new model similar to a two-part daily rainfall model to generate the monthly rainfall. The correlation between monthly rainfalls is modeled by using a lag one autoregressive model. The annual model is driven by the noise term derived from the sum of the 12 monthly rainfalls in a year and this enables the adjustments made to monthly rainfall small. The model developed is referred to as the nested monthly model.

3.1. *Monthly rainfall model*

The nested monthly model consists of a monthly model to generate monthly data, which in turn drives an annual model. Since monthly data is highly variable and may have a considerable number of months of zero rainfall in certain locations, a new model is developed in this paper. To model the appropriate number of months of zero rainfall, the monthly data is generated into two parts. In the first part, it is decided whether the month has zero or nonzero rainfall. From the historical data, the probability of no rainfall in each of the 12 months is calculated. By generating a uniformly distributed random number between 0 and 1, the state of the following month is determined. If the state is dry, then zero rainfall is assigned to the month. For months with nonzero rainfall, the amount of rainfall is generated as follows. Standard normally distributed random numbers (η) with zero mean and unit variance are first generated. Monthly correlations are imparted into the numbers by using a first-order autoregressive model,

$$\varepsilon_j = r_j \varepsilon_{j-1} + (1 - r_j^2)^{1/2} \eta, \tag{1}$$

where ε_j are standardized normally distributed numbers with the monthly correlation r_j for month j and η is a normally distributed random number. The monthly correlation is estimated from the historical monthly rainfall data. If the monthly data is normally distributed, then it can be obtained by imparting the mean and standard deviation.

$$x_j = \bar{x}_j + s_j \varepsilon_j, \tag{2}$$

where $\bar{x}_j$ and s_j are the monthly mean and standard deviation of historical monthly data for month j. In practice, this is rarely the case. The monthly data in general will be skewed. To take skewness into account, a Gamma distribution was used. The shape and scale parameters of the Gamma distribution are estimated by the method of moments. To generate monthly rainfall, the normally distributed number ε_j in Eq. (1) is transformed into a uniformly distributed number u_j (0,1), and a Gamma variate is obtained corresponding to u_j. Alternatively, one can use a Box–Cox transformation[12] to get skewed monthly values.

3.2. *Annual rainfall model*

Once the monthly values for the 12 months in a year have been generated, they are aggregated to form the annual rainfall. The aggregated annual total ($\tilde{Z}_i$) is standardized to have zero mean and unit variance,

$$\tilde{z}_i = \frac{\tilde{Z}_i - \mu'}{\sigma'}, \tag{3}$$

where $\tilde{z}_i$ is the standardized value for year i, and μ' and σ' are the mean and standard deviation of the aggregated annual values. The mean (μ') and standard deviation (σ') of the aggregated annual values can be calculated from the generated annual values.

The standardized annual value is then adjusted by using an annual model. In this case a first-order autoregressive model is used:

$$z_i = r z_{i-1} + (1 - r^2)^{1/2} \tilde{z}_i, \tag{4}$$

where z_i is the adjusted standardized annual value and r is the lag one autocorrelation coefficient estimated from the historical annual rainfall. This is a new approach in which the random noise term in an autoregressive model is replaced by the standardized annual value.[13] The annual rainfall is then obtained from

$$Z_i = \mu + z_i \sigma, \tag{5}$$

where μ and σ are respectively the mean and standard deviation of the historical annual rainfall data. The generated monthly data is then adjusted by the ratio $Z_i/\tilde{Z}_i$. Since the annual rainfall is driven by the monthly model, the adjustment is usually small and hence the impact on the monthly characteristics is small.

4. Model Evaluation

One hundred replicate, each of the length equal to the historical record, were generated for each site. The model evaluation was carried out at annual and monthly time periods. The statistics used to evaluate the annual level are the annual mean, standard deviation, coefficient of skewness, lag one autocorrelation coefficient, extreme events, adjusted range, and low rainfall sums. The maximum and minimum monthly rainfall depths occurring in the historic record or in each of the generated sequences are taken as the extreme events. Rank 1-, 2-, 3-, 5-, 7-, and 10-year low rainfall sums are used. The adjusted range (R) is obtained from

$$R = \max\{D_k\} - \min\{D_k\}, \quad k = 1, 2, \ldots, n, \tag{6}$$

where

$$D_k = \sum_{i=1}^{k} (Z_i - \bar{Z}) \quad \text{and} \quad \bar{Z} = \frac{1}{n}\sum_{i=1}^{n} Z_i.$$

At the monthly level, the following six statistics are used:

(1) monthly standard deviations;
(2) monthly coefficients of skewness;
(3) serial correlation coefficients between successive months;
(4) maximum monthly rainfall;
(5) minimum monthly rainfall;
(6) relative frequency of no rainfall months.

The serial correlation coefficient corresponding to month j is the correlation between the j and the $(j-1)$ monthly pairs.

5. Discussion of Results

The annual and monthly statistics mentioned in the above section were estimated from each of the 100 replicates for both the models. For each statistic, the mean, median, 2.5-, 25-, 75-, and 97.5-percentile values are obtained for comparison. Only the mean values are presented along with the corresponding historical values in this paper.

5.1. *Annual statistics*

The average values of annual mean, standard deviation, coefficient of skewness, and lag one autocorrelation coefficient from 100 replicates are presented in Table 2. It can be seen from this table that the model preserves all the four statistics satisfactorily except the skewness in a few cases. Table 3 shows that the extreme events are reproduced, however the adjusted range is not reproduced satisfactorily. The low-rainfall sums are also reproduced (Table 4).

Table 2. Comparison of historical and generated annual mean, standard deviation, skewness, and lag one autocorrelation coefficient.

Station name	Mean (mm)		Std dev. (mm)		Skewness		Lag 1 autocorrel.	
	Hist	Gen	Hist	Gen	Hist	Gen	Hist	Gen
Adelaide	530	529	108	108	0.057	0.344	−0.010	−0.030
Brisbane	1154	1154	358	354	0.592	0.573	0.016	0.019
Katherine	975	977	251	253	0.046	0.514	0.069	0.044
Melbourne	663	664	116	115	0.098	0.414	0.098	0.094
Perth	868	870	162	163	0.103	0.341	−0.051	−0.037
Sanford	576	577	131	131	0.432	0.384	0.011	0.019
Sydney	1225	1232	331	331	0.607	0.511	0.101	0.092

Table 3. Comparison of historical and generated annual maximum, minimum, and adjusted range.

Station name	Maximum (mm)		Minimum (mm)		Range (mm)	
	Hist	Gen	Hist	Gen	Hist	Gen
Adelaide	786	851	258	285	1205	1477
Brisbane	2243	2291	412	435	7487	4801
Katherine	1575	1739	440	460	4236	3209
Melbourne	927	1013	405	411	687	1721
Perth	1339	1339	509	510	3101	1992
Sanford	923	955	326	294	1639	1645
Sydney	2194	2283	583	544	7089	4819

Table 4. Comparison of historical and generated 2- 5-, and 10-year low rainfall sums.

Station name	2-year (mm)		5-year (mm)		10-year (mm)	
	Hist	Gen	Hist	Gen	Hist	Gen
Adelaide	735	705	2129	2103	4431	4565
Brisbane	1126	1218	4161	4027	8920	9165
Katherine	1144	1168	3516	3611	8028	8055
Melbourne	976	930	2723	2684	6214	5811
Perth	1250	1218	3484	3558	7645	7659
Sanford	762	729	2254	2224	5029	4908
Sydney	1547	1373	4473	4392	9737	9866

5.2. *Monthly statistics*

The mean of each of the seven monthly statistics estimated from the 100 replicates was compared with the corresponding historical values in Figs. 2–4. The monthly mean and standard deviation are reproduced satisfactorily as shown in Fig. 2. Figure 3 shows that the historical and generated coefficients of skewness and correlation between months match well. The extreme rainfall and the number of months of no rainfall are also preserved satisfactorily as shown in Fig. 4.

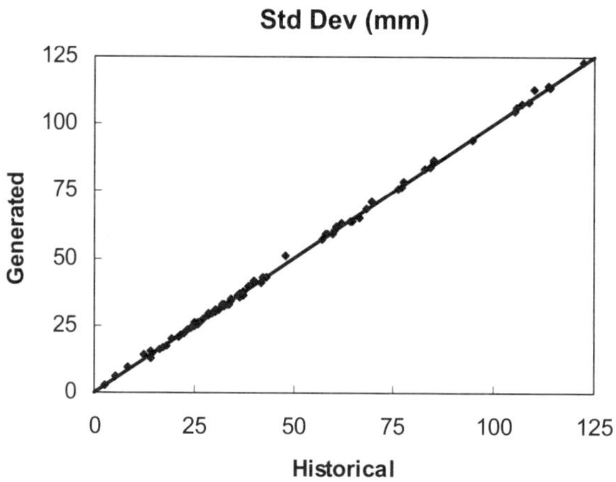

Fig. 2. Comparison of historical and generated monthly standard deviation.

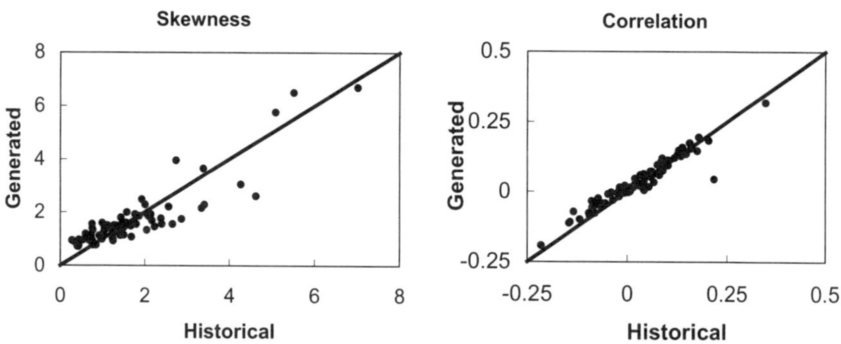

Fig. 3. Comparison of historical and generated coefficient of skewness and correlation between monthly rainfall.

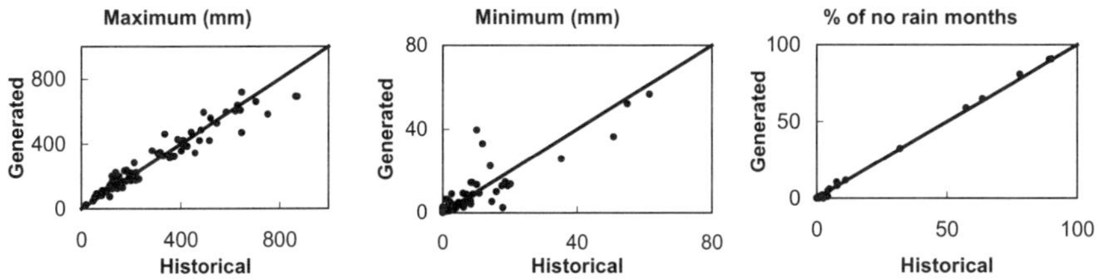

Fig. 4. Comparison of historical and generated extreme rainfall and number of months of no rainfall.

6. Conclusions

A nested monthly model is developed for the generation of monthly rainfall data. Monthly rainfall is first generated using a Gamma distribution, taking into account months with no rainfall. The monthly correlation is modeled by a first-order Markov model. The generated monthly rainfalls are used to drive an annual model to preserve the annual characteristics. The annual rainfall resulting from the annual model is used to adjust the monthly rainfall. The nesting procedure ensures that both the monthly and annual characteristics are preserved. The developed model is applied to seven rainfall stations located in various parts of Australia and was found to perform satisfactorily. The model is simple, performs satisfactorily, and is also easy to implement, and therefore, it is recommended for the generation of monthly rainfall data.

References

1. H. A. Thomas and M. P. Fiering, Mathematical synthesis of streamflow sequences for the analysis of rive basins by simulation, in *Design of Water Resource Systems*, eds. A. Maas *et al.* (Harvard University Press, Cambridge, 1962), Chap. 12.
2. A. A. Harms and T. H. Campbell, An extension to Thomas–Fiering model for the sequential generation of streamflows, *Water Resour. Res.* **3**(3) (1967) 653–651.
3. R. Srikanthan, Stochastic generation of annual and monthly flow volumes, PhD thesis, Unpublished, Monash University, Clayton, Australia (1979).
4. G. G. Svanidze, Osnovy rascheta regulirovaniya rechnogo stoka metodom Monto-Karlo, *Izd. AN GSSR Tbilisi* (1964).
5. D. Valencia and J. C. Schaake Jr., Disaggregation process in stochastic hydrology, *Water Resour. Res.* **9**(3) (1973) 580–585.
6. R. Srikanthan and T. A. McMahon, Stochastic generation of rainfall and evaporation data, *AWRC Technical Paper No. 84* (1985), p. 301.

7. S. Maheepala and C. J. C. Perera, Monthly hydrologic data generation by disaggregation, *J. Hydrol.* **178** (1996) 277–291.
8. J. W. Porter and B. J. Pink, A method of synthetic fragments for disaggregation in stochastic data generation, in *Hydrology and Water Res. Symp.* (Institution of Engineers, Australia, 1991), pp. 187–191.
9. A. Sharma and R. O'Neill, A nonparametric approach for representing interannual dependence in monthly streamflow, *Water Resour. Res.* **138**(7) (2002) 1–10.
10. M. Bayazit and H. Aksoy, Using wavelets for data generation, *J. Appl. Stat.* **28**(2) (2001) 157–166.
11. R. Srikanthan, T. A. McMahon and A. Sharma, Comparison of two nonparametric alternatives for stochastic generation of monthly rainfall, *J. Hydrol. Eng., ASCE* **11**(3) (2006) 222–229.
12. G. E. P. Box and D. R. Cox, An analysis of transformations, *J. Royal Stat. Soc., Ser. B* **26** (1964) 211–243.
13. R. Srikanthan, Stochastic generation of daily rainfall data using a nested model, in 57th Canadian Water Resour. Assoc. Ann. Cong. 16–18 June 2004, Montreal, Canada.

Advances in Geosciences
Vol. 11: Hydrological Science (2007)
Eds. Namsik Park *et al.*

HYDROLOGIC TIME SERIES DATA MODELING USING MULTIPLICATIVE ARIMA

CHAKKRAPONG TAEWICHIT
Office of Engineering and Architecture, Royal Irrigation Department
Ministry of Agriculture and Cooperatives, Bangkok, Thailand
chakkrapong@hotmail.com

SUWATANA CHITTALADAKORN
Department of Water Resources and Engineering
Kasetsart University, Bangkok, Thailand
fengswc@ku.ac.th

This paper presents a methodology on modeling of historical hydrologic data as monthly inflow time series. The stochastic model known as time series model or Box–Jenkins multiplicative ARIMA (autoregressive integrated moving average) model was used for simulating monthly stream flow time series. The parametric and the non-parametric tests were preceded for checking trend and seasonal effects in time series before selecting the model. Diagnostic checks were analyzed for the suitable model that is selected from autocorrelation function and partial autocorrelation function which provided minimum sum square of residual error (SSE). The fittest model was used in order for short-term forecasting of future stream flow for 12 months. For further application, the forecasted inflow was inputted into Genetic Algorithm (GA) model for optimal reservoir operational rule curve, and applied to the studied case of Thap Salao Reservoir Project. The results indicate that monthly stream flow time series is identified as non-stationary, and Box–Jenkins multiplicative ARIMA modeling is suitable for monthly stream flow time series forecasting. The time series pattern of ARIMA$(2,0,2)(2,1,2)_{12}$ was found as an appropriate model for inflow forecasting of this case. And the optimal reservoir operational rule curve corresponding to the changes in hydrological environment could be found by the GA model appropriately.

1. Introduction

A mathematical model representing a stochastic process is called a "stochastic model" or "time series model." In the stochastic approach, a type of model is assumed aimed to represent the most relevant statistical characteristics of the historic series.[9] There are several types of stochastic models which can be used for representing a time series such as the Markov chain model and the autoregressive model. In this case, Box–Jenkins multiplicative ARIMA (autoregressive integrated moving average) model is an efficient method used in order to identify suitable model of time series data using autocorrelation function (ACF) or correlogram r_k and partial autocorrelation (PACF) or correlogram r_{kk}. The forecasted values can be estimated depending upon previous observed data and previous forecasted residual error. Box–Jenkins technique consists of four major stages. They are model selection and identification, the estimation of the model parameters, diagnostic checking of model reliability, and model forecasting.[1] Generally, different kinds of effects such as trend, seasonal, and cycle effects are concealed in time series data. These effects cause non-stationary in time series data, and cannot be analyzed using Box–Jenkins technique. The decomposition analysis has to be proceeded first before following the Box–Jenkins process by using time series transformation in which the original time series is transformed using natural logarithms, and first or second differences.[2] The backward difference was practiced in this paper. Actually, the decomposition of those effects in time series is very difficult to practice. Hence, the statistical analysis, the well-known method, was applied to prove whether it has trend and seasonal effects in the observed time series or not. Various methods were used. In fact, there are parametric test, non-parametric test, and Box–Ljung in which the use of testing hypothesis always has to be applied.

This paper presents the application of time series modeling for hydrologic time series data in water resources engineering focusing on monthly short-term stream flow time series forecasting. This was used in order to identify the short-term optimal reservoir operational rule curve based on assumption that the forecasted stream flow time series is likely to be a key role in the changes of short-term optimal reservoir operational rule curve. Since water resources are limited, the conventional reservoir operational policy is difficult to define to meet the highest efficient management. Hence, the accurate forecasted stream flow time series and the optimal operational policy are the efficient ways to meet that goal.

Optimal reservoir operational rule curve can be identified by using genetic algorithms simulating reservoir water release based on hedging rule. The hedging rule has a unique buffer storage that is preserved to reduce water shortage risk.

The genetic algorithm (GA) uses a search procedure that is generally able to find global optima in complex task domains (Schaffer, 1987 cited in Ref. 3). The search procedures are based on the processes and mechanisms of natural genetic and evolutionary process.[3] GA uses probabilistic and solution is simultaneously search from search space using objective function.[6] The three major processes of GA consist of reproduction, crossover, and mutation depending upon probability.[6] In this paper, Tournament selection with pair of chromosomes or decision variables or lower and upper rule curves were randomly selected for evaluating better fitness,[7] uniform crossover, and modified uniform mutation[8] were used to improve the better solution.

2. The Methodology

- The Thab Saloa irrigation water supply and maintenance project was the pilot project for this study.
- Collecting climatological data, water demand, and monthly stream flow time series data of the pilot project. Fourteen years (Jan. 1989 to Dec. 2002) time series data (stream flow and precipitation) were used for this study.
- Checking consistency and reliability of time series data using double mass curve and single mass curve analysis.
- Non-consistency data and missing data were collected and estimated by slope-ratio method, HEC-4, average and normal ratio methods.[11]
- Identifying suitable multiplicative ARIMA model using Box–Jenkins technique; then, forecasting future stream flow of 12 months (Jan. 2002 to Dec. 2002) using the Box–Jenkins technique. After that, evaluating observed stream flow and forecasted stream flow that entered into the reservoir.
- Identifying the optimal reservoir operational rule curve using GA simulating reservoir water release of 13 years without forecasting future stream flow, together with the studied case of 14 years that includes forecasted future stream flow, and then observing the changes.

3. Pilot Area

The Thab Saloa irrigation water supply and maintenance project is located at Ban Rabum, Rabum sub district, Lansuk district, Authaithani province, Thailand. This project has the responsibility for regulating water release from Thab Saloa reservoir. The Thab Saloa reservoir has a watershed area of about 534 square kilometers. Almost all areas are tropical evergreen forest.

The retention storage and dead storage are 160 MCM (+155.00 mMSL) and 8 MCM (+139.80 mMSL), respectively.

4. Stream Flow Forecasting Using Multiplicative ARIMA Model

Box–Jenkins Technique was used in order to check the stationary of stream flow time series. The effects often caused by non-stationary are trend and seasonal effects. For instance, ACF and PACF were used in order to check trend and seasonal effects of time series. Furthermore, the non-parametric test and the parametric test, together with testing of hypothesis were also applied for trend and seasonal effect analyses. There are runs, turning point, Sign, Daniel's, Kendall's tau, Von Neuman's ratio, rule of thumb, and Box–Ljung tests for trend effect analysis. There are Kruskal and Wallis, and rule of thumb for seasonal effect analysis.[10] Those effects have to be eliminated before selecting a suitable model.

The comparison of correlogram of ACF and PACF of stream flow time series to standard pattern were proceeded for selecting a suitable model. The possible model may have autoregressive, moving average, or mixed autoregressive and moving average.[5] The estimation of model parameters was proceeded using the least square method. Diagnostic checks for suitable model was proceeded by checking minimum sum square of residual error (SSE) and correlogram of residual error based on Box–Ljung test.

5. Results and Discussion

Results of stationary checking, which have to be proceeded first, are described below.

5.1. *Trend effect analysis using non-parametric test*

Runs and Turn point tests show that Z values equal to -5.301 and -4.902, respectively. Both methods provide $|Z| > Z_{\alpha/2}(1.96)$ at 0.5% significant level. Hence, both methods rejected H_0. This means that the time series has the trend effect.

Sign, Daniel, and Kendall tests show that Z values equal to 0, -0.149, and -0.00828, respectively. Those methods provide $|Z| < Z_{\alpha/2}(1.96)$ at 0.5% significant level. Hence, those methods accepted H_0. This means that the time series does not have any trend effect.

5.2. *Trend effect analysis using parametric test*

Von Neumann test shows that M value equals to 1.178. This test provides $M < M_{1-\alpha/2}$ (1.738) at 1% significant level. Hence, H_0 is rejected. This means that the time series has seasonal effect.

Rule of thumb test shows that almost all of 24 values of lag k provide $|r_k| < Z_{\alpha/2}/\sqrt{n}$ (0.1569), whereas only at lag k 1, 12, 13, and 24, with make up 0.4103, 0.4562, 0.2398, and 0.1715, respectively, provide $|r_k| > Z_{\alpha/2}/\sqrt{n}$ (0.1569) at 0.5% significant level. Hence, H_0 is accepted. This means that the time series has no trend effect, but there is seasonal effect.

Box–Ljung test shows that overall lag k provides $Q'_m > \chi^2_{\alpha,m}$ at 0.5% significant level, where m is k. Hence, H_0 is rejected. This means that the time series has the seasonal effect.

5.3. *Seasonal effect analysis using non-parametric test*

The Kruskall and Wallis test was used for checking both positive seasonal effect and multiple seasonal effect. The results show that the positive effect checking provides H value equal to $67.50 > \chi^2_{\alpha,L-1}$ (19.675) at 0.5% significant level. Hence, H_0 is rejected. This means that the time series has positive seasonal effect, whereas negative effect checking provides H value equal to $-156 > \chi^2_{\alpha,L-1}$ (19.675) at 0.5% significant level. Hence, H_0 is accepted. This means that the time series does not have multiple seasonal effects.

5.4. *Seasonal effect analysis based on parametric test*

Rule of thumb test was used for checking both positive seasonal effect and multiple seasonal effect by considering 24 values of lag k. The positive seasonal checking shows that almost all of lag k provide $|r_k| < Z_\alpha/\sqrt{n}(0.137)$, whereas only at lag k 1, 12, 13, and 24, provide $|r_k| > Z_\alpha/\sqrt{n}$ (0.137) at 0.5% significant level, with make up 0.336, 0.493, 0.264, and 0.232, respectively. Hence, H_0 is rejected. This means that the time series has positive seasonal effect.

The multiple seasonal checking shows that almost all lag k provide $|r_k| < Z_\alpha/\sqrt{n}$ (0.137), whereas only at lag k 1, 12, 13, and 24, provide $|r_k| > Z_\alpha/\sqrt{n}$ (0.137) at 0.5% significant level, with make up 0.208, 0.800, 0.198, and 0.648, respectively. Hence, H_0 is rejected. This means that the time

Table 1. Trend effect analysis using non-parametric and parametric test.

Type of testing	Results	Type of testing	Results
Non-parametric		*Parametric*	
Runs	Trend effect	Von Neumann	Trend effect
Turning point	Trend effect	Rule of thumb	No trend effect, but it has seasonal effect
Sign	No trend effect		
Daniel's	No trend effect	Box–Ljung	Seasonal effect
Kendall's tau	No trend effect		

Table 2. Seasonal effect analysis using non-parametric and parametric test.

	Results	
Type of testing	Positive	Multiple
Non-parametric		
Kruskall and Wallis	Seasonal effect	No seasonal effect
Parametric		
Rule of thumb	Seasonal effect	Seasonal effect

series also has multiple seasonal effect. Those results can be summarized as shown in Tables 1 and 2, respectively.

5.5. *Identifying ARMA or multiplicative ARIMA*

As shown in Figs. 1 and 2, the correlogram of ACF and PACF of 13 years of stream flow time series (Jan. 1989 to Dec. 2001) were calculated until lag k equals to 120.

The characteristics of time series that has no trend effect, in which correlogram of ACF and PACF is equal to zero, were checked by using a t-test. The critical t value in which degree of freedom (DF) equal to 155 at 0.5% significant level is 1.96, if calculated $|t| < 1.96$, this means that ACF is zero. As shown in Fig. 1, ACF is high at lag 1; it also has a rapid decrease, and converges into zero (cut off) at lag k 2. This means that the time series has no trend effect; the checking was continuously proceeded by considering ACF at lag k 12, 24, 36, up to 120. This indicates that every lag k equals to 12 has the same characteristic, it means that the time series has seasonal effect. In addition, both correlogram ACF and PACF have a gradual decrease, and converged into zero or dies down exponentially. The general term of the model can be identified as ARIMA(p, d, q)x$(P, D, Q)_{12}$ which

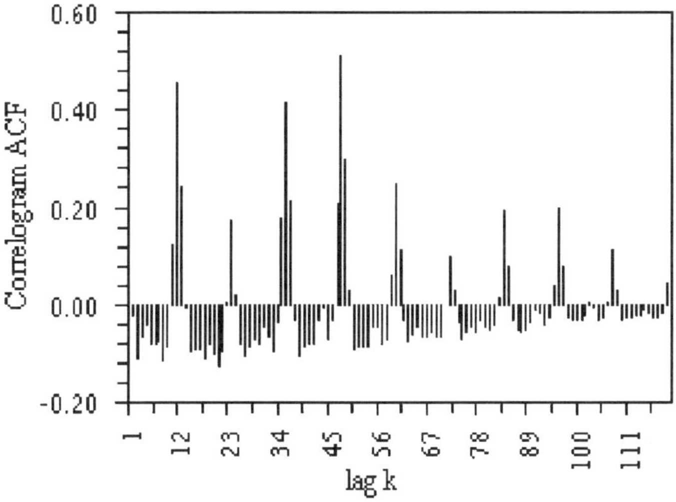

Fig. 1. Correlogram ACF (r_k) of stream flow time series before transform.

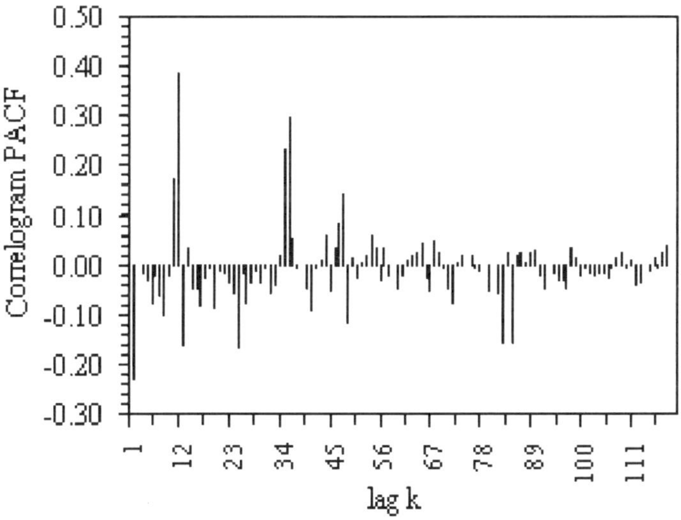

Fig. 2. Correlogram PACF (r_{kk}) of stream flow time series before transform.

includes seasonal effect, parameters d and D are numbers or order of time series transformation for eliminating trend and seasonal effects. The results already mentioned above indicate that it has no trend effect. Hence, d equals zero.

The results as shown in Tables 1 and 2 indicate: three cases have trend effect, four cases have no trend effect, five cases have seasonal effect, and one case has no seasonal effect, together with the results from considering correlogram of ACF and PACF in which the time series has no trend, but has seasonal effect. This can be summarized that the stream flow time series is stationary in trend effect, but non-stationary in seasonal effect. Therefore, the transformation method should be applied to make it become the stationary time series before determining the forecast equation based on Box–Jenkins technique. The transformation was proceeded using first-order backward difference $Zt = Y_t - Y_{t-12}$. After that, ACF and PACF of new time series Zt were checked. As shown in Figs. 3 and 4, it does not have the same characteristic of ACF as every lag k equals 12 ($|t| < 1.96$), the time series has become a stationary time series. Hence, the modified model becomes ARIMA$(p, 0, q) \times (P, 1, Q)_{12}$.

Considering correlogram ACF and PACF characteristics; as mentioned above, term of ARIMA$(p, 0, q)$ has already been identified. The correlogram ACF in Fig. 3 has a rapid decrease, and converges into zero at lag k equal to 1 and 2. As shown in Fig. 4, correlogram PACF has a rapid decrease, and converged into zero at lag k equals 2. The model is most likely to be ARIMA(1,0,2) or ARIMA(2,0,2). Moreover, both ACF and

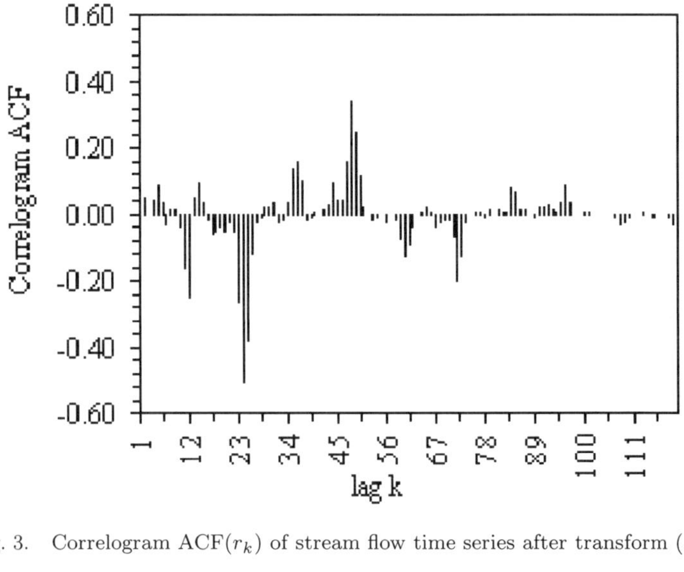

Fig. 3. Correlogram ACF(r_k) of stream flow time series after transform (Zt).

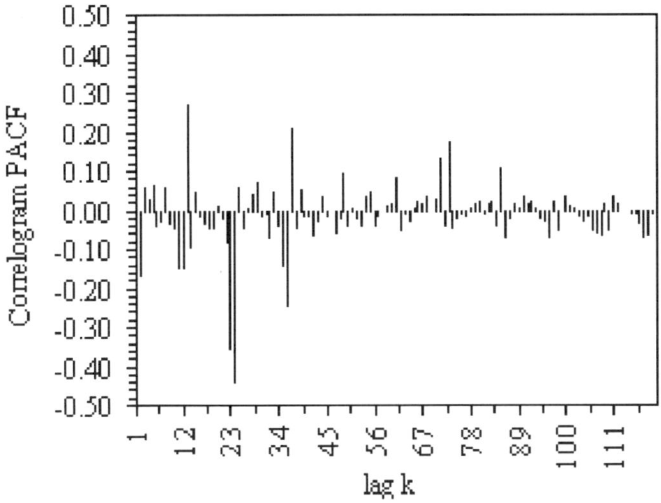

Fig. 4. Correlogram PACF(r_{kk}) of stream flow time series after transform (Zt).

PACF have demonstrated a rapid decrease, and converged into zero from lag k equals to 12. Hence, term of ARIMA$(P, 1, Q)_{12}$ is most likely to be identified as follows: ARIMA$(2,1,2)_{12}$, ARIMA$(2,1,1)_{12}$, ARIMA$(1,1,2)_{12}$, and ARIMA$(1,1,1)_{12}$.

The suitable model is a mix of both terms which provides minimum sum square of residual error (SSE). There is ARIMA$(2,0,2) \times (2,1,2)_{12}$, which provides minimum SSE and variance of residuals is equal to 30,537 and 226.20, respectively. The formula can be written as in Eq. (1).

$$\begin{aligned}(1 + 0.1006B + 0.5519B^2)^{12}(1 - 0.5157B + 0.2786B^2)(1 - B^{12})Y_t \\ = 0.8730 + (1 - 0.1503B + 0.1004B^2)(1 - 0.936B + 0.0473B^2)\varepsilon_t. \quad (1)\end{aligned}$$

Residual checks were proceeded to make sure that ACF of residual error $r_k(e_t)$ of model is simultaneously zero. Box–Ljung test was applied for this purpose. As shown in Fig. 5, overall $Q'_m < \chi^2_{\alpha,m-a}$ is at 0.5% significant level, where a is the number of parameters in Eq. (1), $m = k$, and $m \geq$ zero. This can be summarized that the model fitted with stream flow time series. As shown in Fig. 6, the model has been fitted to observe stream flow time series.

Finally, the model was used in order for short-term forecasting of monthly future stream flow of 12 months (Jan. 2002 to Dec. 2002). The

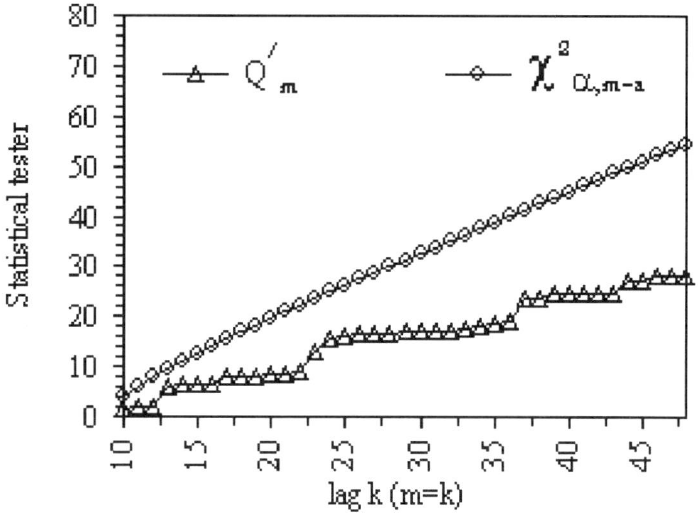

Fig. 5. Residual testing using Box–Ljung criterion.

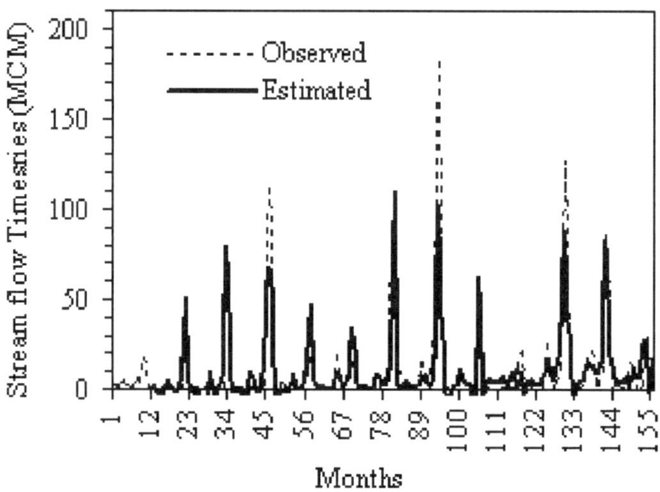

Fig. 6. Observed and estimated stream flow time series (Jan. 1989 to Dec. 2001) with multiplicative ARIMA$(2, 0, 2) \times (2, 1, 2)_{12}$.

formula that was used for point and interval forecasting at $(1 - \alpha)$ 100% confidence limit[4] is illustrated in Eq. (2).

$$\hat{Z}_{t+1}(\ell)_{(1-\alpha)100\%} = \hat{Z}_{t+1}(\ell) \pm Z\left(1 - \frac{\alpha}{2}\right)\sqrt{\sum_{j=0}^{\ell-1} \psi_j^2 \hat{\sigma}_\varepsilon^2}, \qquad (2)$$

where

$\hat{Z}_{t+1}(\ell)_{(1-\alpha)100\%}$ = interval forecast value of time ℓ at confidence level $(1-\alpha)$ 100%.

$\hat{Z}_{t+1}(\ell)$ = point forecast value at time ℓ from the model in Eq. (1).

$Z(1-\alpha/2)$ = standard normal distribution at significant α.

σ_{ε} = variance of residuals or SSE/DF in which $DF = (n-d) - p - q - 1$.

n = number of point data in time series.

ε_t = residual error at time t.

ψ_j = weights of parameter j, which is provided from Eq. (1).

$= -\theta_j + \phi_p\psi_{1-p} + \cdots + \phi_1\psi_{j-1}$ when $j > 0, \psi_0 = 1$ when $j = 0$.

θ_j, ϕ = parameters that provide minimum of SSE.

ℓ = time or period of forecasting.

As shown in Fig. 7, there is a comparison of forecasted and observed stream flow time series.

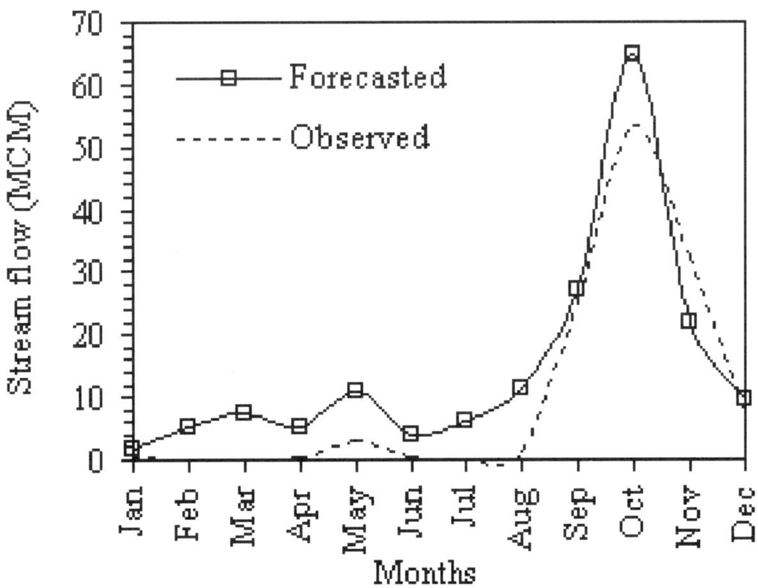

Fig. 7. Observed and forecasted stream flow time series (Jan. 2002 to Dec. 2002).

Table 3. Forecasting accuracy measurement of forecasted stream flow time series.

MAE	MAPE	MSE	RMSE	R	R^2	n
Forecasting accuracy measurement						
6.028	16.708	47.877	6.919	0.946	0.895	12

6. Evaluating Stream Flow Time Series Forecasting

The forecasting accuracy measurement is shown in Table 3.

The results indicate that forecasted inflow has a little positive error. It means that the furcated stream flow time series shifts higher than the observed time series, but there is still good correlation. This may affect irregular effect of stream flow during the period of October 1996 that has a magnitude higher than others. The use of Multiplicative ARIMA Intervention could be applied for this effect but it is a somewhat complicated model.

7. The Changes in Optimal Reservoir Operational Rule Curve

The results are as shown in Figs. 8 and 9, there are differences between two rule curves, which simulate water release without and including forecasted

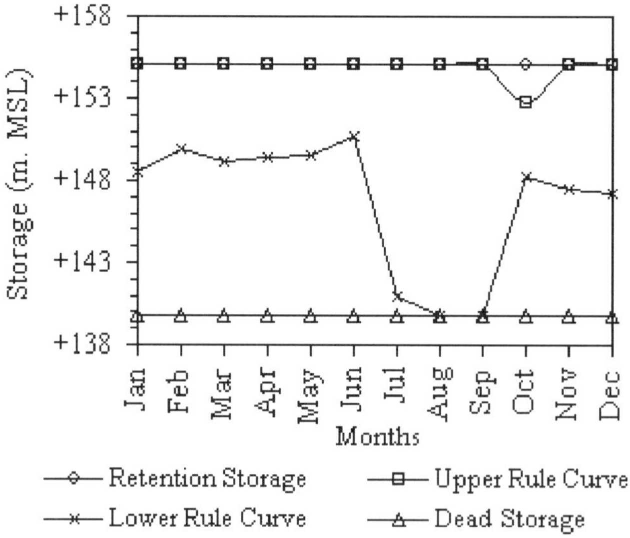

Fig. 8. The use of GA simulating reservoir water release 13 years without forecasted future stream flow for rule curve year 2001.

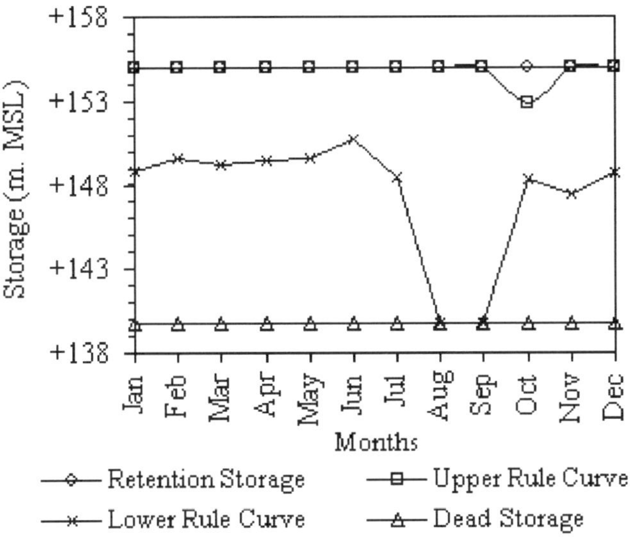

Fig. 9. The use of GA simulating reservoir water release 14 years with forecasted future stream flow for rule curve year 2002.

future stream flow, that the lower rule curve during July and December has been risen so as to preserve buffer storage, and to minimize water shortage corresponding to the changes in hydrological environment based on hedging rule.

8. Conclusion

The results indicate that monthly stream flow time series is identified as non-stationary, and Box–Jenkins multiplicative ARIMA modeling can be suitable for monthly stream flow time series forecasting, and the adoption of this method together with a GA model can appropriately be applied to water resources problem corresponding to the changes in hydrological environment.

References

1. G. E. P. Box, G. M. Jenkins and G. C. Reincel, *Time Series Analysis Forecasting and Control*, 3rd Edn. (Prentice-Hall Inc., USA, 1994).
2. L. Bruce and T. Richard, *Forecasting and Time Series* (Wadsworth, Inc., Belmont, California, 1979).

3. S. Chittaladakorn, Classifier system and genetic algorithm for rule-based global operation of canal gates, Utah State Univ., Logan, Utah, 1996.
4. J. D. Cryer, *Time Series Analysis* (PWS Publisher, Inc., USA, 1986).
5. P. E. Gaynor and R. C. Kirkpatrick, *Introduction to Time-Series Modeling and Forecasting in Business and Economics* (USA, 1994).
6. D. E. Goldberg, *Genetic Algorithms in Search Optimization and Machine Learning* (Addison-Wesley, Reading, Mass, USA, 1989).
7. D. E. Goldberg and K. Deb, A comparative analysis of selection schemes used in genetic algorithms, in *Foundations of Genetic Algorithms* (Morgan Kaufmann, San Mateo, California, 1991), pp. 69–93.
8. Z. Michalewicz, *Genetic Algorithms + Data Structures = Evolution Programs* (Springer-Verlag, New York, Inc., New York, 1992).
9. J. D. Delleur, J. W. Delleur, V. Yevjevich and W. L. Lane, *Applied Modeling of Hydrologic Time Series* (Water Resources Publication, Chelsea, Michigane, USA, 1995).
10. S. Taesombut, *Quantitative Forecasting Techniques* (Physics Center Publisher, Inc., Bangkok, Thailand, 1996).
11. V. Taesombut, *Hydrology* (Physics Center Publisher, Inc., Bangkok, Thailand, 1995).

Advances in Geosciences
Vol. 11: Hydrological Science (2007)
Eds. Namsik Park *et al.*

EVAPOTRANSPIRATION ESTIMATES FOR A STOCHASTIC SOIL-MOISTURE MODEL

CHAVALIT CHALEERAKTRAKOON* and SOMRIT SOMSAKUN
Department of Civil Engineering
Thammasat University, Klong Luang, Pathumthani, 12121, Thailand
**cchava@engr.tu.ac.th*

Potential evapotranspiration is information that is necessary for applying a widely used stochastic model of soil moisture (I. Rodriguez Iturbe, A. Porporato, L. Ridolfi, V. Isham and D. R. Cox, Probabilistic modelling of water balance at a point: The role of climate, soil and vegetation, *Proc. Roy. Soc. London* **A455** (1999) 3789–3805). An objective of the present paper is thus to find a proper estimate of the evapotranspiration for the stochastic model. This estimate is obtained by comparing the calculated soil-moisture distribution resulting from various techniques, such as Thornthwaite, Makkink, Jensen–Haise, FAO Modified Penman, and Blaney-Criddle, with an observed one. The comparison results using five sequences of daily soil-moisture for a dry season from November 2003 to April 2004 (Udornthani Province, Thailand) have indicated that all methods can be used if the weather information required is available. This is because their soil-moisture distributions are alike. In addition, the model is shown to have its ability in approximately describing the phenomenon at a weekly or biweekly time scale which is desirable for agricultural engineering applications.

1. Introduction

The probability distribution of soil moisture at a site is usually necessary for the agribusiness sector. For example, it is critical to the schedule of supplying water to irrigation fields, a basis for surface water management decision, and a crucial input to crop yield models. Several stochastic models (see e.g. Refs. 2, 6, and 11) that are based mainly on the concept of soil-moisture balance are available for determining the probability distribution. Among these statistical models, Rodriguez-Iturbe *et al.*[11] is popular because the model considers the stochastic and intermittent characteristics of rainfall, and uses an accepted relationship between water losses (i.e. infiltration, evapotranspiration, and leakage) and soil moisture which has been supported by Milly[7] and Salvucci.[12]

To define the water loss function, the estimate of the evapotranspiration is usually necessary. Empirical techniques (Thornthwaite, Makkink, Jensen–Haise, FAO Modified Penman, and Blaney–Criddle) can be used for calculating the estimate from readily available climatic data at a given location. Since there are many empirical techniques, the search for a suitable one is a crucial question.

Therefore, an objective of the present study is to find an appropriate evapotranspiration formula for the stochastic model[11] by comparing the soil-moisture probability distribution of all techniques with those of observations. Five series of soil-moisture data from November 2003 to April 2004 (Udornthani Province, Thailand) were used for the comparison. Results from this study have indicated that all evapotranspiration equations are possible to use in the statistical model because their soil-moisture distributions are alike.

2. Stochastic Soil Moisture Model

Let s be the relative soil moisture content. Rodriguez-Iturbe *et al.*[11] gives the probability density function $\mathbf{p}(s)$ of the soil moisture at a site under seasonally fixed condition as

$$\mathbf{p}(s) = \frac{a}{\rho(s)} \exp\left[-\gamma s + \lambda \int \frac{du}{\rho(u)}\right], \tag{1}$$

where λ is the rate of arriving rainfall based on a Poisson process, $\gamma = \alpha n z$ where $1.0/\alpha$ is the mean depth of rain events assuming to follow an exponential distribution function, n is the soil porosity, z is the soil depth, $\rho(s)$ is the total normalized water loss, and a is determined from

$$\int_0^1 \mathbf{p}(s)ds = 1.0. \tag{2}$$

Figure 1 shows the schematic representation of the loss function $\rho(s)$ against the soil moisture s. When $0 \leq s \leq s^*$ (s^* is the soil moisture at wilting point), the total loss $\rho(s)$ depends mainly on evapotranspiration $E(s)$. It varies linearly from zero to a normalized maximum evapotranspiration η, $\eta = E(s^*)/(nz)$. For $s^* \leq s \leq 1.0$, it is the sum of the evapotranspiration η and a corresponding leakage loss $z(s)$ of $L(s)$ where

$$z(s) = ks^c. \tag{3}$$

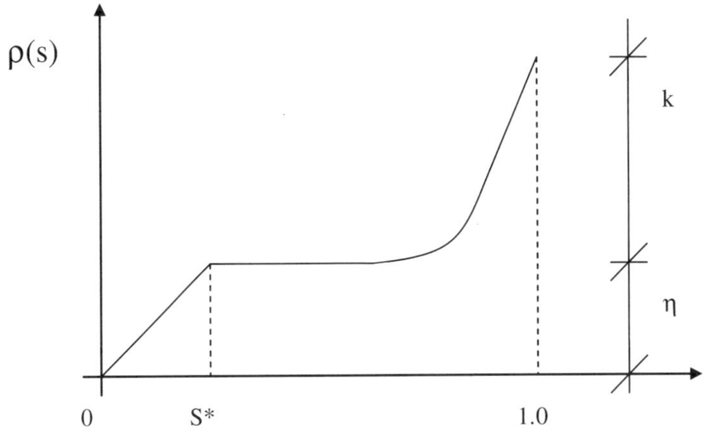

Fig. 1. Schematic representation of the loss function against the soil moisture.

Note, in (3), that k is the normalized hydraulic conductivity of saturated soil, and c is a constant depending on soil type.[1] Mathematically, the function $\rho(s)$ as presented in Fig. 1 can be written as

$$\rho(s) = \frac{E(s)}{nz} + \frac{L(s)}{nz} = \begin{cases} \frac{\eta}{s^*} s & 0 \le s \le s^*, \\ \eta + ks^c - k(s^*)^c & s^* < s \le 1.0. \end{cases} \tag{4}$$

Substitute (4) in (1), the probability distribution function $\mathbf{p}(s)$ of the soil moisture becomes

$$\mathbf{p}(s) = \begin{cases} \frac{as^*}{\eta s} \exp\left(-\gamma s + \frac{\lambda s^*}{\eta} \int_0^s \frac{\mathrm{d}u}{u}\right) & 0 < s \le s^*, \\ \frac{a}{\eta + ks^c - k(s^*)^c} & s^* < s \le 1.0. \\ \quad \times \exp\left[-\gamma s + \lambda \int_{s^*}^s \frac{\mathrm{d}u}{\eta + ku^c - k(s^*)^c}\right] & \end{cases} \tag{5}$$

3. Potential Evapotranspiration Estimates

Five evapotranspiration formulae, namely Thornwaite, Makkink, Jensen–Haise, FAO modified Penman, and Blaney–Criddle, are considered for defining the water loss function of the soil-moisture model. Thornwaite[13] developed an equation to calculate the referred evapotranspiration from mean temperature and latitude data as

$$E(s^*) = 1.6m\ell \left(\frac{10T}{I}\right)^b, \tag{6}$$

where m is the crop coefficient, ℓ is the number of 12-h durations of possible sunshine, T is the mean monthly temperature in °C, I is the heat index, and b is a constant given as follows:

$$b = 6.75 \times 10^{-7} I^3 - 7.71 \times 10^{-5} I^2 + 1.792 \times 10^{-2} I + 0.49239. \tag{7}$$

Makkink considered solar radiation for calculating the water loss by[9]

$$E(s^*) = m\left(0.61Q\frac{\Delta}{\Delta + \nu} - 0.12\right), \tag{8}$$

where Q is the incoming solar radiation, Δ is the slope of saturated vapor pressure–temperature curve, and ν is a psychrometric constant ($\nu = 0.49$ or 0.27 for temperature in °C or °F, respectively). If an observation of the solar radiation is unavailable, it may be obtained from

$$Q = Q'\left(0.26 + 0.50\frac{n}{N}\right), \tag{9}$$

where Q' is the mean solar radiation for cloudless sky, n is the duration of actual sunshine, and N is the longest period of possible sunshine.

The Jensen–Haise method also used the solar radiation to incorporate the recognized link between a source of energy and evapotranspiration as[4]

$$E(s^*) = m[C_T(T - T_X)Q] \tag{10}$$

in which

$$C_T = \frac{1}{38 - (2/305)h + [380/(e_1 - e_2)]} \tag{11}$$

and

$$T_X = -2.5 - 0.14(e_2 - e_1) - \frac{h}{550}, \tag{12}$$

where h is the elevation above sea level in meters, and e_2 and e_1 are, respectively, saturation vapor pressures in kPa at mean maximum and mean minimum temperatures of the warmest month in the considered area.

Penman[10] first combined factors to account for a supply of energy and a mechanism to remove the water vapor from the immediate vicinity of the evaporating surface. Later, Doorenbos and Pruitt[3] modified the Penman method to

$$E(s^*) = m\frac{\Delta Q_n + \gamma E_a}{\Delta + \gamma} = m\frac{\Delta Q_n + \gamma(0.27 + 0.233U_2)(e_s - e_{\text{d}})}{\Delta + \gamma}, \qquad (13)$$

where Q_n is the net solar radiation, γ is a psychrometric constant, E_a is the rate of evaporation, U_2 is wind speed at 2 m above ground and e_s is the vapor pressure at mean air temperature and e_{d} is the pressure at dew point. If the solar radiation is not measured, it will be computed from

$$Q_n = Q'(1 - r)\left(0.26 + 0.5\frac{n}{N}\right) - \sigma T^4(0.56 + 0.0797\sqrt{e_{\text{d}}})\left(0.1 + 0.9\frac{n}{N}\right), \qquad (14)$$

where r is the reflection constant and σ is the Stefan–Boltzmann constant.

The last formula considered is the SCS Blaney–Criddle method. It is the standard method recommended by the USDA-Soil Conservation Service (SCS), is well known in the Western US, and is used extensively throughout the world.[5] The equation is given as

$$E(s^*) = m\frac{T'p}{100}, \qquad (15)$$

where T' is the average air temperature in °F and p is the percent daytime during a cultivation period.

4. Assessment of the Stochastic Model with the Evapotranspiration Estimates

The observations of daily soil moisture at five locations in Amphoe Muang, Amphoe Bandung, Amphoe Nong-wao-sor, Amphoe Pen, and Amphoe Kudjub (Udornthani Province, Thailand) were measured (see their sites in Fig. 2) by members of our research team because there is no governmental related agency responsible for the data collection. The measurement was taken by field density test (sand cone method). It is observed during a dry season from November 2003 to April 2004 which is critical for managing limited available water. The total period of the measured data is short because of the constraints in budget and time available for this study.

Table 1 presents the soil type for each site. The table also shows the nearby weather gauging stations used (see their locations in Fig. 2) to supply necessary climate data. The total length of selected rainfall data is 10 years (1991–2000).

Fig. 2. Location of weather and soil moisture stations used.

The evapotranspiration estimation methods as described earlier[6–15] were applied to these data. Note that the estimation using mean values is adequate because the studied stochastic soil-moisture model has been proposed under seasonally fixed condition, as mentioned earlier. Table 2 shows the results of potential evapotranspiration estimates for all sites. The simple Thornthwaite technique gives its results that agree well with the FAO modified Penman and the Jensen–Haise equations. The estimates of the Makkink method are close to those of the standard Blaney–Criddle formula. They were used to define a water loss function $\rho(s)$ by (4).

Table 1. Location and soil type of sites used for measuring soil moisture in Udornthani Province, Thailand and nearby weather gauging stations.

Soil-moisture observation site		Nearby weather gauging station		
Location	Soil type (Class)	Station no.	Latitude	Longtitude
Muang	L	68170	17°24′35″	102°47′45″
Bandung	SL	68212	17°41′55″	103°15′60″
Nong-wao-sor	CL	68232	17°09′26″	102°34′43″
Pen	SiCL	68022	17°41′40″	102°54′55″
Kudjub	C	68110	17°25′17″	102°36′11″

Note: L = loam, SL = sandy loam, CL = clay loam, SiCL = silty clay loam, and C = clay.

Table 2. Parameters and Kolmogorov–Smirnov goodness-of-fit test of rainfall hypotheses based on 0.05 significant level.

Station	Poisson hypothesis			Exponential hypothesis		
	Parameter, λ (time/day)	Kolmogorov–Smirnov statistics		Parameter, α^{-1} (mm)	Kolmogorov–Smirnov statistics	
		Calculated	Critical		Calculated	Critical
Muang	0.093	0.252	0.262	8.4	0.056	0.084
Bandung	0.067	0.260	0.262	11.0	0.031	0.096
Nong-wao-sor	0.097	0.240	0.262	8.4	0.046	0.084
Pen	0.093	0.261	0.262	7.9	0.049	0.084
Kudjub	0.092	0.260	0.262	8.8	0.065	0.084

Table 3. Potential evapotranspiration estimate of each individual site and method.

Site	Potential evapotranspiration estimate (mm/day)				
	Thornthwaite	Makkink	Jensen–Haise	Modified Penman	Blaney–Criddle
Muang	4.17	3.65	4.28	4.4	3.7
Bandung	4.2	3.7	4.28	4.39	3.75
Nong-wao-sor	4.0	3.5	4.15	4.27	3.55
Pen	4.35	3.8	4.37	4.44	3.85
Kudjub	4.08	3.55	4.2	4.33	3.6

Next, the Poisson parameter λ of rainfall arrival rate and exponential parameter α^{-1} of rainfall depth were estimated by method of moment.[8] Then, the Poisson and exponential assumptions were investigated using Kolmogorov–Smirnov goodness-of-fit test at 0.05 significant level. Table 3

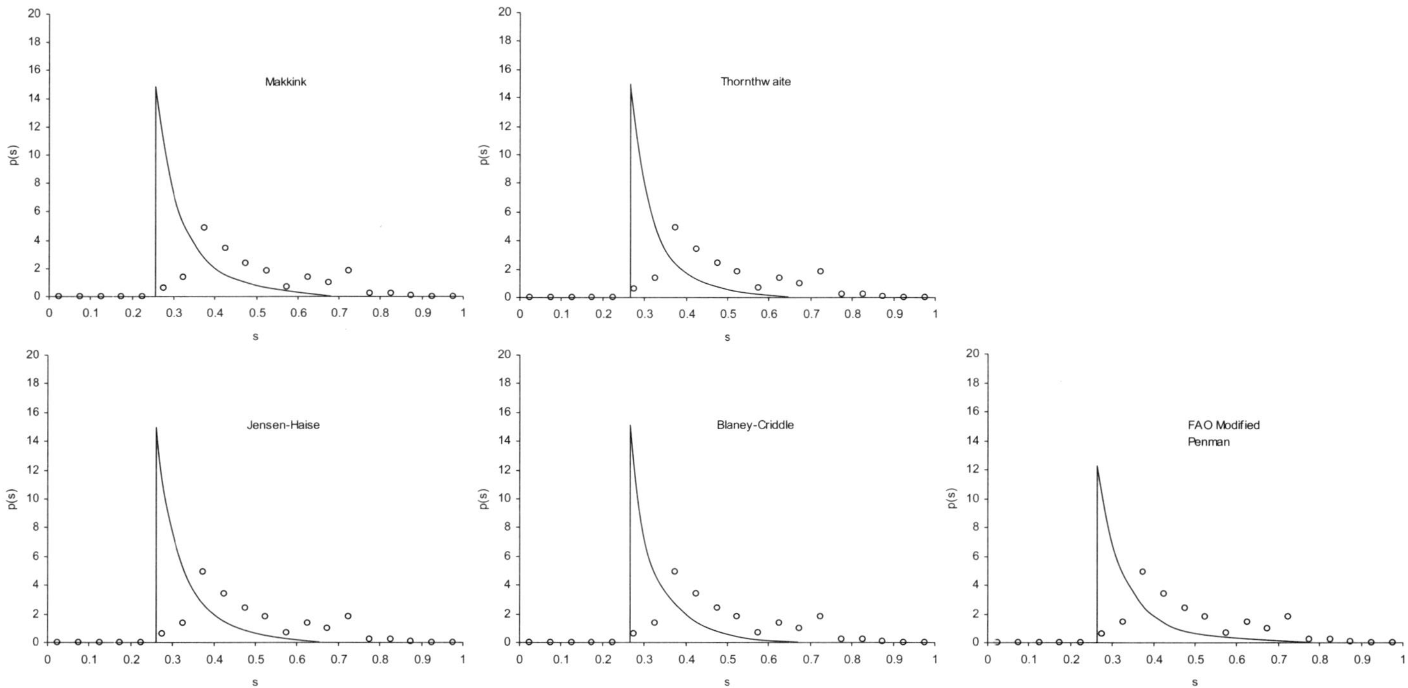

Fig. 3. Soil-moisture probability density of the various evapotranspiration techniques for Amphoe Nong-wao-sor, Udornthani Province ($\lambda = 0.097$ time/day, $\alpha^{-1} = 8.4\,\text{mm}$).

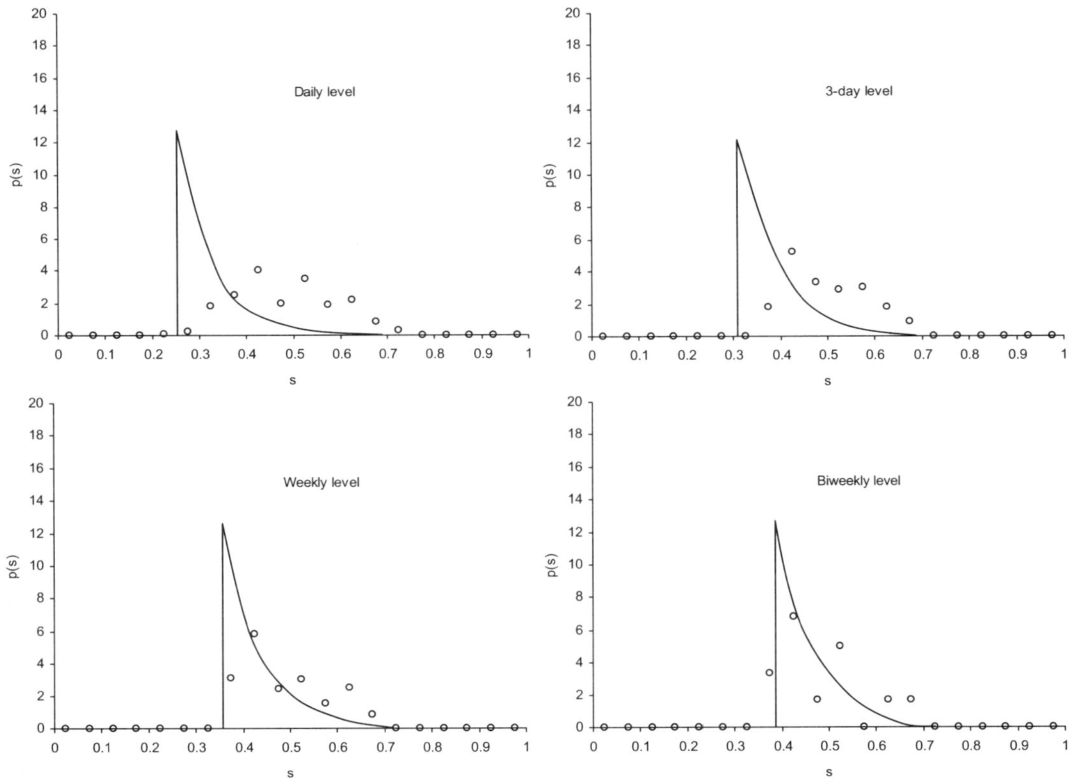

Fig. 4. Observed and calculated probability density of soil moisture at daily, 3-day, weekly and biweekly time intervals for Amphoe Muang, Udornthani Province ($\lambda = 0.093$ time/day, $\alpha^{-1} = 8.4\,\text{mm}$).

shows that the assumptions can be accepted. All calculated Kolmogorov–Smirnov statistics are less than their critical counterparts.

Now, a soil-moisture probability distribution at daily time interval was determined with (5). Figure 3 presents the soil-moisture probability density of the evapotranspiration techniques for Amphoe Nong-wao-sor, Udornthani Province (Thailand). All distributions of considered techniques are similar. Any formula can be used provided that the weather information required is available. The figure also compares them with an observed one. The results demonstrate that the stochastic model considered, (5), does not describe the historic daily soil-moisture satisfactorily because the calculated densities deviate largely from the empirical one. This deviation may be due to the lack of adequate soil-moisture data.

The model was further investigated for its ability to describe the frequency of observed soil moisture at longer levels that are necessary for water managers in the agri-business sector. For instance, they usually make decisions in supplying available water at weekly interval. Figure 4 compares the empirical with calculated probability density of soil moisture at daily, 3-day, weekly and biweekly time intervals (Amphoe Muang, Udornthani Province). The model fits the weekly and biweekly phenomena better than the daily and 3-day ones.

5. Summary and Conclusions

Potential evapotranspiration is fundamental information to success in applying a popular stochastic soil-moisture model.[11] Since several estimation methods (Thornthwaite, Makkink, Jensen–Haise, FAO Modified Penman, and Blaney–Criddle) are available, an objective of this paper is therefore to find the one that is suitable for the stochastic model. The appropriate technique is achieved by comparing the modeled soil-moisture distribution of all referred methods with a corresponding observed one.

The sequence of daily soil-moisture observations measured at five sties — Amphoe Muang, Amphoe Bandung, Amphoe Nong-wao-sor, Amphoe Pen and Amphoe Kudjub of Udornthani Province (Thailand) — during a dry period (November 2003 to April 2004) and the 10-year (1991–2000) series of climate data of nearby weather gauging stations were used for the comparison. Results have shown that every considered technique is feasible for calculating the evapotranspiration estimate used in the stochastic model because their soil-moisture distributions are quite similar. Based on the limited soil-moisture data in the study area, the model describes reasonably the process at a weekly or biweekly time scale that is essential for the applications in agricultural engineering. Further research work to cover a wide enough range of climate and soil type is necessary for the results to be generally applicable.

References

1. R. B. Clap and G. N. Hornberger, Empirical equation for some soil hydraulic properties, *Water Resources Research* **14** (1978) 601–604.
2. D. R. Cox and V. Isham, The virtual wait-time and related processes, *Appl. Prob.* **18** (1986) 558–573.
3. J. Doorenbos and W. O. Pruitt, Guidelines for prediction of crop water requirements, *FAO Irrigation and Drainage Paper No. 24*, 2nd Edn., (FAO Rome, Italy, 1977), 156 pp.
4. M. E. Jensen and H. R. Haise, Estimating evapotranspiration from solar radiation, *Journal of the Irrigation and Drainage Division, ASCE*, **89** (1963) 15–41.
5. M. E. Jensen, R. D. Burman and R. G. Allen, *Evapotranspiration and Irrigation Water Requirements*, (ASCE, New York, 1990), 332 pp.
6. P. C. D. Milly, An analytic solution of the stochastic storage problem applicable to soil water, *Water Resources Research* **29** (1993) 3755–3758.
7. P. C. D. Milly, A minimalist probabilistic description of root zone soil water, *Water Resources Research* **37**(3) (2001) 457–463.
8. D. C. Montgomery and G. C. Runger, *Applied Statistics and Probability for Engineers*, 2nd Edn. (John Wiley and Sons Inc., Singapore, 1999).
9. P. Palayasoot, Estimation of pan evaporation and potential eva potranspiration of rice in the central plain of Thailand by using various formulas based on climatological data, M.S. Thesis, Utah State University, 1965.
10. H. L. Penman, Natural evaporation from open water, bare soil and grass, *Proceedings of the Royal Society of London*, **A193** (1948) 120–146.
11. I. Rodriguez Iturbe, A. Porporato, L. Ridolfi, V. Isham and D. R. Cox, Probabilistic modelling of water balance at a point: The role of climate, soil and vegetation, *Proceedings of the Royal Society of London*, **A455** (1999) 3789–3805.
12. G. D. Salvucci, Estimating the moisture dependence of root zone water loss using conditionally averaged precipitation, *Water Resources Research* **37**(5) (2001) 1357–1365.
13. C. W. Thornthwaite, An approach toward a rational classification of climate, *Geographical Review* **38** (1948) 55–94.

Advances in Geosciences
Vol. 11: Hydrological Science (2007)
Eds. Namsik Park *et al.*

ISOTOPIC COMPOSITIONS OF RAINFALL DEPEND ON THE DIURNAL VARIATIONS OF HUMIDITY AND CLOUD SYSTEM IN THAILAND*

KIMPEI ICHIYANAGI
Institute of Observational Research for Global Change (IORGC)
Japan Agency for Marine-Earth Sciences and Technology (JAMSTEC)
2-15 Natsushima-Cho, Yokosuka-city, Kanagawa 237-0061, Japan
kimpei@jamstec.go.jp
and
Graduate School of Science and Technology, Kumamoto University
2-39-1 Kurokami, Kumamoto 860-8555, Japan
kimpei@sci.kumamoto-u.ac.jp

KEI YOSHIMURA
Institute of Industrial Science, The University of Tokyo
4-6-1 Komaba, Meguro-ku, 153-8505, Tokyo, Japan

MANABU D. YAMANAKA
IORGC, JAMSTEC, 2-15 Natsushima-Cho, Yokosuka-city, Kanagawa 237-0061
Japan/Also at Graduate School of Science and Technology, Kobe University
1-1 Rokkodai-cho, Nada-ku, Kobe 657-8501, Japan

The majority of past precipitation studies investigating stable isotopes and the amount effect have used monthly data. Only a few such studies have incorporated daily observation data. Daily rainfall was sampled at Bangkok and Phuket in Thailand, between August and November in 2001. The isotopic compositions and rainfall amounts at both stations differed with respect to daily variability. The δ^{18}O decreased as rainfall amounts increased when the rainfall rate was less than 30 mm only at Phuket; however, no clear trend was apparent when the rainfall rate exceeded 30 mm at both stations. The diurnal variations in rainfall intensity, frequency of rainy days, and relative humidity were composed of light (<30 mm) and heavy (>30 mm) rain at both stations. Rainfall characteristics and relative humidity for heavy rain events were higher than for light rain events at both stations. There were obvious diurnal variations in relative humidity for light and heavy rain events at Bangkok, while no variation for heavy rain events at Phuket. Low relative humidity in the daytime enhanced the evaporation of rain drops during their descent from the cloud-base to the ground, therefore the isotopic composition in rainfall

*This work was supported by IORGC for the project research in FY2001–2002.

became higher at Bangkok. Black-body temperature (T_{BB}) data from the Geostationary Meteorological Satellite (GMS-5) were used to create composite maps of convective activity at Phuket. Clouds developed along the coastal area of the Malay Peninsula only in the evening for light rain events. In contrast, clouds associated with heavy rain events extended from the Indian Ocean to Indochina throughout the day. The isotopic compositions in heavy rain events at Phuket remained constant, because those were not so decreasing from the initial value of the Indian Ocean air mass caused by the partial condensation of continuously supplied humid air mass by the large-scale cloud system.

1. Introduction

Stable isotopes (δ^2H, δ^{18}O) in precipitation are widely used to infer temperature and precipitation within the global hydrological cycle. Dansgaard[1] analyzed monthly observational datasets provided by the Global Network for Isotopes in Precipitation (GNIP) project, of the International Atomic Energy Agency (IAEA) and the World Meteorological Organization (WMO). That study revealed a linear relationship between precipitation amount and isotopic composition, or the so-called precipitation amount effect, at low-latitude stations. The precipitation amount effect can be explained by Rayleigh condensation (i.e. enrichment of the atmosphere by evaporation from falling raindrops) and isotopic exchange between falling drops and surrounding vapor. Araguás-Araguás and Froehlich[2] revealed the spatial and temporal variability of long-term monthly means of the stable isotopic composition of precipitation over the Asia-Pacific region. They reported that seasonal changes in precipitation isotopes in the tropical Pacific were controlled by rain-out processes in deep convective clouds and by isotopic exchanges beneath the cloud. At Bangkok, Ichiyanagi and Yamanaka[3] noted a strong negative correlation between long-term monthly means of precipitation amounts and δ^{18}O in May, July, August, October, November and December. However, no correlation was apparent in the other six months. Lawrence *et al.*[4] analyzed isotopic variations of rainfall and water vapor in detail. They showed that isotope ratios are related to the intensity and organization of storm systems upwind of the collection site; ratios decrease as storm activity and organization increase. Processes controlling rainfall isotopes include rainwater origin, transport processes in the air mass, and formation processes at the collection site. To investigate the controlling factors determining the isotopic composition of rainfall, further analyses using daily or other short-term observations are warranted.

2. Data and Methods

Daily rainfall isotopic data were collected at Bang Na Agromet (13.44°N, 100.34°E) in Bangkok and at Phuket Airport (8.07°N, 98.19°E) from August to November 2001, and were subsequently analyzed. Both δ^2H and δ^{18}O of precipitation were measured by a MAT-252 mass spectrometer in the Ecological Research Center of Kyoto University, Japan. Analytical errors for the entire procedure were 1.5‰ and 0.1‰ for δ^2H and δ^{18}O, respectively.

Diurnal variability was calculated using three-hourly rainfall and relative humidity data provided by the Thailand Meteorological Department (TMD). There were no three-hourly data at Bang Na Agromet in Bangkok, so data from Bangkok Metropolis Station (13.73°N, 100.56°E) were used instead. Convective activity was derived from equivalent black-body temperatures (T_{BB}) from infrared channel 1 (11 μm) T_{BB} (IR1) of the Geostationary Meteorological Satellite (GMS-5). The satellite provides a resolution of 0.2° × 0.2° in the area of interest (10°–20°N and 85°–130°E).

3. Results and Discussion

3.1. *Daily rainfall variability*

Figure 1 shows daily rainfall and its isotopic compositions at Bangkok and Phuket. Light rain with high δ^{18}O occurred in August and November, while heavy rain with low δ^{18}O was common in September and October. Short-term (daily) variability, however, showed a different pattern between two stations. Two rainfall peaks occurred on 18 September and 7 October at Bangkok. A large peak at the end of September was followed by intermittent peaks of decreasing magnitude into early November at Phuket. Although Phuket had more rainy days than Bangkok, both stations had approximately the same number of heavy rain days (>30 mm/day). The δ^{18}O value was near −5‰ until early September at Phuket and until mid-September at Bangkok. Between mid-September and mid-October, sudden δ^{18}O drops to less than −10‰ appeared more frequently at Bangkok than at Phuket. Subsequently, δ^{18}O at Bangkok remained at a higher level than at Phuket. Yoshimura *et al.*[5] reported similar temporal variability at three stations in Thailand (Bangkok and two stations in northern Thailand) during the 1998 rainy season. In that study it was found that large-scale moisture transport influenced the daily variability of rainfall isotopes at all stations. In this study, however, the daily rainfall isotope trends at Bangkok

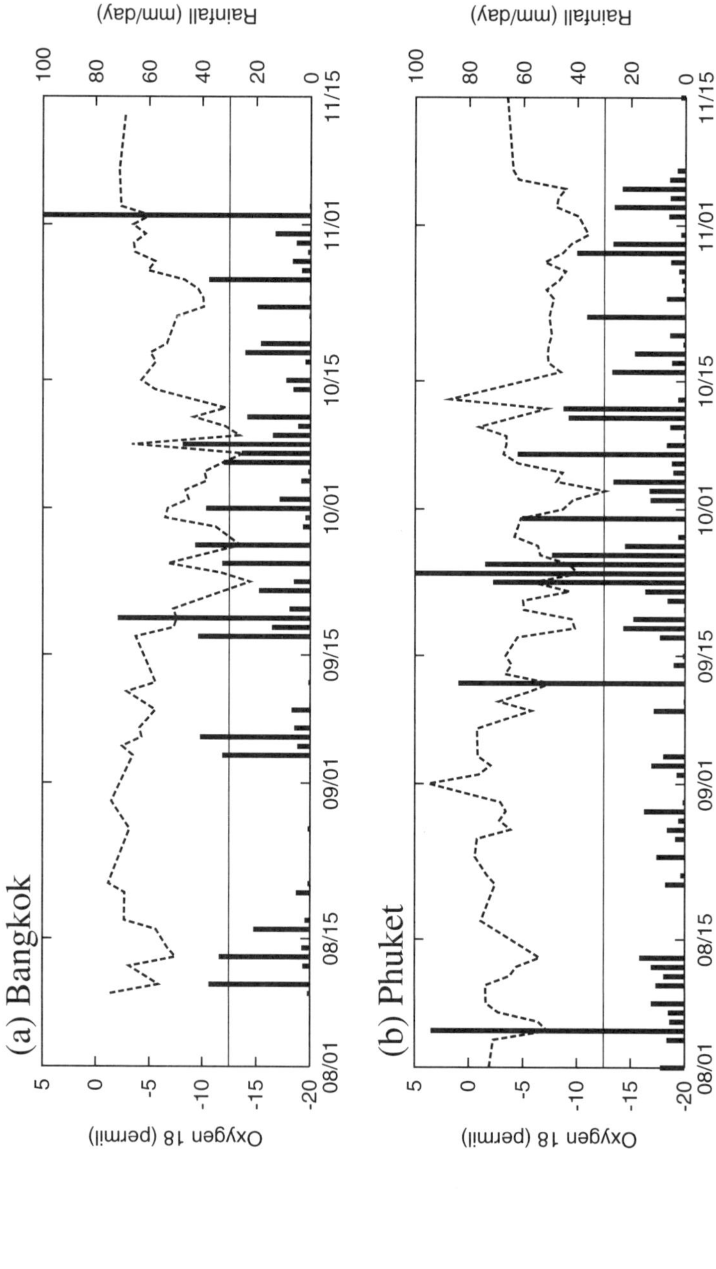

Fig. 1. Time series of daily rainfall amount (bar graph) and $\delta^{18}O$ (dot line graph) at Bangkok and Phuket from August to November 2001. The solid line indicates 30 mm/day in rainfall amount.

differed from those at Phuket. It is located in the southern part of Thailand surrounded by the Indian Ocean, the isotopic variability at Phuket differs from that of other stations in the inland of Indochina.

Figure 2 shows the relationship between $\delta^{18}O$ and rainfall amount at two stations. The $\delta^{18}O$ in rainfall ranged from $-15‰$ to $0‰$ and from $-10‰$ to $+5‰$ at Bangkok and Phuket, respectively. In general, the $\delta^{18}O$ values were decreasing with rainfall amounts increasing, so-called the "amount effect".[1] There was negative correlation between $\delta^{18}O$ and rainfall amount when daily rainfall amounts were less than 30 mm at Phuket. The amount effect was not so clear at Bangkok. However, light rain (<10 mm) with low $\delta^{18}O$ (<$-5‰$) samples was common at both stations. Initial water vapor, which constituted the rainwater source, might be characterized by

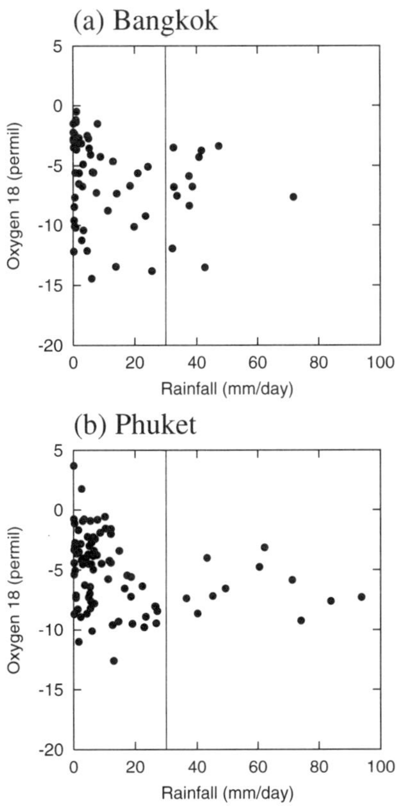

Fig. 2. Scatter plots between $\delta^{18}O$ and rainfall amount at Bangkok and Phuket. Solid line indicates 30 mm/day in rainfall amount.

low δ^{18}O values. Also heavy rain (>30 mm) with high δ^{18}O (> − 10‰) samples was common at both stations. In the next section, the diurnal variations in rainfall and relative humidity are examined in relation to light (<30 mm) and heavy rain (>30 mm) events.

3.2. *Diurnal variation*

Diurnal variations in rainfall intensity, frequency of rain days, and relative humidity were investigated using three-hourly routine observations at Bangkok and Phuket. The frequency was defined as the ratio of rain days at each time to total days from 1 August to 15 November in 2001. Figure 3 shows the diurnal variations in rainfall intensity, frequency of rain days, and relative humidity composites for light (<30 mm) and heavy

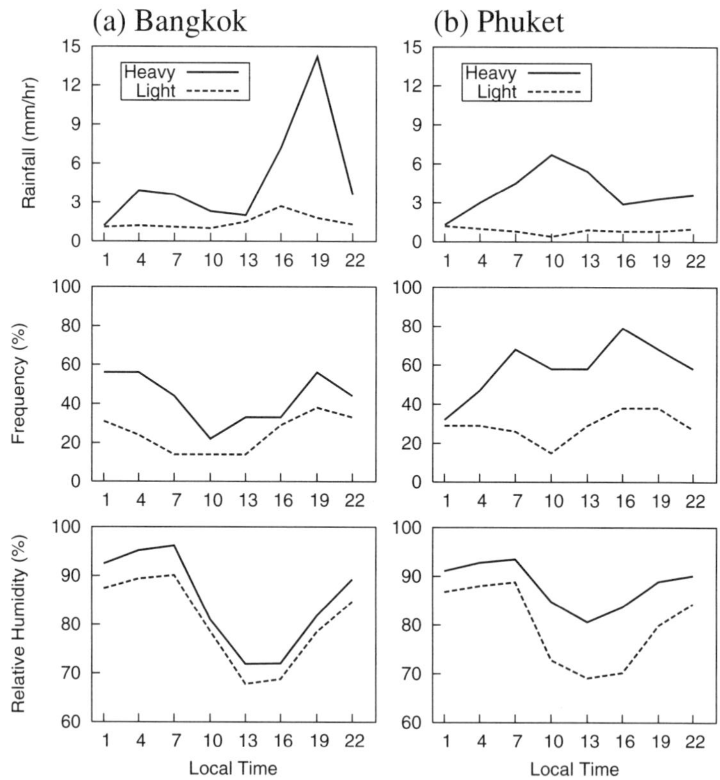

Fig. 3. Diurnal variations in rainfall intensity, frequency of rainy days, and relative humidity for light and heavy rain events at Bangkok and Phuket.

(>30 mm) rain at both stations. Rainfall intensity at Bangkok shows two peaks (4 mm in the early morning and 12 mm in the evening) for heavy rain events and one peak (3 mm in the evening) for light rain events. The frequency of rain days also shows two peaks (more than 50% in the early morning and evening) for heavy rain events and one peak (more than 30% in the evening) for light rain events. Relative humidity shows a clear diurnal variation, high in the morning and evening and low in the daytime. Also, relative humidity for heavy rain events is higher than that for light rain events throughout the day.

Rainfall intensity at Phuket is less than 1 mm throughout the day for light rain events and more than 3 mm almost throughout the day. There are two peaks (6 mm in the morning and 4 mm in the night) for heavy rain events. The frequency of rain days is less than 40% for light rain events throughout the day and more than 50% for heavy rain events except in the early morning. There are two peaks (60% in the morning and 80% in the evening) for heavy rain events. Relative humidity shows a clear diurnal variation, high in the morning and evening and low in the daytime. But, relative humidity for heavy rain events is higher than that for light rain events throughout the day. It is less than 70% in the daytime for light rain events, and more than 80% throughout the day for heavy rain events. Low relative humidity in the atmosphere enhances secondary evaporation of rain drops during their descent from cloud-base to ground.[6]

Comparing between Bangkok and Phuket, the differences in rainfall characteristics and relative humidity between for heavy and light rain events in Phuket were much larger than those in Bangkok. Ichiyanagi *et al.*[7] estimated the origin of water vapor at Bangkok and Phuket using isotope circulation model.[8] During the summer monsoon season, the water vapor at Phuket mainly comes from the Indian Ocean origin, while the Indian Ocean and Land origins were mixed at Bangkok. Therefore, the large differences in rainfall characteristics and relative humidity between for heavy and light rain events at Phuket might be caused by the single air mass origin. In the next section, the diurnal variation of convective activity is considered only at Phuket.

3.3. *Convective activity*

Diurnal variation in convective activity for Phuket was investigated using hourly equivalent black-body temperatures ($T_{\rm BB}$) from GMS-5. Figures 4 and 5 are composites of GMS $T_{\rm BB}$ data at 00, 06, 12, and 18 GMT (07, 13,

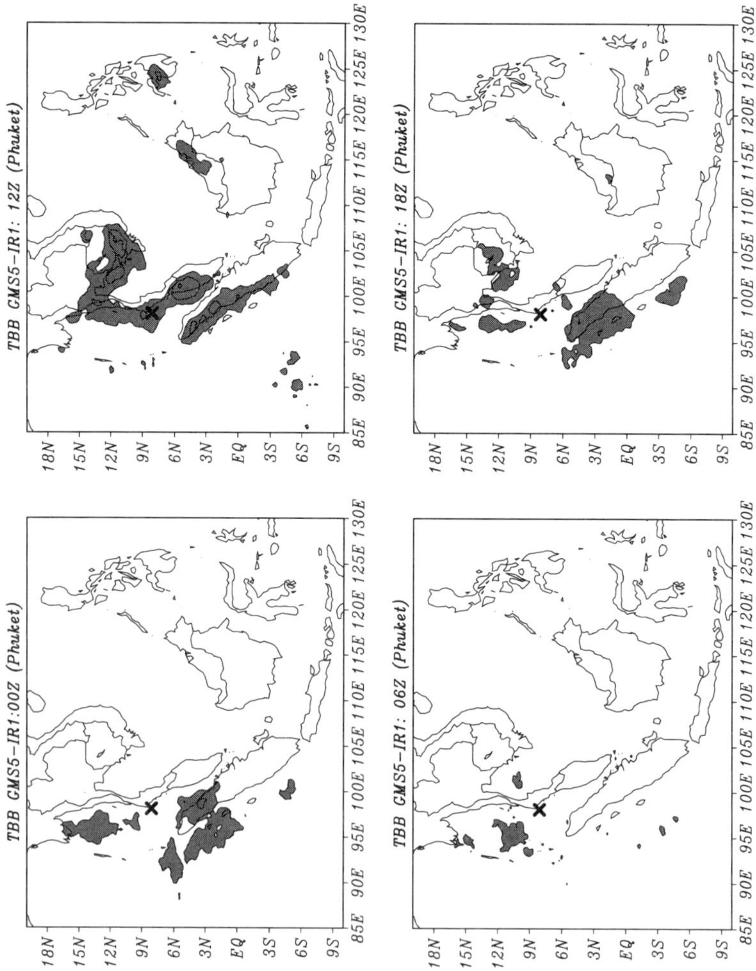

Fig. 4. Composites of GMS T_{BB} data for light rain events at Phuket. Panels show 00, 06, 12, and 18 GMT. The contour interval is 10 K. Shading indicates areas colder than 250 K.

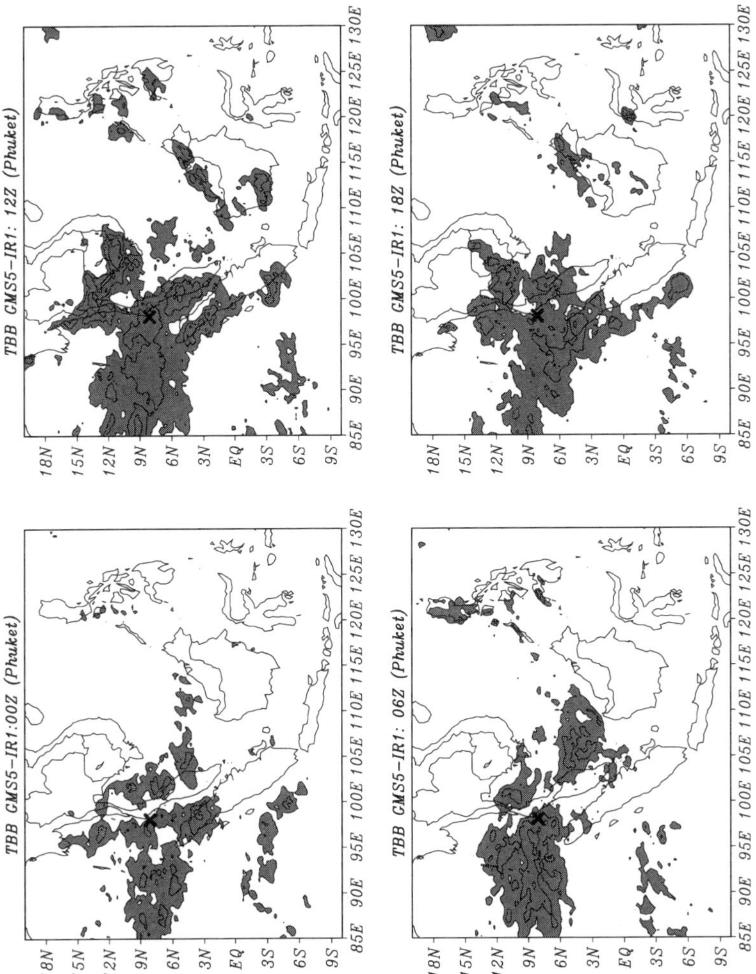

Fig. 5. As in Fig. 4, but for heavy rain events.

19, and 01 LST, respectively) at Phuket for light and heavy rain events, respectively. In Fig. 4, there were few clouds over Thailand for light rain events at 00, 06, and 18 GMT. Only in the evening (12 GMT), clouds covered over the western coast of the Malay Peninsula. On the other hand, clouds were common from the Indian Ocean to the southern Indochina throughout the day (00, 06, 12, and 18 GMT) for heavy rainfall as shown in Fig. 5.

Distributions of clouds show noticeable differences between light and heavy rain events. Clouds around Phuket show a diurnal variation for light rain events, while much more active throughout the day for heavy rain events. Geographical conditions and cloud distributions suggest that light rain resulted mostly from local-scale cloud systems forced by land-sea breezes.[9] Heavy rain, in contrast, was accompanied by large-scale cloud systems from the Indian Ocean to Indochina. Light and heavy rain events at Phuket might be caused by local- and large-scale cloud systems, respectively. Rindsberger *et al.*[10] pointed out that the amount effect was controlled primarily by the large-scale motion of the air masses in particular, by the degree of vertical mixing of the air layers above the ocean.

4. Conclusions

Daily rainfall samples were collected at two stations in Thailand to examine whether diurnal variation and convective activity affected rainfall isotopes. Light rain with low $\delta^{18}O$, which were caused either by the condensation of residual vapor in a cloud system or by condensation at high altitudes, were common at both stations. Heavy rain with high $\delta^{18}O$ was also common at both stations. Except for these samples, there is an obvious decreasing trend when daily rainfall amounts were less than 30 mm at Phuket. Diurnal variations of rainfall and relative humidity were analyzed for light (<30 mm) and heavy (>30 mm) rain events. There are two peaks of rainfall for heavy rain events at Bangkok in the early morning and the evening. Rainfall amounts during the heavy rain events at Phuket are relatively high throughout the day, except in the early morning. Large diurnal variations in relative humidity occur for both light and heavy rain events at Bangkok, but for only light rain events at Phuket. The relative humidity during the daytime at Phuket for light rain events is much lower than for heavy rain events. This low relative humidity enhanced the secondary evaporation of rain drops during their ascent from the cloud base to the ground, then isotopic compositions in rainfall is high even if the rainfall amount is larger.

Composite analyses of convective activity from T_{BB} data showed a clear diurnal variation over Thailand for light rain events at Phuket. In light rain events, clouds were common along the coastal region during the evening (12 GMT); few clouds were present at 00, 06, and 18 GMT. These clouds were associated with local-scale cloud systems related to land/sea breeze convergence systems. In heavy rain events, however, clouds were common throughout the day. These large-scale cloud systems formed over the Indian Ocean, which provided continuous moisture flux to Indochina. The decrease range in isotopic composition of rainfall is smaller due to the partial condensation of vapor continuously supplied by the convective ascent of humid air masses.[11] Therefore, rainfall isotopes were not so decreasing caused by the partial condensation of continuously supplied humid air mass by the large-scale cloud system. These results show that the diurnal variation of humidity and cloud system are important for determining the isotopic compositions of rainfall.

Acknowledgments

Meteorological data used in this article was from GEWEX Asia Monsoon Experiment (GAME) — Tropics observation. The authors thank Prof. Michio Hashizume of Chulalongkorn University and the staff of the Thailand Meteorological Department (TMD) for their cooperation in sampling rainfall.

References

1. W. Dansgaard, *Tellus* **16** (1964) 436.
2. L. Araguás-Araguás and K. Froehlich, *Journal of Geophysics Research* **133** (1998) 28721.
3. K. Ichiyanagi and M. D. Yamanaka, *Hydrological Processes* **19** (2005) doi: 10.1002/hyp.5978 3413.
4. J. R. Lawrence, S. D. Gedzelman, D. Dexheimer, H.-K. Cho, G. D. Carrie, R. Gasparini, C. R. Anderson, K. P. Bowman and M. I. Biggerstaff, *Journal of Geophysics Research* **109** (2004), doi: 10.1029/2003JD004046.
5. K. Yoshimura, T. Oki, N. Ohte and S. Kanae, *Journal of Geophysics Research* **108** (2003), doi: 10.1029/2003JD003477.
6. H. Peng, B. Mayer, S. Harris and R. Krouse, *Tellus* **59** (2007) 698.
7. K. Ichiyanagi, K. Yoshimura and M. D. Yamanaka, *SOLA* **1** (2005) 113.
8. K. Yoshimura, T. Oki and K. Ichiyanagi, *Journal of Geophysics Research* **109** (2004) doi: 10.1029/2004JD004764.

9. T. Ohsawa, H. Ueda, T. Hayashi, A. Watanabe and J. Matsumoto, *Journal of Meteorological Society of Japan* **79** (2001) 333.
10. M. Rindsberger, Sh. Jakke, Sh. Rahamim and J. R. Gat, *Tellus* **42** (1990) 263.
11. H. Celle-Jeanton, R. Gonfiantini, Y. Travi and B. Sol, *Journal of Hydrology* **289** (2004) 165.

Advances in Geosciences
Vol. 11: Hydrological Science (2007)
Eds. Namsik Park *et al.*

DIFFERENCED SERIES FOR BETTER ACCURACY IN PHASE-SPACE PREDICTION OF HYDROLOGICAL TIME SERIES

DULAKSHI S. K. KARUNASINGHA
Department of Engineering Mathematics, Faculty of Engineering
University of Peradeniya, Peradeniya, Sri Lanka
dulakshi@yahoo.com

SHIE-YUI LIONG
Tropical Marine Science Institute, National University of Singapore
No. 14, Kent Ridge Road, Singapore 119223, Singapore
tmslsy@nus.edu.sg

Since the superiority of the dynamical systems approach over the conventional methods is revealed, techniques to further improve the accuracy of phase-space prediction are on the lookout. Yu *et al.* (*J. Hydroinformatics* **6**(3) (2004) 209–223) reported that the first differenced series can give better prediction accuracy than the original time series, and it was demonstrated for one-step ahead prediction. However, direct extension of the method for longer prediction horizons does not provide better prediction performance. This study proposes a technique to longer prediction horizons. The technique with first differenced series is shown to produce better prediction accuracy than the original series. However, as the prediction horizon increases, the percentage improvement in prediction accuracy over the original series decreases. A disadvantage of the proposed technique is that one has to build T prediction models to make T-step ahead predictions. The technique was tested on a chaotic Lorenz series and two daily river flow time series, with local prediction models up to five-step ahead predictions.

1. Introduction

Prediction of hydrological and meteorological time series is an important task in understanding the hydrological and meteorological systems. In the past, linear stochastic approaches such as ARMA were widely used in the prediction of hydrological time series. However, since most hydrological systems are complex and nonlinear, in the recent past, the nonlinear chaotic dynamical systems approach, also referred to as the phase-space prediction approach, has been tested on many hydrological problems.[1–3] Particularly due to the method's potential shown in short-term prediction, the approach is now gaining popularity in many other diverse fields (e.g. physics, chemistry, biology, meteorology, etc) too. The major applications in hydrology have been reported in river flow analysis[2,4,5] and in rainfall analysis.[3,6–8] River flow time series analysis with chaotic dynamical systems approach has shown a lot of advancement than the rainfall analysis; the chaos-based techniques have shown good potential in short-term prediction.[2,4,5,9] This has triggered researchers to experiment more on chaos applications. As a result, current research shows lots of enthusiasm in finding the ways to improve the accuracy of phase-space prediction. For example, several methods have been proposed from different data pre-processing methods[5,9,10] to radical approaches such as inverse approaches[11,12] to get better prediction performances from phase-space prediction. This study also attempts to use first differencing, a data pre-processing technique, to improve the prediction accuracy. Yu *et al.*[10] reported that the first differenced series can give better prediction accuracy than the original time series in river flow prediction, which was demonstrated for one-step ahead prediction. However, direct extension of the method for longer prediction horizons does not provide better prediction performance.

This study proposes a technique to longer prediction horizons. In this paper, a brief introduction to chaotic time series analysis and phase-space prediction is given first and then the methodology to use first differenced series for prediction is explained. This is followed by the application results on a chaotic Lorenz series and two river flow time series and the conclusions of the study.

2. Chaotic Time Series Analysis

Analysis of chaotic time series has three main phases: (1) system characterization, (2) determining phase-space parameters for prediction, and (3) forecasting the time series. Basically, system characterization investigates whether a time series is chaotic. System characterization may also include determining whether the system is deterministic or stochastic, whether the dynamics are linear or nonlinear, the number of degrees of freedom of the system, the extent of predictability of the system, etc. The aim of forecasting is to predict the future with the assistance of observed past patterns of the system. The key to the chaotic time series analysis is state-space reconstruction.

A state space is defined as the multi-dimensional space whose axes consist of variables of a dynamical system. When the state space is reconstructed using observed time series data, it is called a phase-space. Although there are many methods to reconstruct the state space, the time delay coordinate method is currently the most popular choice. Packard *et al.*[13] and Takens[14] described the time delay coordinate method to approximate the state space from a scalar time series. According to the method, the state vector X_i can be expressed as

$$X_i(x_{i-(m-1)\tau}, x_{i-(m-2)\tau}, \ldots, x_i), \tag{1}$$

where x_i is the observed value at time $t_i = i\Delta t$, Δt is the sampling interval, m is called the embedding dimension, τ is a suitable multiple of the sampling interval, and Δt is called the time delay. Two approaches are used to determine these parameters (m, τ) needed for phase-space prediction: (1) standard approach and (2) inverse approaches.

The standard approach determines phase-space parameters using the criteria available in literature. In the inverse approaches (e.g. Babovic *et al.*[11]; Phoon *et al.*[12]) an optimal set of phase-space parameters are generally chosen from a predetermined range of values of the parameters such that the prediction accuracy is maximized.

2.1. *Phase-space prediction*

Chaotic time series have a short-term predictability. Therefore, when the phase-space parameters are determined for a chaotic time series, one can use the phase-space prediction methods to forecast the time series. In phase-space prediction, the basic idea is to set a functional relationship between the current state X_i and future state X_{i+T} in the form

$$X_{i+T} = f_T(X_i), \tag{2}$$

where T is referred to as lead time. For a chaotic system, the predictor F_T which approximates f_T is necessarily nonlinear. There are two strategies to obtain F_T: (1) local approximation and (2) global approximation. In global approximation a function F_T valid over the entire state space is considered. Neural networks, polynomial, and rational functions, for example, can be used as global approximators. On the other hand, the local approximation approach subdivides the domain of the attractor into many subsets each represented by an approximation F_T^i valid only for that subset. The set of all F_T^i functions constitutes the F_T for the case of local approximation. Local averaging technique and local polynomial technique are the widely used local approximators. Local averaging method is used in this study and is explained below.

2.2. *Local averaging method*

In local approximation, only the states near the current state are used to make prediction. To predict a future state X_{i+T}, an Euclidean metric is imposed on the phase-space to find the k-nearest neighbors of the current state X_i. Once the nearest neighbors are found, one can project each of these states X_n to their respective future states X_{n+T}, and construct a local predictor using this group of future states. A local predictor can be constructed in several ways. Among them, the averaging technique is the most popular way. Here, the estimate to future state $\hat{X}_{i+T}$ is calculated as

$$\hat{X}_{i+T} = \left(\sum_{n=1}^{k} X_{n+T}\right) \Big/ k. \tag{3}$$

When the local averaging method is used, the optimal number of nearest neighbors has to be determined. The inverse approach of Phoon *et al.*[12] suggested the phase-space parameters m, τ, and the number of nearest neighbors (k) to be determined such that the prediction accuracy is optimal.

3. Prediction with Differenced Time Series Data

For an original (observed) time series: $x_1, x_2, \ldots, x_N$, the first differenced time series, with one time lag, is obtained as $dx_1, dx_2, \ldots, dx_{N-1}$, where $dx_i = x_i - x_{i+1}$. This is shown in Table 1. The method proposed by Yu *et al.*[10] is to predict this first differenced time series instead of the original time series. In one-step prediction, first differenced time series produces $d\hat{x}_N$, where ^ denotes that it is an estimated (predicted) value. Using $d\hat{x}_N$ one may estimate the corresponding one-step ahead original series value as

$$\hat{x}_{N+1} = x_N - d\hat{x}_N. \tag{4}$$

However, for other prediction horizons, the values produced by first differenced series is of little use. For example, for the original time

Table 1. Original and first differenced time series.

Original time series	First differenced time series
x_1	$dx_1 = x_1 - x_2$
x_2	$dx_2 = x_2 - x_3$
—	—
—	—
x_{N-2}	$dx_{N-2} = x_{N-2} - x_{N-1}$
x_{N-1}	$dx_{N-1} = x_{N-1} - x_N$
x_N	

series, the three lead-time ahead prediction of x_N is x_{N+3}, whereas for the differenced time series the three lead-time ahead prediction of dx_{N-1} is $dx_{N+2} = (x_{N+2} - x_{N+3})$, which is of little interest when one's interest is in corresponding raw values (e.g. for a river flow time series, dx_{N+2} may be of little interest while x_{N+3} is the desired value). A direct extension of the first differenced series to predict meaningful higher prediction horizons (say T step ahead) may be to use a differenced series with several time lags. That is, a series $dx_1^T, dx_2^T, \ldots, dx_{N-T}^T$, where $dx_i^T = x_i - x_{i+T}$. Then, using the T-step ahead prediction of the series, $d\hat{x}_N^T$, one can estimate the T-step ahead value of the original series as

$$\hat{x}_{N+T} = x_N - d\hat{x}_N^T. \tag{5}$$

However, this study noted that these results are not satisfactory.

4. First Differenced Time Series Data for Higher Prediction Horizons ($T > 1$)

This study suggests the following technique to incorporate first differenced series for higher lead time prediction. Consider Eq. (6).

$$x_{N+T} = x_N - dx_N - dx_{N+1} \cdots - dx_{N+T-1}, \tag{6}$$

where x_N is the last observed value and x_{N+T} is the value that has to be predicted. T is the prediction horizon (note that when the differenced values are replaced by the corresponding original series values, the equation is balanced, and therefore, the equation is valid). However, when the time series is known only up to x_i, the instance where the need of prediction arises, the values $dx_i(= x_i - x_{i+1})$, $dx_{i+1}(= x_{i+1} - x_{i+2}), \ldots,$ are unknown. Therefore, the predicted values of dx_i, $dx_{i+1}, \ldots,$ using the first differenced time series are used as estimates to those values. Thus, the prediction formula can be written as

$$\hat{x}_{i+T} = x_i - d\hat{x}_i - d\hat{x}_{i+1} \cdots - d\hat{x}_{i+T-1}, \tag{7}$$

where $\hat{x}_{i+T}$ is the predicted value at time $i + T$ and $d\hat{x}_i$, $d\hat{x}_{i+1}, \ldots,$ are the one-step, two-step,... T-step ahead predicted values using the first differenced time series. One may devise several methods to predict the values, $d\hat{x}_i$, $d\hat{x}_{i+1}, \ldots,$ using first differenced series. For example, making iterative prediction or the prediction using different lead-times are two such possible methods.

4.1. *Iterative prediction*

One of the several possible ways to estimate $d\hat{x}_i, d\hat{x}_{i+1}, \ldots, d\hat{x}_{i+T-1}$ is an iterative prediction. First, one lead-time ahead prediction of dx_{i-1}, that is, $d\hat{x}_i$, is conducted. $d\hat{x}_i$ is then considered a value in the series (i.e. dx_i) and the one lead-time ahead prediction of it, $d\hat{x}_{i+1}$, is estimated. The procedure is repeated up to the required prediction horizon. Although this method is simple, the prediction error grows with increasing lead-times.

This study suggests the following way.

4.2. *Prediction using different lead-times*

When the time series is known only up to x_i, the estimates of dx_i, $dx_{i+1}, \ldots, dx_{i+T-1}$ can be made using differenced time series for the desired lead-times, as shown in Table 2. The $1, 2, \ldots, T$ lead-time predictions of the point dx_{i-1}, as shown in Eq. (7), predict the value $\hat{x}_{i+T}$ in the original time series. The study adopted this method for higher prediction horizons, and the methodology followed and the results obtained are discussed in the following sections.

5. Data and Methodology

A theoretically chaotic Lorenz series and two daily river flow series are used in this study. The Lorenz model is given by the following three ordinary differential equations:

$$\begin{aligned} \dot{x} &= \sigma(y - x), \\ \dot{y} &= -xz + \gamma x - y, \\ \dot{z} &= xy - bz. \end{aligned} \tag{8}$$

Table 2. Prediction using different lead-times.

Original time series	First differenced time series
x_1	$dx_1 = x_1 - x_2$
x_2	$dx_2 = x_2 - x_3$
—	—
—	—
x_{i-1}	$dx_{i-1} = x_{i-1} - x_i$
x_i	$d\hat{x}_i$
$\hat{x}_{i+1}$	$d\hat{x}_{i+1}$
—	—
$\hat{x}_{i+T-1}$	$d\hat{x}_{i+T-1}$
$\hat{x}_{i+T}$	

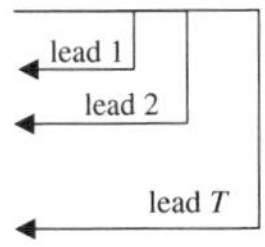

When standard parameter values $\sigma = 16$, $b = 4$, and $\gamma = 45.92$ are used, the orbits of the Lorenz system reside on a geometric object of dimension 2.06 (approximately) and exhibit nonperiodic, chaotic motion.[15] The $x(t)$ component is solved from the above equations by the fourth order Runge–Kutta method with a time step of $\Delta t = 0.01$. Seven thousand five hundred values of this $x(t)$ time series is used in this study.

Two river flows with very different flow characteristics are chosen: (1) Mississippi river at Vicksburg and (2) Wabash river at Mt Carmel. The Mississippi river is characterized with large flow rates (mean around 18,500 m^3/s) while the Wabash river is of moderate flow rates (mean around 750 m^3/s). These daily flow data are available in the US Geological Survey web site (http://water.usgs.gov/), and are used in the study. These two river flows have been shown to demonstrate chaotic behavior.[10,16,17]

The Mississippi river is one of the world's largest river systems with about 3,705 km in length and a river basin of around 3.2 million km^2. The average amount of water discharged to the Gulf of Mexico is about 18,500 m^3/s and a standard deviation of 9,728 m^3/s. The minimum and maximum flow values are about 3,900 m^3/s and 52,100 m^3/s, respectively. The daily flow time series of the Mississippi river measured at Vicksburg, Station No. 07289000 (Hydrologic Region 08 of USGS) for the period from 01 January 1975 to 31 December 1993 is used in this study.

The Wabash river is 765 km long flowing southwest from northwest Ohio. The basin area is approximately 85,750 km^2. The Wabash river has a mean flow rate of about 750 m^3/s and a standard deviation of around 792 m^3/s. The minimum and maximum flow values are about 48 m^3/s and 7,023 m^3/s, respectively. The daily river flow measured at Mt Carmel

Station, Station No. 03377500 (Hydrologic Region 5 of USGS), for the period from 01 January 1960 to 31 December 1978 is used in this study.

The analysis performed can be briefed as follows. First, the first differenced series of each time series is constructed as shown in Table 1. Each series is divided into three parts: training, test set, and validation sets. Then the optimal phase-space parameters (m, τ) and the optimal number of nearest neighbors (k) for a certain prediction horizon is sought for each time series considering the prediction accuracy on the test set. Then, with these optimal parameter sets the unseen data set, i.e. validation set, is predicted and the prediction accuracy of the first differenced and the original series are compared. Two error indicators, a relative error measure and an absolute error measure, are used to evaluate the model prediction performance. The relative error measure is the Normalized Root Mean Square Error (NRMSE) as given below:

$$\mathrm{NRMSE} = \sqrt{\frac{\sum_{i=1}^{N}(x_i - \hat{x}_i)^2}{\sum_{i=1}^{N}(x_i - \bar{x})^2}}, \tag{9}$$

where $\hat{x}_i$ is the predicted value of x_i and $\bar{x}$ is the average value of the time series. A zero value in NRMSE indicates a perfect prediction, while a value larger than 1 indicates that the predictions are no better than using the average value of the time series ($\bar{x}$). The absolute error measure, mean absolute error (MAE), is given as

$$\mathrm{MAE} = \frac{\sum_{i=1}^{N}\|x_i - \hat{x}_i\|}{N}, \tag{10}$$

where x_i is the desired value and $\hat{x}_i$ is the predicted value. Some details of the analysis are explained in the next paragraph.

Like in the inverse approach of Phoon *et al.*,[12] in this study the parameter set (m, τ, k) is optimized simultaneously with the least prediction error as the objective function. The data set is divided into three parts: training, test set, and validation sets. For the Lorenz time series, the first 6,000 points are used for training set, the next 750 for test set, and the last 750 for validation set. The first 15 years (approximately 5,480 data points) are used for training; the next 2 years (approximately 730 data points) for testing; and the last 2 years (approximately 730 data points) for validation.

Then, for a certain combination of embedding dimension (m), time delay (τ), and nearest neighbors (k), the phase-space is reconstructed and the test set is used to check the performance of the trained model. The combination of m, τ, and k which gives the least prediction error (NRMSE) on the test set is selected as the optimal parameter set for prediction on the validation set. The Genetic-Algorithm-based method proposed by Liong *et al.*[16] is used for this purpose. The optimal parameters sets are sought from the Genetic-Algorithm-based search from the combinations of the following ranges of parameters. The same ranges are considered for the original and the first differenced series of the Lorenz series and the two river flow time series. The range for the embedding dimension considered is 1–10. The range for time delay values is from 1–100, and the range for the number of nearest neighbors (k) is 1–100. Simulations are conducted for three different lead-times 1, 3, and 5 for all the time series considered.

6. Results

The results obtained for the chaotic Lorenz series are shown in Table 3 and those for the river flow series are shown in Tables 4 and 5, respectively.

Table 3. Prediction errors on validation set for original and first differenced Lorenz series with local averaging prediction method.

		Prediction error on validation set	
Lead time	Time series	NRMSE	MAE
Lead 1	Original (x)	0.01442 (2, 4, 9)	0.1202
	Differenced (dx)	0.00294 (4, 5, 7) **(80%)**[a]	0.0223 **(81%)**[a]
Lead 3	Original (x)	0.01919 (3, 6, 9)	0.1738
	Differenced (dx)	0.01012 (5, 4, 7) **(47%)**[a]	0.0795 **(54%)**[a]
Lead 5	Original (x)	0.02268 (6, 3, 6)	0.2071
	Differenced (dx)	0.01896 (6, 3, 3) **(16%)**[a]	0.1393 **(33%)**[a]

( )Values in parenthesis are optimal m, τ, and k, respectively.
[a]The percentage reduction of prediction error compared to original time series.

Table 4. Prediction errors on validation set for original and first differenced Mississippi series with local averaging prediction method.

Lead time	Time series	Prediction error on validation set NRMSE	MAE
Lead 1	Original (x)	0.0454 (2, 1, 5)	259.6
	Differenced (dx)	0.0396 (2, 8, 18) **(13%)**[a]	214.6 **(17%)**[a]
Lead 3	Original (x)	0.1437 (2, 1, 9)	836.9
	Differenced (dx)	0.1336 (2, 1, 60) **(7%)**[a]	767.82 **(8%)**[a]
Lead 5	Original (x)	0.2644 (2, 1, 8)	1589.4
	Differenced (dx)	0.2543 (5, 5, 71) **(4%)**[a]	1533.2 **(4%)**[a]

( ) Values in parenthesis are optimal m, τ, and k, respectively.
[a]The percentage reduction of prediction error compared to original time series.

It should be noted that for ease of computation, the optimal (m, τ, k) set found for some T-step ahead prediction is used for 1-step, 2-step, ..., $(T-1)$-step ahead prediction of first differenced series. For example, 3-step ahead prediction with first differenced series requires 1-step and 2-step ahead predicted values too (see Eq. (7)). In this study, these values are also predicted using the optimal set of 3-step ahead prediction. Therefore, it can be expected that if the optimal set for each time step was used, the prediction performance can be even better than that is reported here. Table 3 shows that the first differenced series with the proposed technique produces significantly better predictions than the original series. The percentage improvement, however, has decreased with the increasing lead-time. The percentage improvement over original series at a lead-time of 5 is still very significant (16% in terms of NRMSE and 33% in terms of MAE) suggesting that the technique can be effective even at higher prediction horizons than 5. Tables 4 and 5 show that the method is effective on river flow series too. In these series too the percentage improvement decreases as the prediction horizon increases. In the river flow series, although the

Table 5. Prediction errors on validation set for original and first differenced Wabash series with local averaging prediction method.

		Prediction error on validation set	
Lead time	Time series	NRMSE	MAE
Lead 1	Original (x)	0.1163 (2, 6, 16)	51.0
	Differenced (dx)	0.0641 (2, 1, 15) **(45%)**[a]	27.4 **(46%)**[a]
Lead 3	Original (x)	0.2883 (3, 1, 21)	124.8
	Differenced (dx)	0.2519 (5, 2, 33) **(13%)**[a]	116.4 **(7%)**[a]
Lead 5	Original (x)	0.4390 (2, 1, 25)	199.8
	Differenced (dx)	0.4147 (4, 2, 49) **(6%)**[a]	192.3 **(4%)**[a]

( ) Values in parenthesis are optimal m, τ, and k, respectively.
[a]The percentage reduction of prediction error compared to original time series.

percentage improvement at lead-time 5 is not as significant as in the Lorenz series the predictions are still better than using the original series.

7. Conclusion

The study proposed a technique to predict higher prediction horizons ($T > 1$) with the first differenced series of a given time series. First differenced time series is shown to produce better prediction accuracy than the original time series in phase-space prediction with local averaging models up to several lead-times. The percentage improvement with the proposed method over original series prediction, however, decreases with the increasing lead time. One disadvantage with the proposed method is that one may have to build T prediction models to make T-step ahead predictions. Note, however, that this study has used only one model (one parameter set) and the results are still satisfactory. For the synthetic Lorenz series, the predictions that the proposed technique produces at 5-step ahead are significantly better than the original series predictions, suggesting that it could be effective even at longer prediction horizons than 5. For the river flow series too, the proposed technique is effective up to about 5 lead times. Research is on the way for finding possible theoretical reasoning for better prediction with first differenced series.

References

1. E. N. Lorenz, Deterministic non-periodic flow, *J. Atmos. Sci.* **20** (1963) 130–141.
2. A. W. Jayawardena and F. Lai, Analysis and prediction of chaos in rainfall and streamflow time series, *J. Hydrol.* **153** (1994) 23–52.
3. I. Rodriguez-Iturbe, B. F. De Power, M. B. Sharifi and K. P. Georgakakos, Chaos in rainfall, *Water Res. Res.* **25**(7) (1989) 1667–1775.
4. A. Porporato and L. Ridolfi, Clues to the existence of deterministic chaos in river flow, *Int. J. Mod. Phys. B* **10**(15) (1996) 1821–1862.
5. A. Porporato and L. Ridolfi, Nonlinear analysis of river flow time sequences, *Water Res. Res.* **33**(6) (1997) 1353–1367.
6. M. B. Sharifi, K. P. Georgakakos and I. Rodriguez-Iturbe, Evidence of deterministic chaos in the pulse of storm rainfall, *J. Atmos. Sci.* **47**(7) (1990) 888–893.
7. B. Sivakumar, S. Y. Liong and C. Y. Liaw, Evidence of chaotic behaviour in Singapore rainfall. *J. Am. Water Res. Ass.* **34**(2) (1998) 301–310
8. B. Sivakumar, S. Y. Liong, C. Y. Liaw and K. K. Phoon, Singapore rainfall behaviour: Chaotic? *J. Hydrol. Eng.* **4**(1) (1999a) 38–48.
9. A. W. Jayawardena and A. B. Gurung, Noise reduction and prediction of hydro-meteorological time series: Dynamical systems approach vs stochastic approach, *J. Hydrology* **228** (2000) 242–264.
10. X. Y. Yu, S. Y. Liong and V. Babovic, *J. Hydroinformatics* **6**(3) (2004) 209–223.
11. V. Babovic, M. Keijzer and M. Stefansson, Optimal embedding using evolutionary algorithms, *Proc. 4th Int. Conf. Hydroinformatics*, Iowa city, July 2000.
12. K. K. Phoon, M. N. Islam, C. Y. Liaw and S. Y. Liong, Practical inverse approach for forecasting nonlinear hydrological time series, *J. Hydrol. Eng. (ASCE)* **7**(2) (2002) 116–128.
13. N. H. Packard, J. P. Crutchfield, J. D. Farmer and R. S. Shaw, Geometry from a time series, *Phys. Rev. Lett.* **45**(9) (1980) 712–716.
14. F. Takens, Detecting strange attractors in turbulence, in *Dynamical Systems and Turbulence*, Lecture Notes in Mathematics, (eds.), D. A. Rand and L. S. Young, (Springer-Verlag, Berlin, 1981), Vol. 898, pp. 366–381.
15. H. D. I. Abarbanel, Analysis of Observed Chaotic Data (Springer, New York NY, 1996).

16. S. Y. Liong, K. K. Phoon, M. F. K. Pasha and C. D. Doan, Efficient implementation of inverse approach for forecasting hydrological time series using micro GA, *J. Hydroinformat.* **7** (2005) 151–163.
17. D. S. K. Karunasinghe and S. Y. Liong, Chaotic time series prediction with a global model: Artificial Neural Network, *J. Hydrol* **323** (2006) 92–105.

Advances in Geosciences
Vol. 11: Hydrological Science (2007)
Eds. Namsik Park *et al.*

INTENSE CONVECTIVE RAIN ESTIMATION USING GEOSTATIONARY METEOROLOGICAL SATELLITE

WARDAH TAHIR* and SAHOL HAMID ABU BAKAR
Faculty of Civil Engineering, Universiti Tekonologi MARA
Shah Alam, 40450, Selangor, Malaysia
**warda053@salam.uitm.edu.my; wardah_tahir@yahoo.com*

MAZNORIZAN MOHAMAD
Satellite Divison, Malaysian Meteorological Department
Jalan Sultan, 466667, Petaling Jaya, Selangor, Malaysia

In this study, a quantitative rainfall estimates model is developed using the information from the Geostationary Meteorological Satellite-5 (GMS-5) infrared (IR) image. The model is to estimate intense convective rainfalls which are responsible for the majority of flash flooding in urban areas of Klang River Basin, Malaysia. Information from records of GMS-5 IR images have been retrieved for selected convective cell clouds associated with intense rain to be trained with the radar rain rate in a back-propagation neural network. The model has been validated against the actual convective rainfalls observed over upper Klang River Basin. Results show a satisfactory performance of the model in estimating a real averaged total rainfall with a correlation coefficient of r as 0.91 between the estimated values with the Thiessen areal averaged total rainfall.

1. Introduction

Flash flood has been the most frequent natural hazard in the urban areas of Malaysia, causing massive damages to properties and infrastructures and in several cases, loss of life. It has been widely known that storms of convective origins are responsible for the majority of flash flood events. The rain measuring system, whether the conventional rain gauges or the more advanced Remote Sensing and Transmission Unit (RSTU) panel, can only be sparsely installed at suitable locations; hence it is considered as point rain measurement. The use of meteorological satellite images in estimating rainfall has become an attractive option in improving the performance of flood forecasting and warning systems. Although the estimates are indirect, meteorological satellites with fine temporal and spatial resolution cover broader area including rugged mountains (radar beam blockage problem) and areas inaccessible to rain gages such as deep forests and oceans, hence should be taken as complementary to radar and rain gauge measurements. In addition, the rain estimation from the observation of cloud development and movement would enable earlier forecast of critical storm events.

2. Objective of Study

The research objective is to develop a Quantitative Rainfall Estimate (QRE) model for intense convective rain using the information from the geostationary meteorological satellite infrared (IR) images, weather radar information, and Numerical Weather Prediction (NWP) model products. Artificial Neural Network (ANN) technique is applied as the main tool in the model development.

3. Background Information

3.1. *Deep convection*

The presence of thick cumulonimbus clouds that reach to high cold altitudes is an indication of deep convective activity. The temperatures of cloud tops for severe storms are very low, sometimes below the undisturbed tropopause temperature. Intense rains and thunderstorms are the most frequent representations of deep convection. Griffith *et al.*[1] had found that a convective system is more active and produces the greatest rainfall rates when the tops are becoming colder and expanding.

Hees[8] discovered that convective overshoot is visible in high resolution infrared images of very cold region of about 10 km diameter and becomes the only direct evidence for active deep convection. The cirrus anvils associated with well-developed convection can reach to 10^5 km^2, and can last long after the actual convection has ceased while the actual deep convection takes place in smaller regions.

3.2. *Geostationary meteorological satellite*

Geostationary meteorological satellites have high altitudes around 36,000 km, maintain a fixed geographic location, and orbit at the same speed with which the earth rotates; so they stay above the same spot on the earth. The satellites make observations at around 20–30 min intervals throughout each day, enabling meteorologists to observe clouds at all levels of the atmosphere over both the land and the oceans.

Radiation can be measured at specific wavelengths by using a filter placed in front of the satellite sensor. In satellite meteorology, the larger the radiance from an element the darker the pixel of an infrared image is. Lower clouds are warmer and thus emit more radiation than higher clouds, which are much colder. Low clouds appear gray on an infrared satellite picture while high clouds appear white. By closely analyzing satellite images, meteorologists determine cloud heights and thicknesses.

4. Methodology

4.1. *Data collection — geostationary meteorological satellite images*

Meteorological satellite images have been acquired from the Malaysian Meteorological Department which receives the images from the Japanese National Space Development Agency (NASDA). The GMS-5 is positioned at 140° east and has a spin scan radiometer that returns visible and infrared images. The images utilized for this study are hourly infrared GMS-5 IR images channel 10.3–11.3 μm with spatial resolution of 7 km × 5 km. Images have been collected for the year 2003 and for the wet months (March, April, October, and November) of 2004 and 2005. Validation data used are the latest GMS series that is MTSAT (Multifunctional Transport Satellite) IR images for the rainy days of the year 2006.

4.1.1. *Geolocation of pixel*

An important aspect of meteorological satellite image interpretation is the accurate geographic location of image pixels. The process of geolocating pixels is called *navigation*. The process requires current satellite orbit and altitude parameters, and also the algorithms that use these parameters to calculate the latitude and longitude of each pixel. The meteorological geostationary satellite images received by the Malaysian Meteorological Department are provided with an algorithm which performs the navigation; hence the border lines representing countries can be placed correctly as illustrated in Fig. 1(a).

To locate a specific area, the image is first cropped from 100° to 110°E and 0° to 10°N, as shown in Fig. 1(b) to produce a 10° × 10° image. Since the number of pixels per image of 10° × 10° can be determined by image

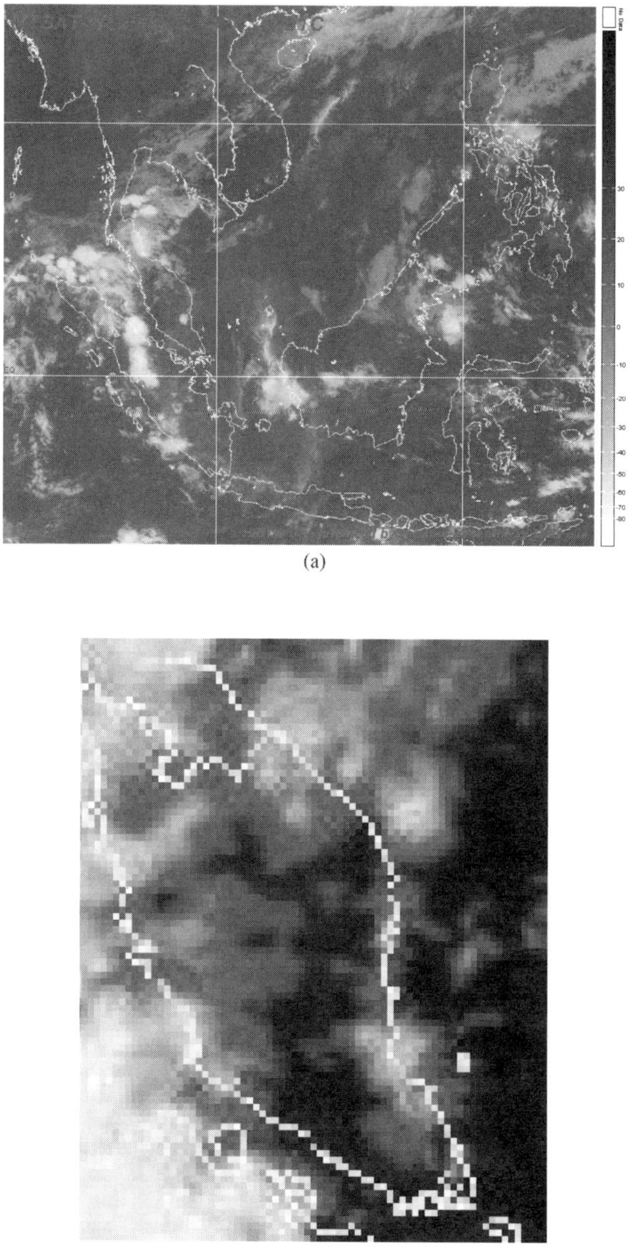

Fig. 1. (a) An example of MTSAT image; (b) Cropped GMS-5 image 100° to 110°E and 0° to 10°N.

processing software like MATLAB^{TM}, the geolocation of each pixel or which pixel is closest to a particular latitude–longitude point on earth can be calculated.

4.1.2. *Reading pixel cloud top brightness temperature*

An infrared image is composed of pixels with discrete digital values which correspond directly to an equivalent blackbody radiation. The 8-bit image display system allows a range of value from 0–255 for a pixel. A color or monochromatic gray shade can be assigned to each digital value to produce an enhanced image. In a gray scale infrared (IR) image, cold clouds are high clouds with pixel brightness values indicated by the corresponding gray scale color tone. Using the pixel brightness value, the temperature can be estimated from the following formulas produced by the National Oceanic and Atmospheric Administration (NOAA):[6]
For brightness value, $B > 176$

$$T = 418 - B$$

or when $B \Leftarrow 176$

$$T = 330 - (B/2) \qquad (1)$$

Note that the resulting temperatures, T, are in Kelvin and the highest pixel values correspond to the coldest temperatures. The cloud position on earth is then estimated after some parallax correction as suggested by Kidder and Vonder Haar.[11]

4.2. *ANN model development*

Rainfall is an outcome of a complex nonlinear atmospheric process involving many parameters. Even complicated physically based numerical weather prediction model cannot precisely estimate rainfall at the required temporal and spatial resolution, especially for flash flood simulation. An alternative to the physically based model is the data-driven approaches, which are based on machine learning. Artificial Neural Network (ANN) is a widely used artificial intelligence technique which provides a powerful means of performing nonlinear regression between combinations of prime factors to the outcome of interest. Numerous works have been done by other researchers on the use of ANN in the rainfall and atmospheric processes modeling as described among others by Kim and Barros[3] on quantitative flood forecasting using multisensor data and ANN, Sooroshian *et al.*[10] on application of ANN in satellite-based estimates of tropical rainfall, Kuligowski and Barros[7] on short-term precipitation forecasting using ANN, and Bankert[8] on cloud classification of a AVHHR imagery in maritime regions using a probabilistic ANN.

Neural networks, which are inspired by biological nervous systems are composed of simple elements operating in parallel. It is the connections between elements which determine the network function. The multilayer back-propagation neural network consists of layers of neurons, with each layer being fully connected to the next layer by interconnection strengths or weights w. Initial estimated weight values are progressively corrected during a training process that compares predicted outputs to target outputs, and back-propagates any error to determine the appropriate weight adjustments necessary to minimize the error. The algorithm used here for adjusting the weights is based on the *generalized delta rule.* The simplest version of gradient descent algorithm in minimizing error is to change each weight by an amount proportional to the accumulated $\partial E/\partial w$:

$$\Delta w = - \in \frac{\partial E}{\partial w}, \tag{2}$$

where $\in$ is the learning rate, w is weights, and E is total error.[4]

4.2.1. *ANN input variables*

Following Kidder and Vonder Haar,[11] there are two ways of verifying rainfall estimates from satellite image that are (i) comparing with rain gauge values and (ii) comparing with radar rain estimates. Comparison of satellite rain estimates with radar is considered more reasonable, since it samples an area comparable with the size of satellite pixels, even though, generally radar rain estimation is less accurate than the rain gauge measurement. Since rain gauge samples area roughly 0.1 m in diameter, comparison of satellite rain estimates with rain gauge would be poor due to the high spatial variability of rain. Since radar technique is also an indirect estimation of rain rate, a verification procedure of the technique using the radar network system available in Peninsular Malaysia had been performed beforehand by comparing radar rain estimation with point gage measurement. The results had shown a satisfactory performance of the radar rain rate to be applied further in the ANN satellite rainfall estimation model development.

Peninsular Malaysia has a network of six Weather Surveillance Radars (WSR), and one Doppler type radar. The radar data used for the research have been taken from Subang WSR81S B and S radar located at Lat 3.12°N and Long 101.55°E and altitude of 32 m. The acquired data are hourly image of radar displays and 15-min interval of radar reflectivity values. Figure 2 shows a sample of radar display and the Look-Up Table (in set).

The variables selected as inputs to the neural network and their coefficients of correlation are given in Table 1. As an attempt to represent the overshoot of convective clouds, Sobel operator is used to estimate the cloud top brightness temperature (*Tb*) gradient surrounding a pixel of interest. The Precipitable Water Content (PW), Relative Humidity (RH), and Vertical Wind (VW) are the three variables included to represent the fundamental meteorological elements in convective rain formation, as described by Fox and Collier.[5] These variables are obtained from the Numerical Weather Prediction (NWP) model products. The Malaysian Meteorological Department (MMD) adopts the Fifth Generation Mesoscale Model (MM5) as its NWP model. For the purpose of NWP historical data retrieval, NCEP Reanalysis Data have been recommended by the MMD to

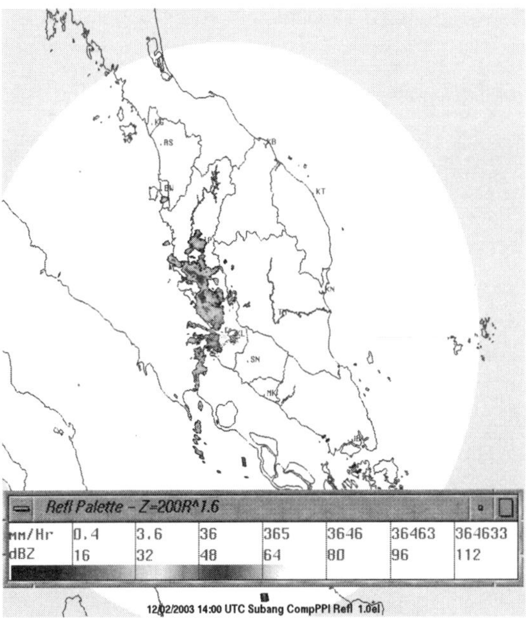

Fig. 2. Sample of a radar display and the Look-Up Table (in set).

Table 1. Correlation coefficient between parameters involved.

Variables	Correlation with	Spearman's rho	Significant (2-tailed)
Pixel temperature	Radar rain rate	−0.734	0.000
Mean 5 × 5 pixel	Radar rain rate	−0.517	0.000
Std deviation 5 × 5 pixel	Radar rain rate	−0.283	0.000
Temperature change (K/h)	Radar rain rate	0.880	0.000
Maximum sobel	Radar rain rate	0.413	0.000
Precipitable water content	Mean 5 × 5 pixel temperature	−0.239	0.001
Relative humidity	Mean 5 × 5 pixel temperature	−0.165	0.020
Relative humidity	Precipitable water content	0.728	0.000
Vertical wind	Relative humidity	0.301	0.000

Note: Correlation is significant at the 0.01 level.

be used in this research. The NCEP Reanalysis data are provided by the NOAA/OAR/ESRL PSD, Boulder, Colorado, USA, from their web site at *http://www.cdc.noaa.gov/*. An example of an image of PW downloaded from the web site is given in Fig. 3.

4.2.2. *Multilayer back-propagation ANN*

The study employs the multilayer feed-forward back-propagation type network in the satellite rainfall estimation model development. The input variables are entered as an array to each neuron in the input layer. The values are then transmitted to the next layer (hidden layer) through links, after being multiplied with the weights, w, associated with each link. At each neuron in the hidden layer, the weighted transmitted values are summed up together with a weighted bias, b. A suitable transfer function is next used by the results to generate a level activity for the neuron. Then, the output layer will receive these activation levels through links, and the steps are repeated as before. This training process requires a set of examples of proper network behavior — network inputs p and target outputs t. During training, the weights and biases of the network are iteratively adjusted to minimize the average squared error between the network outputs a, and the target outputs t.

About 204 sets of data have been retrieved for convective rain events and used in the training. The final network consists of an input layer with

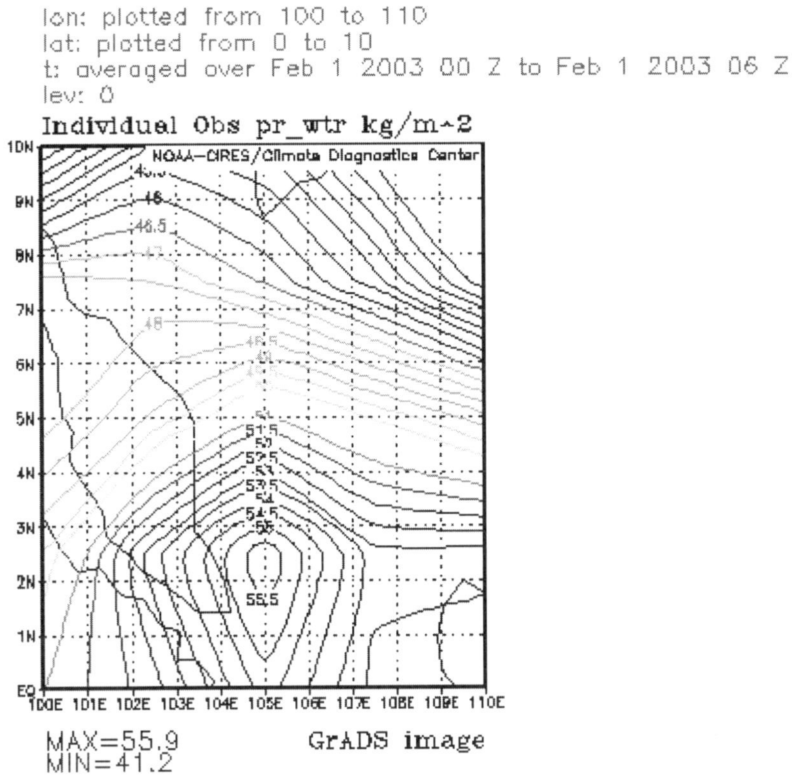

Fig. 3. Sample of PW data retrieved from the NCEP web page.

eight input variables, one hidden layer with three neurons, and one output layer. The transfer functions are log-sigmoid function for hidden layer and linear function for the output layer as indicated in Fig. 4. The output of the model is the rain rate (mm/h) at the collocated time of the satellite images.

5. Results and Analysis

5.1. *Case study area*

The Klang River Basin is located toward the west coast of Peninsular Malaysia, and consists of the most urbanized and still rapidly urbanizing regions including the Federal Capital of Kuala Lumpur. Most of the important areas in the basin have been frequently affected by recurrent

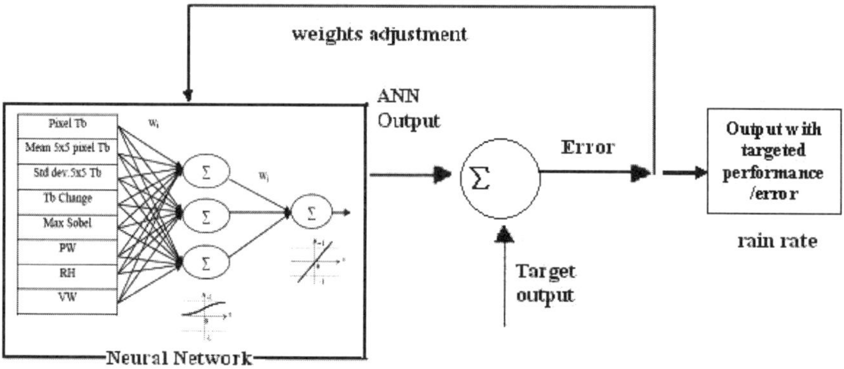

Fig. 4. Multilayer back-propagation ANN adopted in the study.

flooding of Klang River and its tributaries. The Klang River has a total stream length of about 120 km and has a total catchment area of about 1200 km^2 at the mouth of Port Klang. The main river has headwaters at the mountainous areas to the north-east of Kuala Lumpur where the maximum elevation rises to above 1400 m. Most of the catchments are hilly, and rivers are generally fast-flowing. The study will focus on the upper Klang River Basin as shown in Fig. 5, which comprises of an area of 468 km^2.

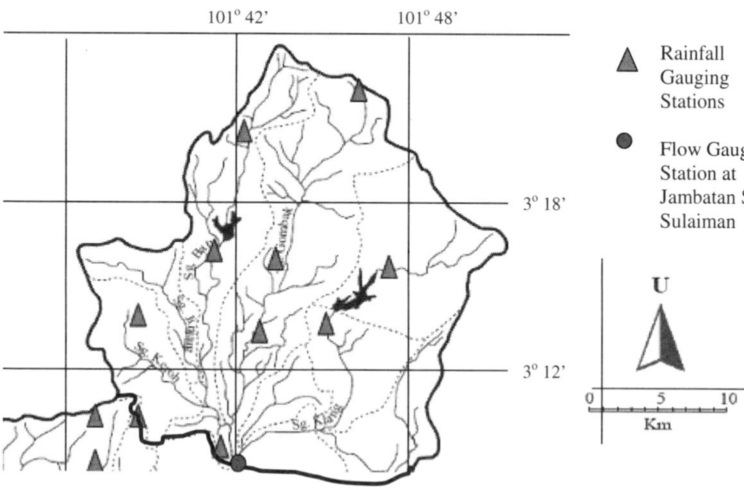

Fig. 5. Upper Klang River basin.

5.2. *Model application*

The ANN satellite rainfall estimation model is applied to estimate the real rainfall for the upper Klang River Basin by repeated runs on every pixel covering the area. Rain is considered zero, for infrared image brightness temperature is greater than 250 K following Stout *et al.* in the work of Kidder and Vonder Haar[11] who had adopted a threshold count of 250 K to define clouds in infrared channel. An example is given below for rainfall estimation during a flash flood event dated 10 June 2003 over the catchment. The first row consists of three pixel representations over the catchment arranged in time sequence with cloud top brightness temperature assigned. The second row is the estimated rainfall rate for each pixel for duration (d) between 14:23 and 15:23 (e) between 15:23 and 16:23. Note that one of the input variables to the ANN model is the temperature change per hour. Since the output from the model is the rain rate for each pixel in mm/hr, the areal average rain depth for the hour can then be estimated. The results are shown in Fig. 6.

The procedures are repeated for the whole storm event duration occurring on the date. Figure 7 shows the hourly estimates for areal averaged rain for upper Klang River Basin using the model and plotted with the Thiessen areal averaged rain and the recorded flow at the outlet station. The graph indicates a close approximation between the estimated rain and the recorded rain during the flash flood event.

5.3. *Model validation*

The study advocates the use of gauged rainfall measurements as the "truth," and hence uses them for validation purposes. Hourly interval rainfall data have been collected from the Drainage and Irrigation Department (DID) of Malaysia for 10 stations in the upper Klang River Basin as shown in Fig. 5. Convective rainfall events of category heavy (>30 mm/h) for at least one station are selected from the year 2006 and wet months of 2005 for validation purposes. A total of 107 hourly rainfall data sets from 33 storm events have been selected. To measure the closeness of the ANN satellite rainfall estimation values to the gauged rainfall values, the coefficient correlation, r, has been determined. Figure 8(a) is the plot of ANN hourly areal averaged rain estimation versus gauge measured Thiessen areal averaged rain. An r value of 0.63 is obtained which indicate not a very strong correlation between these two set of values. The results are very

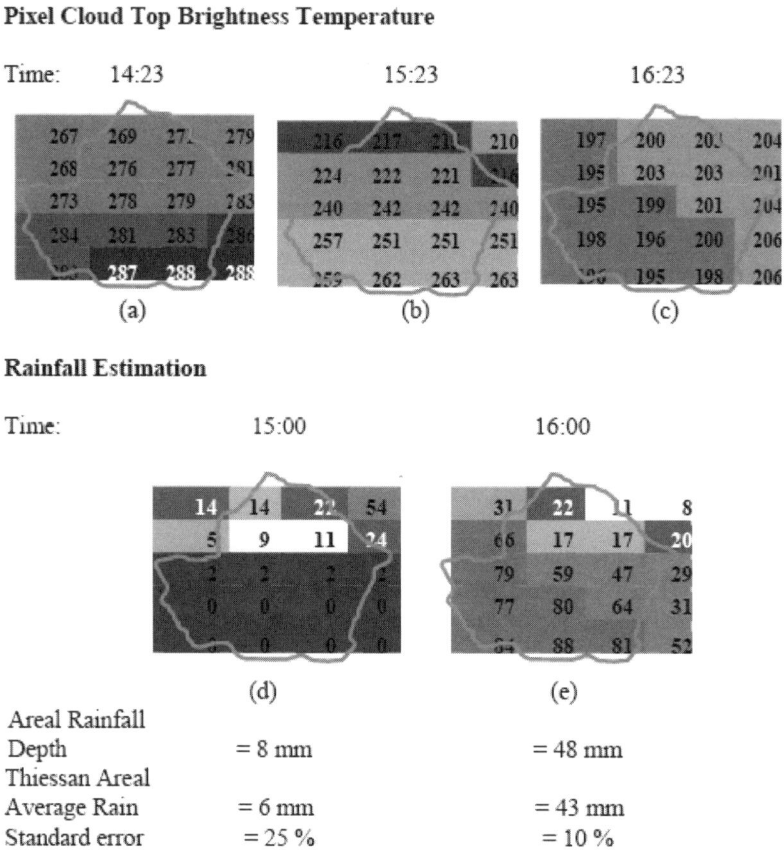

Fig. 6. Areal rainfall estimation over the catchment.

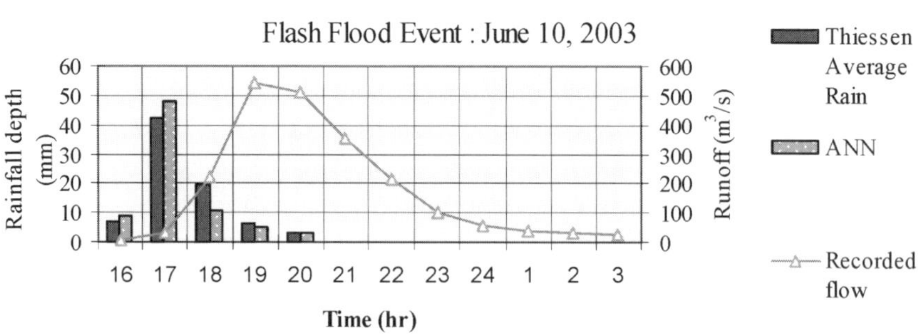

Fig. 7. ANN satellite areal average rain estimation plotted along with Thiessen areal average rain versus flow measured at the catchment outlet.

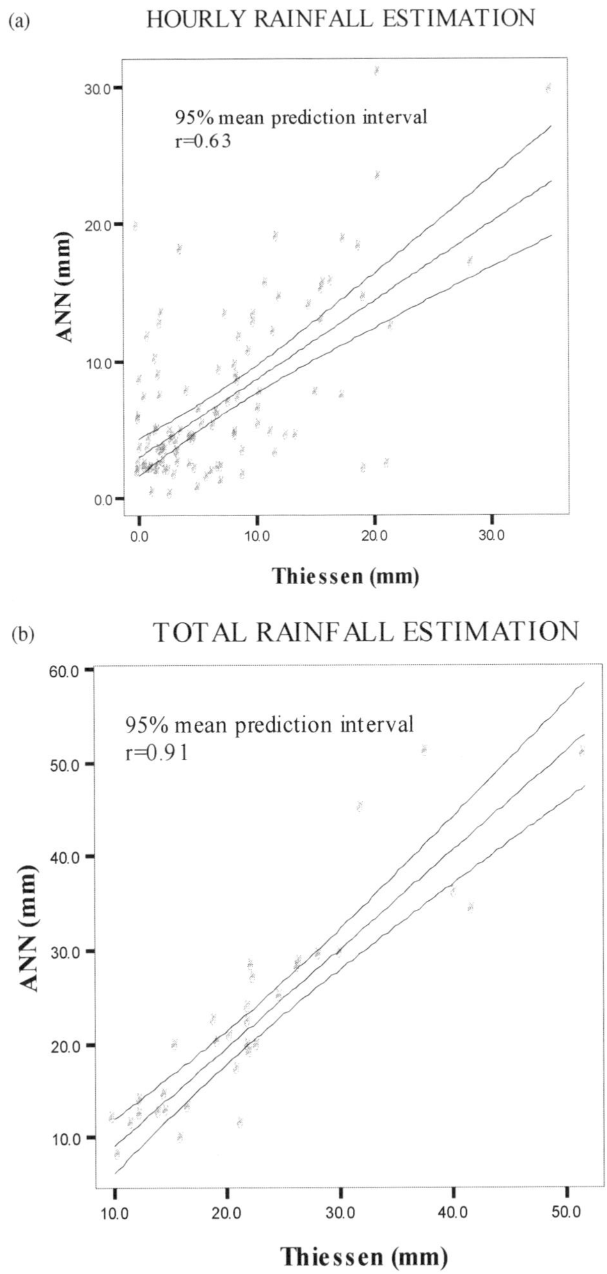

Fig. 8. Validation of ANN areal averaged rainfall estimation against gauge measured Thiessen areal averaged rain; (a) hourly (b) total storm duration.

Table 2. Accuracy measures (ANN versus Thiessen).

	Hourly	Total
Mean Absolute Error (MAE)	3.8	3.3
Mean Squared Error (MSE)	30.9	22.3
Root-Mean-Squared Error (RMSE)	5.6	4.7

much improved for the total areal averaged rain depth estimation. Total here means the cumulative rainfall within the storm duration. The storm events included here are 2-h, 3-h, and 4-h rainfall. Figure 8(b) displays the plot of total areal averaged rain estimated by ANN against the gauge measured. The more convergence of the points to the linear line in the figure and the r value of 0.91 indicate that the model performs better for longer duration of estimation. This is in conformity with the statement by Kidder and Vonder Haar[11] that satellite-based rainfall estimation performs better for larger area and longer duration.

5.4. *Other accuracy measures*

Other accuracy measures performed in the study are the determination of Mean Absolute Error (MAE) and Root Mean Squared Error (RMSE). Table 2 shows the values of MAE and RMSE for hourly estimation as 3.8 and 5.6, respectively, while for total rainfall estimation as 3.3 and 4.7, respectively. The values are considered satisfactory with a point that better estimation is expected for longer time duration.

6. Conclusion

The paper presents a QRE model development for intense convective rainfall estimation. The results indicate a very encouraging performance by the model. The model can then be applied as an input to a flood forecasting and warning system or any available flood modeling software which requires rainfall input. In addition, a cross-correlation technique can be applied to provide a tracking of the movement of the estimated convective cell for quantitative rainfall forecasting.

The main contribution of this work is the exploratory and correlation analysis between the geostationary meteorological satellite infrared images and intense convective rain for an equatorial country like Malaysia and the development of a quantitative rainfall estimates model incorporating the knowledge from the observation.

Acknowledgments

The authors would like to thank the Institute of Research, Development and Commercialization of Universiti Teknologi MARA for funding the project and to many helpful individuals from the Malaysian Meteorological Department and the Drainage and Irrigation Department of Malaysia.

References

1. C. G. Griffith, W. L. Woodley, P. G. Grube, D. W. Martin, J. Stout and D. N. Sikdar, *Mon. Wea. Rev.* **106** (1978) 1153–1171.
2. G. A. Vicente, Visiting Scientist's report, NASA/GSFC Distributed Active Archive Center DAAC, USA (2001).
3. G. Kim and A. P. Barros, *J. Hydrology* **246** (2001) 45–62.
4. M. T. Hagan, H. B. Demuth and M. Beale, *Neural Network Design* (Thomson Publishing, 1996), pp. 19–21.
5. N. I. Fox and C. G. Collier, *J. Hydrology* **237** (2000) 1–16.
6. NOAA, Latest retrieved on June 12, 2007 from. http://www.goes.noaa.gov/ECIR2.html
7. R. J. Kuligowski and A. P. Barros, *Mon. Wea. Rev.* **126** (1998) 470–482.
8. R. L. Bankert, *J. Appl. Meteorol.* **33** (1994) 909–918.
9. R. M. Hees, J. Lelieveld and W. D. Collins, *J. Geophys. Res.* (1999) 9213–9228.
10. S. Sorooshian, K. L. Hsu, X. Gao, H. V. Gupta, B. Imam and D. Braithwaite, *Bull. Am. Meteorol. Soc.* **81** (2000) 2035–2046.
11. S. Q. Kidder and T. H. Vonder Haar, *T. H. Satellite Meteorology, An Introduction* (Academic Press, 1995).

Advances in Geosciences
Vol. 11: Hydrological Science (2007)
Eds. Namsik Park *et al.*

ESTIMATING GLOBAL DISTRIBUTION OF SPATIAL AND TEMPORAL CORRELATION LENGTH OF POINT RAINFALL INTENSITY USING LOW FREQUENT OBSERVATIONS FROM SPACE

EIICHI NAKAKITA
Disaster Prevention Research Institute, Kyoto University
Gokasho Uji, Kyoto 611-0011, Japan

LISAKO KONOSHIMA
Department of Urban Engineering,
Graduate School of Engineering, Kyoto University
Gokasho Uji, Kyoto 611-0011, Japan
konoshima@hmd.dpri.kyoto-u.ac.jp

The purpose of this research is to develop a method to estimate the global distribution of climatologically important stochastic parameters such as temporal and spatial correlation length of instantaneous rainfall from the low frequent observation from space. A model to estimate population variance from the sample variance of low frequent observation is used, by considering temporal correlation. In this paper, a revised temporal–spatial correlation model already showed in the previous paper (C. Nakakita, S. Okane and L. Konoshima, *Advances in Geosciences 2006*, Vol. 6 (World Scientific, 2007)), which also considers the spatial correlation of the observed area, is corrected and discussed. Also, we focus on the estimation method in lower latitude, and model validation and improvement are shown with the comparison to the ground radar observation in the Kinki area of Japan. An application to the west part of Asian monsoon region during the southwest monsoon season is shown, and temporal correlation length, spatial correlation length, and population variance of the region are estimated.

1. Introduction

The rainfall stochastic properties observed from a satellite in non-sun-synchronous orbit have an outstanding point compared to ground observations in its global coverage of time and space. On the other hand, the observation has limitation in its intermittency. One of such satellite, TRMM, the Tropical Rainfall Measuring Mission loads a first space-borne precipitation radar, PR, and during its mission and broad observation in the area between 35° N and 35° S, provides a detailed data set of rainfall and latent heating over vastly unobserved oceanic and tropical continental regimes. One of the primary purposes of TRMM is estimating the variation of region-average monthly precipitation in 5° square-grid within less than 10% errors, and its feasibility has been examined since before the launch of TRMM as in Bell[1] and Ikebuchi *et al.*[2] Besides that, Nakakita *et al.*[3] have proposed a method to consider the rainfall dependency upon topographic elevation after TRMM was launched. These researches, however, were based on *a priori* estimates of climatologically important stochastic variables using long-term ground observations.

Meanwhile, Nakakita *et al.*[4] developed a method to estimate population variance from an intermittent observation of satellite, based on a relation between the observation frequency and the sample variance of monthly rainfall derived from a continuous observation of ground radar. This method not only corrects the sample variance of monthly rainfall but at the same time estimates temporal correlation length as a stochastic parameter by introducing a temporal correlation function of point instantaneous rainfall.

In this paper, a temporal–spatial correlation model previously introduced in Nakakita *et al.*,[5] which also considers the spatial correlation of the observed area, is corrected and its accuracy is discussed again. Also, we focus on the estimation method in lower latitude, where the observation frequency is not enough, in particular, less than 15 times a month on average.

On the other hand, Global Precipitation Measurement (GPM), a followed-on and expanded mission of the current ongoing TRMM is proposed, and is to be launched in 2013. GPM is composed of a core nonsynchronous orbit satellite and constellation of about eight sub-sun-synchronous orbit satellites and by these plural observations, global rainfall distributions will be available every three hours in GPM era.

The global estimation method shown in this paper would not only improve the estimation accuracy of low frequent observation of TRMM, but would also assess the accuracy in the higher observation frequency of the coming GPM, since the model describes the relation between the observation frequency of TRMM and estimation accuracy of stochastic values obtained from TRMM.

2. Theory/Methodology

2.1. *The model*

As was emphasized by Nakakita *et al.*,[5] the sample variance of monthly rainfall calculated directly from low frequent observation includes sampling error, and tends to estimate larger than the true value. From past research, it is a known fact that as observation frequency increases (or in a continuous observation), the variance of monthly rainfall converges to a certain value as shown in Fig. 1 (see Ref. 3). This relation is formulated by a stochastic model of sample variance, taking in the formula of autocorrelation function (temporal correlation function) of point rainfall intensity.[4]

The model was extended to a temporal–spatial correlation model by Nakakita *et al.*,[5] in order to take the observation footprint into consideration. However, the model derived in the paper included an error in the formulation. Here in this paper, we corrected the derivation and the new spatial model is shown. This new temporal–spatial model formula is derived as below.

Spatial monthly rainfall $H_{j,n}(x)$ of month j at point x in the arbitrary domain B_R (with radius R and area V_{BR}) can be calculated by integrating the point rainfall intensity, $P(x, t_{i_j})$, with the area B_R and the total hour of a month, with the parameter of observation frequency n (number of observations in a month).

With $H_{j,n}(x)$ and its m months sample mean of spatial monthly rainfall $M_{n,m}(x)$, expectation of variance of regional monthly rainfall $E[S^2_{n,m}]$ can

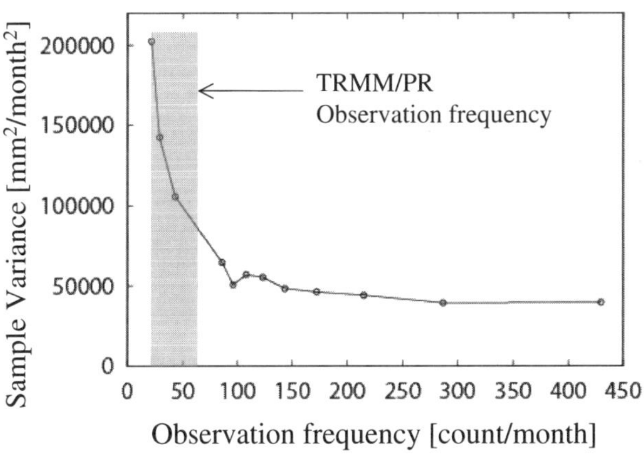

Fig. 1. The relation between sample variance and observation frequency.

be derived with the parameters n and m.

$$E[S_{n,m}^2] = E\left[\frac{1}{m}\sum_{j=1}^{m}\left\{\frac{T}{n}\sum_{i=1}^{n}\frac{1}{V_{B_R}}\int_{B_R} p_j(x,t_{i_j})dx \right.\right.$$
$$\left.\left. - \frac{1}{m}\sum_{j=1}^{m}\frac{T}{n}\sum_{i=1}^{n}\frac{1}{V_{B_R}}\int_{B_R} p_j(x,t_{i_j})dx\right\}^2\right]. \quad (1)$$

Here, by approximating the temporal and spatial correlation function below,

$$c(r,\tau) = \mu_{i2}e^{-\nu|\tau|}\left(1+\frac{c\,|r|}{2}\right)e^{-c|r|} - \mu_{i1}^2, \quad (2)$$

where $1/\nu$ and $1/c$ give the temporal correlation length and the spatial correlation length, respectively. Using the above equation, a relation among the expectation of sample variance of the monthly rainfall $E[S_{n,m}^2]$ and the observation frequency in a month n, and the number of months m (record length) can be modeled as

$$E[S_{n,m}^2] = \mu_{i2}\begin{pmatrix} \frac{1}{m}\frac{T^2}{n^2}\sum_{j=1}^{m}\sum_{i=1}^{n}\sum_{k=1}^{n} e^{-\nu|(k-i)\Delta T|} \\ -\frac{1}{m^2}\frac{T^2}{n^2}\sum_{j=1}^{m}\sum_{i_j=1}^{n}\sum_{l=1}^{m}\sum_{k_l=1}^{n} e^{-\nu|(l-j)T+(k-i)\Delta T|} \end{pmatrix}$$
$$\cdot\frac{1}{V^2}\int_{B_R}\int_{B_R}\left(1+\frac{c\,|r|}{2}\right)e^{-c|r|}dx'dx''. \quad (3)$$

Here, the integrator B_R is the footprint of TRMM, which is the area of about 4 km radius.

Once the parameters ν and c are estimated, by making m and n infinite, the expectation of sample variance (i.e. population variance) of monthly rainfall can be analytically derived.

$$\lim_{n\to\infty,m\to\infty} E[S_{n,m}^2]$$
$$= \frac{\mu_{i2}}{T^2V^2}\int_0^T\int_0^T e^{-\nu|\tau|}\int_{B_R}\int_{B_R}\left(1+\frac{c|r|}{2}\right)e^{-c|r|}dx'dx''dt'dt''. \quad (4)$$

2.2. *Method applied to TRMM/PR observation*

From the formula derived in Sec. 2.1, an arbitrary domain of any size (here, larger than the TRMM footprint) can be regarded as a virtual footprint

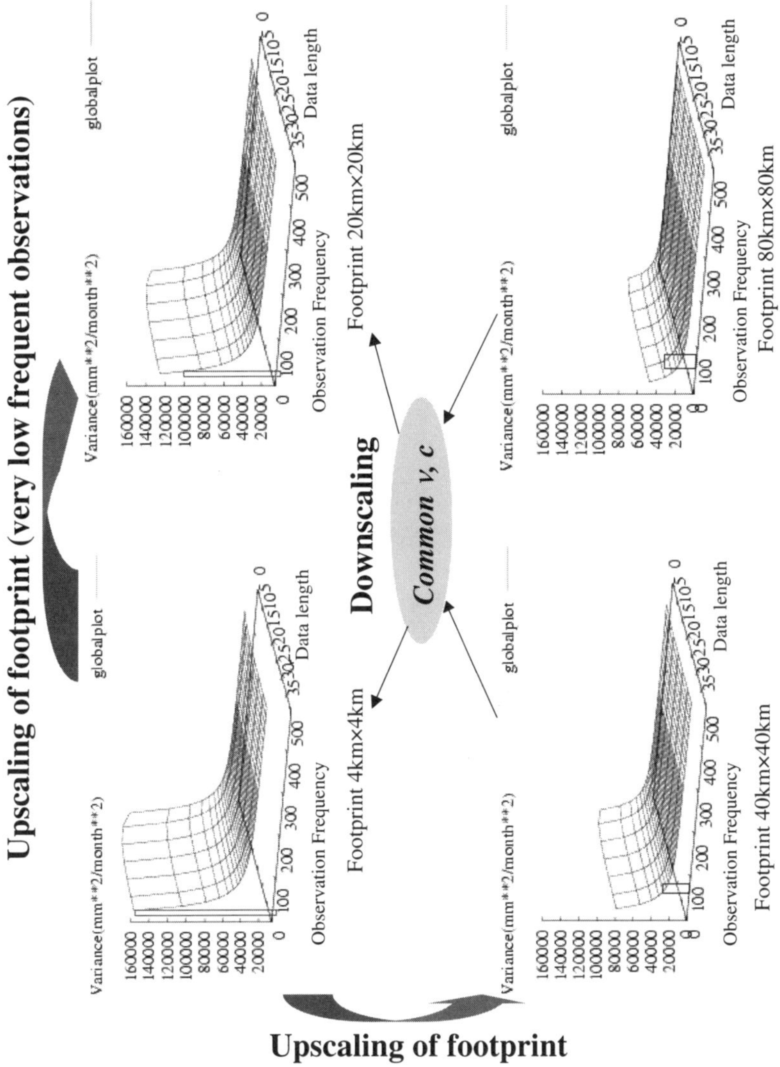

Fig. 2. Image of changing footprints.

and can be calculated in various scales. Since the temporal and spatial correlation lengths are common in all footprints, the parameters ν and c are determined as one through the various footprints. In areas where observation frequency is extremely low, as in latitudes near the equator, parameters can be estimated by changing the domain sizes until it has enough observation frequency i.e., determined from various footprints of different sizes. As shown in Fig. 2, when the footprint is as small as 4 km square-grid as in TRMM/PR, the TRMM observation frequency is too low to determine the parameters, because the model function (Eq. (3)) at that range is unstable, but as the footprint becomes as large as 40 km square-grid, the TRMM observation frequency range becomes higher where the model function is stable, and the parameters can be estimated correctly.

The parameters ν and c are identified by comparing with the sample variance of monthly rainfall calculated directly from satellite observation in the various observation frequencies n and month data m.

As an alternative way to estimate c, since the TRMM/PR has swath width of 247 km, the spatial correlation length is able to estimate directly from instantaneous data, with the spatial correlation function. By determining these parameters, not only the population variance is estimated, but the temporal correlation length and spatial correlation length of the area are also determined.

Here, since the spatial correlation was hard to estimate in the model, the parameter c was estimated directly from TRMM/PR.

3. Application

3.1. *Validation in Kinki, Japan*

The used data set is PR2A25, version 6 data of rainfall intensity at 3 km height. The applied period is during June–October 1998–2006, which is the full observation length of TRMM.

The model was first applied to the Kinki area in Japan within the area of 2.5° square-grid to examine the model validity with the TRMM/PR data, in comparison to the ground radar. The Miyama radar of the MLIT (Ministry of Land, Infrastructure and Transport, Japan) locates at latitude 35°2′18″ North, and longitude 135°22′48″ East. The observation area is in a circular range with a radius of 120 km which is almost of the same size and area of the 2.5° square-grid of TRMM observation. The power of the received echo is recorded sequentially every 5 min in PPI scan at an elevation angle of 0.8°, about 3.5 km height at the circumference. The reflectivity Z is converted into rainfall rate according to a power-law relationship

$$Z = BR^{\beta}, \tag{5}$$

with optimized parameters using rain gauges. The raw data arranged by polar coordinate are converted to 80×80 grid data arranged by Cartesian coordinate, where the size of the grids is 3 × 3 km, as the spatial resolving ability of the radar is 3 km. The 34 years' data of 1988–1994 during the same months as TRMM are used.

The result is shown in Fig. 3. The pink line shows the estimated result considering spatial correlation, while the turquoise line shows the estimated value considering only temporal correlation. The blue line shows the estimation result from continuous data of ground-based radar. The samples were taken by descretizing the continuous observation. From Fig. 3, it is clear that by introducing the spatial correlation function, the estimated population variance from TRMM/PR shows a value nearly close to that from the continuous data of ground radar. The small difference can be thought of as the bias between the ground radar and space-borne radar. This also implies that the larger difference of temporal correlation model with TRMM and ground observation can be regarded as the difference of footprints of the two radars.

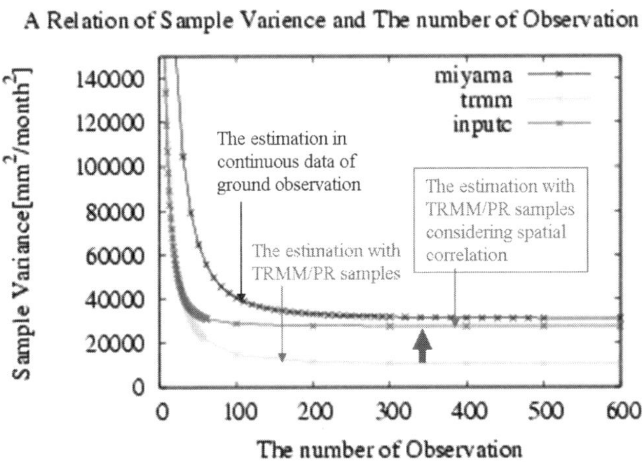

Fig. 3. The population value estimated from ground radar and TRMM.

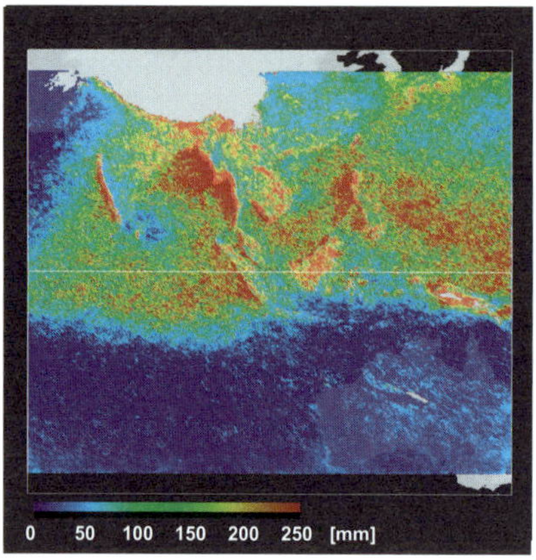

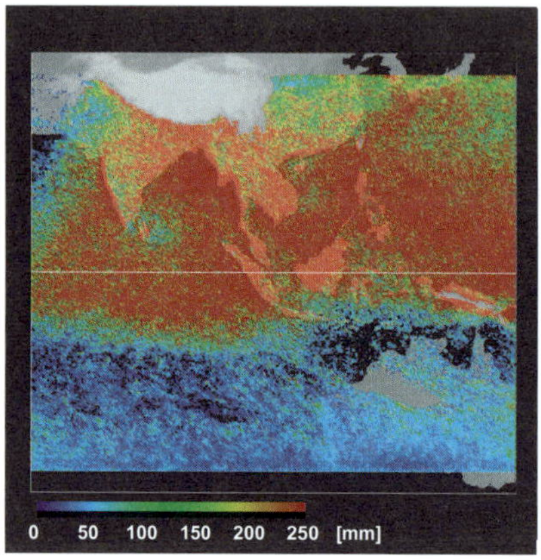

Fig. 4. The average (left) and standard deviation (right) of monthly rainfall calculated directly from TRMM/PR (mm/month).

3.2. *Application to Asian monsoon areas*

As an attempt to global estimation, the model was applied to the east part of Asian monsoon area during the southeast monsoon season, 120°E to 150°E longitude.

Figure 4 shows the average monthly rainfall and the standard deviation of monthly rainfall of the area, which are calculated directly from TRMM/PR.

4. Results

The results of the temporal correlation length, spatial correlation length of instantaneous rainfall, and the estimated population variance of area monthly rainfall were derived from the model, as shown in Fig. 5. The variation of the spatial correlation lengths well expresses the difference between the ocean (including coastal area and Japan islands) and the land area. The estimated population variance shows similar variation with the average monthly rainfall (shown in Fig. 4). However, the estimated temporal correlation lengths in the middle latitude (where the observation frequency is high) shows the same value at all grids. This error can be thought of as the estimation method for low latitude is not suitable in estimating at higher latitude, because there, the observation frequency varies greatly in latitudinal average. Besides that, the lower estimated temporal correlation

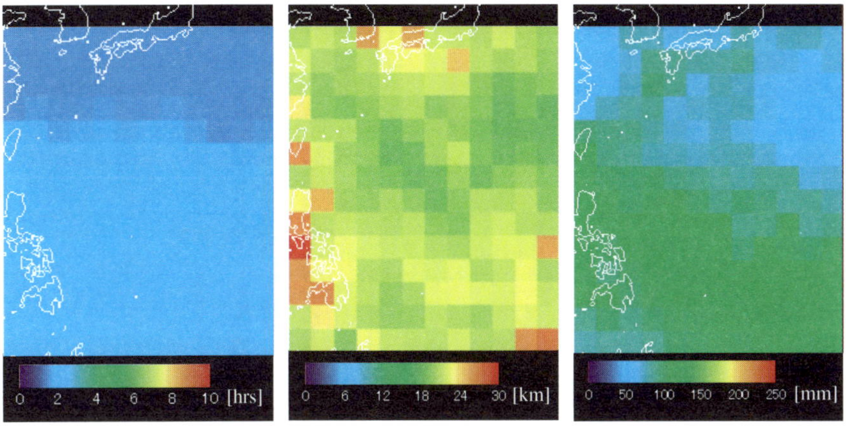

Fig. 5. Temporal correlation length (left) and spatial correlation length (middle) of instantaneous rainfall and standard deviation (right).

lengths, the spatial correlation lengths, and the population variance were successfully estimated.

As is clear in Fig. 6, the estimated population variance is greatly revised, especially in the lower latitude.

5. Conclusions and Further Prosperities

As a conclusion, this formulation made the model accurate by considering the effect of the footprints as shown in the validation with the ground radar. The method also succeeded to estimate in lower latitude, by taking in the method to enlarge the spatial domain sizes.

Also, the climate characteristics during the southeast monsoon were estimated in the Asian region. These estimated values show the same order as the values estimated by the ground rain gauge of the area.

Finally, the possibility of estimating global distribution of temporal and spatial correlation characteristics with TRMM/PR was showed.

For future works, by combining TMI product with PR as an alternative method to increase observation frequency can be studied. By using TMI data, parameters are expected to be estimated with higher accuracy since TMI has higher observation frequency. As a further improvement and possibilities of model, in the low latitude area of Asian monsoon, the diurnal cycle should be taken into consideration, which would make the model performance higher.

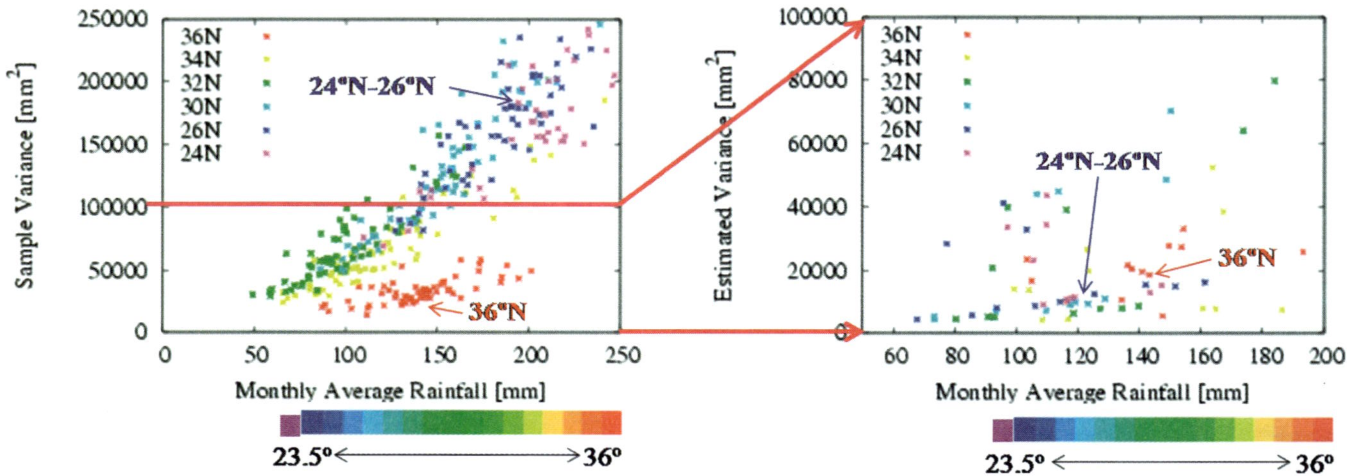

Fig. 6. The average–variance relations of sample variance calculated from TRMM/PR (left) and of revised variance estimated from model (right).

Acknowledgments

The observed data products used here in this research are provided from JAXA, the Japan Aerospace Exploration Agency.

References

1. T. L. Bell, A space–time stochastic model of rainfall for satellite remote sensing study, *J. Geophys. Res.* **92** (1987) 9631–9643.
2. S. Ikebuchi, E. Nakakita, K. Kakimi and T. Adachi, Accuracy of space and time average estimation on precipitation by using satelite data, in *Proc. Int. Symp. HEIFE* (1993), pp. 216–226.
3. E. Nakakita, T. Okimura, Y. Suzuki and S. Ikebuchi, Development of method of estimating spacio-temporal averaged rainfall with TRMM/PR information using a dependency of rainfall on topographic elevation, *Ann. J. Hydraulic Eng., JSCE* **46** (2002) 25–30.
4. E. Nakakita and S. Okane, Method of correcting variance of point monthly rainfall directly estimated with low frequent observations from space, *Ann. J. Hydraulic Eng., JSCE* **50**(CD-ROM) (2006) 6.
5. E. Nakakita, S. Okane and L. Konoshima, Method of correcting variance of point monthly rainfall directly estimated using low frequent observations from space, *Advances in Geosciences 2006*, Vol. 6 (World Scientific Publishing Company Pte Ltd, 2007).
6. I. Rodriguez-Iturbe, N. Marani, P. D'Odorico and A. Rinaldo, On space-time scaling of cumulated rainfall fields, *Water Res. Res.* **34**(12) (1998) 3461–3469.
7. D. R. Cox, F. R. S. and V. Isham, A simple spatial–temporal model of rainfall, in *Proc. Roy Soc. Lond.* **A415** (1988) 317–328.

Advances in Geosciences
Vol. 11: Hydrological Science (2007)
Eds. Namsik Park *et al.*

ANALYSIS OF MONAMI WAVES IN AQUATIC VEGETATION

S. PATIL* and V. P. SINGH†
Department of Biological and Agricultural Engineering
Texas A&M University, College Station 77843, USA
**cesandeeppatil@yahoo.com*
†*vsingh@tamu.edu*

A. K. RASTOGI
Department of Civil Engineering
Indian Institute of Technology Bombay
Powai, Mumbai 400 076
akr@civil.iitb.ac.in

Flow through flexible submerged vegetation exhibits Kelvin–Helmholtz (KH) vortices at the top of the stems causing a periodic movement that is referred to as interfacial monami waves. Using linear instability theory, Patil *et al.* (2006) derived simplified expressions for amplitude and periods of monami wave in terms of easily measurable vegetation parameters. These expressions are used in this paper for validation and found to be matching with the corresponding experimental data of past researchers. The bending angle of the stem as one of the key parameters that is related to monami amplitude and period is emphasized. Assuming KH vortices as a function of overflow depth, equations for monami waves are also proposed using linear wave theory. The proposed models of monami waves presented in this paper can be applied to applicable hydrodynamic situations in inland or coastal wetlands for better prediction of turbulence, nutrients or sediments between obstructed vegetation flow and free overflow flow.

1. Introduction

The uniform flow in an open channel or river is characterized by a vertical logarithmic current and turbulent environment. The presence of vegetation destroys the shear in the velocity and increases the turbulence within vegetation. If the water level is above stem height, then the flow is divided into two parts: the flow through vegetation and the overflow.[1] Both the parts are characterized by a different magnitude of velocity and connected

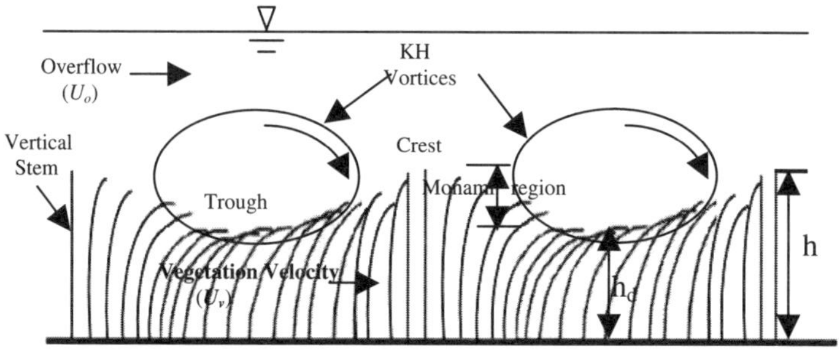

Fig. 1. KH vortex and waving of stems (monami) at interface.

by a strong shear around the interface (i.e. the junction between the two layers).[2] This shear is attributed to the abrupt drop in the overflow velocity below the interface due to stem interference and makes the flow susceptible to Kelvin–Helmholtz (KH) instability.[3] As a result, the large size KH vortices are observed advecting in the flow direction at the interface.[4] In flexible-type vegetation, a direct visible effect of the advecting KH vortices is the wave motion of the upper part of the stems called *monami*[5] (Fig. 1). Monami, owing to its continuous periodic movement, has predominant effect on vertical exchange of momentum,[6] particulate sediment transport,[7] biological transport like larvae[8] or pollen dispersion[2] and also on changing the morphology of vegetation[10] or even its removal.[11]

Physically, the tip of a vertical stem is the crest of a monami wave, whereas its height under maximum bent position is the trough. While moving from trough to crest in the first half ($= T/2$) of wave period T, stems enter in the clear water causing periodic transition of density from ρ_o (water density $= 10^3\,\mathrm{kg/m^3}$) to ρ_v (vegetation density) and vice versa in the second half of T. The amplitude and frequency of the monami wave have been derived by Patil *et al.*[12] from circle geometry for the bending stem (Fig. 2) and linear instability theory respectively that is given in brief in the next section. They also proposed an expression for the Richardson number for the monami instability. This paper describes the validation of derived expression by using experimental data of past researchers. The dependency of monami frequency on the bending angle of stem is also explored.

2. Amplitude and Period of Monami Wave

To design stem geometry, let the stem base A be the origin and AB and AD be the x and z-axes, respectively. Soon after the stem bends, it is

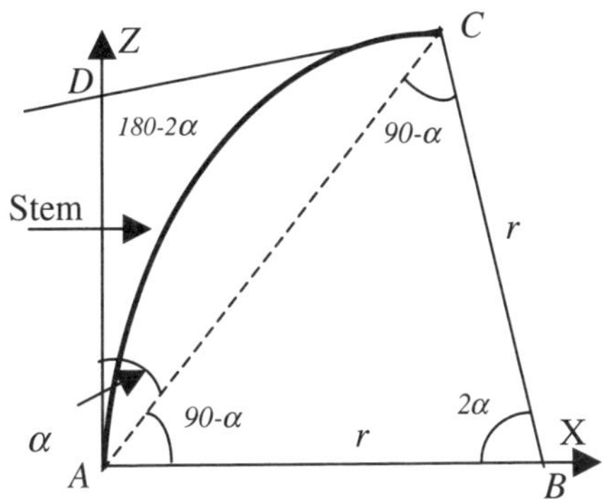

Fig. 2. Geometry of a bent stem.

assumed that it takes a shape of a circular arc with the center (point B) and radius, r appears on x–z plane depending on the magnitude of bending angle α, with z-axis (Fig. 2). In the bent position, arc (AC) $= h$ whereas ℓ(AC) is the linear distance AC. In the figure, $\ell(\mathrm{AD}) = \ell(\mathrm{CD})$ as both are tangent to the arc (AC), hence $\angle\mathrm{DAC} = \angle\mathrm{DCA} = \alpha$. From $\Delta\mathrm{ACD}$, $\angle\mathrm{ADC} = (180 - 2\alpha)$ and therefore from quadrilateral ABCD, $\angle\mathrm{ABC} = 2\alpha$. Now the arc (AC) $= h = r\ (2\alpha)$; i.e.

$$r = \frac{h}{2\,\alpha}, \quad \text{for } \alpha \neq 0. \tag{1}$$

It should be noted that for $\alpha = 0$, arc (AC) will turn into a vertically straight stem for which (i) $\angle\mathrm{ACB} = \angle\mathrm{CAB} = 90°$, (ii) $\ell(\mathrm{BC})//\ell(\mathrm{AB})$ and horizontal. Point B will disappear and the geometry will disintegrate. Therefore, the following analysis is for $\alpha \neq 0$ unless specified. Applying sine rule to $\Delta\mathrm{ABC}$, the co-ordinates of points A, B, and C are (0, 0), $(r, 0)$ and $\{2r\sin^2\alpha,\ r\sin 2\alpha\}$, respectively. Thus, the z-coordinate of C is $r\sin 2\alpha$ which provides the bent height (h_d) of stem. Thus, $h - h_\mathrm{d}$ is the monami wave height which provides monami amplitude, $a = \frac{h-h_\mathrm{d}}{2}$. Note that the circle geometry described in this section is valid for the vegetation in bent condition, i.e. for $\alpha \neq 0$. For $\alpha = 0$, the formulation changes.

Linear instability theory is applied on the linearized governing equations in which the total flow field is $(\bar{U} + u, w, \bar{P} + p, \bar{\rho} + \rho)$ in which values with bars corresponding to the overflow and without bar are the dynamic perturbations at the interface due to the dowel entry. Thus, the values of the perturbations are $u = U_\mathrm{v} - U_\mathrm{o}, w = W_\mathrm{v} - W_\mathrm{o}, p = P_\mathrm{v} - P_\mathrm{o}, \rho = \rho_\mathrm{v} - \rho_\mathrm{o}$ and $\bar{U} = U_\mathrm{o}, \bar{\rho} = \rho_\mathrm{o}$ The continuity and incompressibility equations are, respectively

$$\frac{\partial u}{\partial x} + \frac{\partial w}{\partial z} = 0, \tag{2}$$

$$\frac{\partial \rho}{\partial t} + U_o \frac{\partial \rho}{\partial x} + w \frac{d\rho_o}{dz} = 0. \tag{3}$$

As the density stratification in the present problem is not depth variant but is time-variant within a periodic cycle of monami, $\frac{d\rho_o}{dz}$ is the averaged quantity over a wave period. Momentum conservation equations in x and z directions are

$$\rho_o \left(\frac{\partial u}{\partial t} + U_o \frac{\partial u}{\partial x} + w \frac{\partial U_o}{\partial z} \right) = -\frac{\partial p}{\partial x}, \tag{4}$$

$$\rho_o \left(\frac{\partial w}{\partial t} + U_o \frac{\partial w}{\partial x} \right) = -\frac{\partial p}{\partial z} - \rho a_{br}, \tag{5}$$

where $a_{br} = \frac{Ed^2}{32(\rho_o - \rho_v) g h^2 r \cos^2 \alpha}$.[12] Substituting perturbations such that $\{u, w, p, \rho\} = \{U(z), W(z), P(z), \rho^*\} e^{ik(x-Ct)}$ (C is the celerity), monami wave period as a reciprocal of buoyant frequency is derived as

$$T = \sqrt{\left[\frac{32\, a\, \rho_o h^2 r \cos^2 \alpha}{Ed^2} \right]}. \tag{6}$$

The wave amplitude and period can fully describe the monami wave.

3. Validity of the Model

The proposed theory is applied to the experimental data of three researchers, Ghisalberti and Nepf.[4,6,13] First, $h_d = C_z = r(\sin 2\alpha)$ is substituted in Eq. (1) to get

$$\frac{\sin(2\alpha)}{2\alpha} = \frac{h_d}{h}. \tag{7}$$

Ghisalberti and Nepf[4] used already bent stems to simulate natural Eelgrass meadow. As such, stem height in the already bent condition is assumed as the initial height h. During monami, the amplitude of monami wave is measured and subtracted from h to find h_d. The known values of h and h_d are substituted in Eq. (7) to calculate bending angle α by iterations. Further, Ghisalberti and Nepf[4] used low density polyethylene blades (stem) (material density $\rho_m = 920\,\text{kg/m}^3$ and its modulus of elasticity,

$E = 3 \times 10^8\,\mathrm{N/m^2}$). The blades were 12.7 cm long with rectangular cross-section of thickness = 0.1 mm and width 3 mm. Six blades in each group were fixed in randomly spaced wooden dowels at the bottom. Monami region covers upper part of blades which due to synchronous waving, spreads blades uniformly in the flow. As such, assuming equally spaced blades, center to center spacing of blades is taken as the value for ΔS, which is = 0.02 m for the given number of blades = 1890 blades/m^2. Also, the bending of blades is particularly through the thickness where width has no influence. Therefore, blade thickness = 0.1 mm is considered as the effective stem diameter. Figure 3 shows the agreement of wave periods with Eq. (6) for $\Delta S = 0.017$ m and $d = 0.17$ mm which confirms blade spacing (ΔS) and blade thickness as the scaling for the distribution and bending for a rectangular blade, respectively, for monami region. Figure 3 also includes the data of Ikeda and Kanazawa[6] and Vivoni.[13] Ikeda and Kanazawa[6] used circular nylon stems ($\rho_m = 1400\,\mathrm{kg/m^3}$ and $E = 2 \times 10^9\,\mathrm{N/m^2}$) of 0.24 mm diameter and 5 cm height whereas Vivoni's[13] experiments consist of vinyl plastic stems, 16 mm long blades with rectangular cross-section (similar to Ghisalberti and Nepf,[4] 0.30 cm wide and 0.025 cm thick, and $E = 2 \times 10^9\,\mathrm{N/m^2}$. In Fig. 3, three data points appear to deviate. The

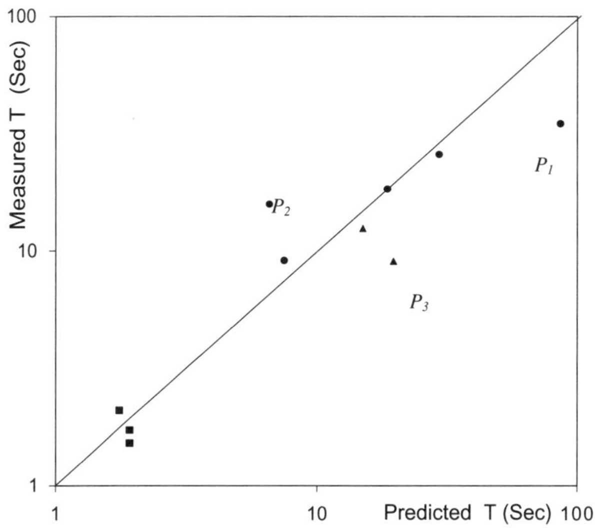

Fig. 3. Monami wave period (predicted and observed).

right side point (P_1) predicts higher T because of the corresponding longest $h = 0.113\,\text{m}$ and lowest values of flow velocity $= 3.4 \times 10^{-4}\,\text{m/s}$ and amplitude $= 7\,\text{mm}$. This configuration provides smaller value of $\alpha = 21.6°$ calculated from Eq. (7) and for which the circular arc assumed for Eq. (1) has lesser validity. The over-prediction of period is because in Eq. (1), r tends to higher value as $\alpha \to$ smaller value. Similarly, data point (P_3) of Vivoni[13] is over predicted because of $h = 0.113\,\text{m}$ with least amplitude $= 5\,\text{mm}$ reflects small $\alpha = 21°$ and shows onset of monami.

Data point (P_2) shows a higher measured period. This is because of the corresponding smaller $h = 0.064\,\text{m}$, higher $a = 32\,\text{mm}$, for which $\alpha = 44$ calculated from Eq. (7) and hence smaller $r = 4.1\,\text{cm}$. Such a flow pattern of smaller h with higher α allows higher depth of overflow. This exhibits a case of terrestrial canopy for which, the forward sweep is much higher than the ejection.[4] The resultant imbalance causes nonlinearity in the movement and increases the wave period in experiments. As Eq. (6) is based on linearized governing equations, it does not capture the nonlinearity and thus under-predicts the period for unconfined (terrestrial) canopies. However, Eq. (6) has significance as most of the monami waves are examined in confined canopies that have overflow depth comparable with stem height. Finally, a simple form of wave equation for the monami wave now can be proposed as

$$\eta = a\cos\frac{2\pi t}{T},$$

$$\eta = \left(\frac{h - r\sin 2\alpha}{2}\right)\cos\left(2\pi t\sqrt{\left[\frac{32a\rho_o h^2 r\cos^2\alpha}{Ed^2}\right]}\right). \tag{8}$$

Figure 4 shows wave period from Eq. (6) as a function of overflow depth (h_o). The spurious points P_1, P_2, and P_3 are excluded due to the unrealistic deviation explained above.

Although the data points are less in numbers, it seems from Fig. 4 that h_o can be a controlling parameter to get required monami wave configuration and desired in-canopy mixing. This can be tentatively judged from the inter-comparison between data points in Fig. 4. The high $\Delta U = 10.49\,\text{cm/s}$ in GN-A causes faster oscillation of KH vortices which reduce T even though h_o is higher compared to GN-E. For similar $Q \sim 40/48\,\text{cm}^3/\text{s}$, reduction in h_o from 20 cm (GN-E) to 10 cm (GN-G) causes increase in $\overline{U}$

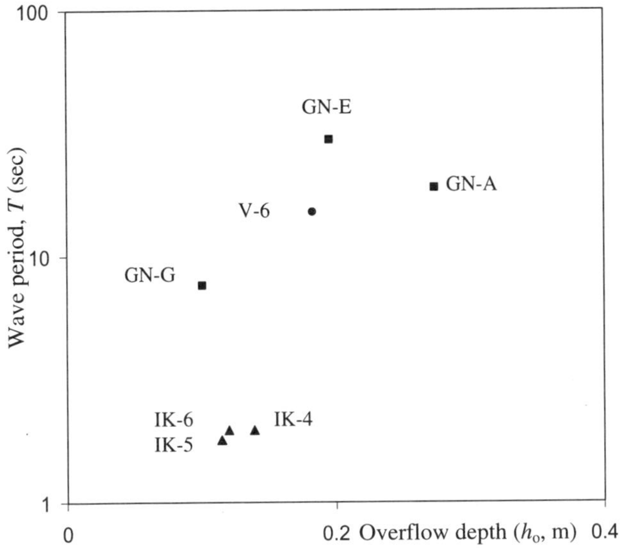

Fig. 4. Wave period of monami as a function of overflow depth (■, Ghisalberti and Nepf; ▲, Ikeda and Kanazawa; •, Vivoni).

and ΔU, hence substantial reduction in T. Same can be observed from IK-4 to IK-5. For the same magnitude of $\overline{U} \sim 7\,\text{cm/s}$, reduction in h_o from 27 cm (GN-A) to 10 cm (GN-G) significantly reduces T even under reduced shear strength. However, IK-4 and IK-6 show that for the slight reduction in h_o from 14 cm (IK-4) to 12.1 cm (IK-6) does not change T if it is accompanied by substantial reduction of discharge that reduces the shear. These trends in particular, GN-A to GN-G suggests that the effect of reduced h_o is stronger than that of ΔU and therefore h_o can be the governing parameter to control the monami wave period and length. For the similar overflow depth, increase in shear obviously reduces T from GN-E to V-6 owing to faster oscillations of KH vortices at interface. Reduction in T increases the monami frequency of plants' motion that can potentially influence the mixing of nutrients, pollens or sediments within the canopy and also the exchange between overflow and vegetation regions. Motivating from the fact that h_o can be a governing parameter of monami wave, linear water wave theory[14] is applied to find the wavelength (L) of monami wave. First of all, vegetation region in Fig. 1 is totally removed and the figure is now viewed reversely (rotate 180°) in the absence of vegetation. Thus, the monami wave is now a wave on water surface. The plane water surface of overflow is viewed as the flow bottom and as being perfectly horizontal, satisfies the necessary no-flux condition of linear wave theory. In this situation, L can be implicitly predicted from the linear dispersion equation as

$$\frac{1}{gT^2} = \frac{\tanh(h_o/L)}{L}$$

Substituting Eq. 6, the implicit expression for wavelength is

$$\frac{\tanh(h_o/L)}{L} = \frac{Ed^2}{32ag\rho_o h^2 r \cos^2\alpha} \tag{9}$$

The magnitudes of L with h_o are shown in Fig. 5 which show similar trend as T in Fig. 4 as expected. In Fig. 6, the dependency of predicted wavelength on the overflow depth and mixing layer thickness is plotted. The mixing layer thickness is substituted as a possible governing scale in place of h_o to calculate L from Eq. (9). The predicted values of wavelength are also compared with the observed wavelengths which are calculated from the measured phase speed. It can be seen that although mixing length can be a probable governing scale, no confirmed dependency can be proposed at this stage and some experimental work is needed to reach to a firm conclusion.

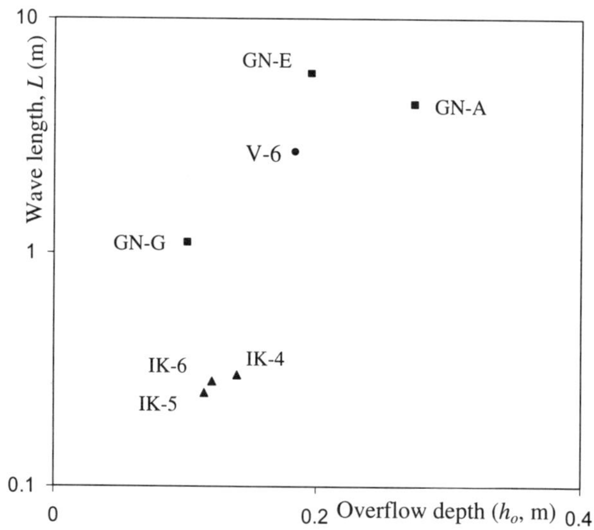

Fig. 5. Wavelength of monami as a function of overflow depth •, Ghisalberti and Nepf; ■, Ikeda and Kanazawa; ▲, Vivoni.

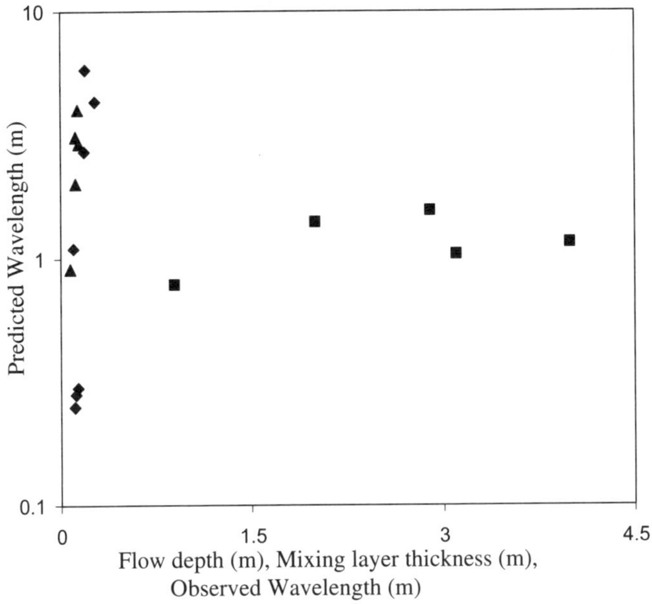

Fig. 6. Wavelength of monami as a function of overflow depth and mixing layer thickness is plotted (x-axis = flow depth for ▲; mixing layer thickness for ♦; observed wave length for ■).

4. Conclusion

Monami effect provides an important mixing layer on the top of seagrass meadow that provides necessary CO_2 and dissolved nutrients to seagrass leaves.[15,16] Patil *et al.*[12] have derived wave amplitude and period for the monami wave that have briefly represented in the paper. During monami movement, flow in the monami region experiences periodic stratification in density synchronous with the periodic movement of stems. The predicted values of wave periods are in agreement with the measured values in the experiments with variable h[4] and constant h.[6] The model as based on linear theory cannot capture the nonlinear effect in monami motion; however, such effects occur at higher bending, i.e. for $\alpha \to \pi/2$ which reflect terrestrial canopies. The wave equation (Eq. (8)) shows that all the vegetation parameters are playing an important part in governing the magnitude of monami motion. Both amplitude and period of monami wave are expressed as a function of easily measurable vegetation parameters (stem material, diameter, height, angle, and radius of bent stem) and hence can be predicted using a scientific calculator. The experimental validation shows that the information of monami wave can be important in studying various wetland hydrodynamics such as pollen dispersion,[9] vegetation removal,[11] and several such environmental applications.[16]

References

1. M. R. Raupach, J. J. Finnigan and R. Brunet, Coherent eddies and turbulence in vegetation canopies: The mixing-layer analogy, *Boundary Layer Meteorol.* **78** (1996) 351–382.
2. F. G. Carollo, V. Ferro and D. Termini, Flow velocity measurements in vegetated channels, *J. Hydraul. Eng.* **128**(7) (2002) 664–673.
3. P. K. Kundu and I. M. Cohen, *Fluid Mechanics* (Acadamic Press, New York, 2002), 730p.
4. M. Ghisalberti and H. M. Nepf, Mixing layers and coherent structures in vegetated aquatic flows, *J. Geophys. Res.* **107**(C2) (2002) 1–11.
5. J. D. Ackerman and A. Okubo, Reduced mixing in a marine macrophyte canopy, *Functional Ecol.* **7** (1993) 305–309.
6. S. Ikeda and M. Kanazawa, Three-dimensional organized vortices above flexible water plants, *J. Hydraul. Eng. ASCE* **122**(11) (1996) 634–640.
7. M. C. Gambi, A. R. M. Nowell and P. A. Jumars, Flume observations on flow dynamics in Zostera marina L. (Eelglass) Beds, *Marine Ecol. Progress Ser.* **61** (1990) 159–169.
8. J. E. Eckman, The role of hydrodynamics in recruitment, growth and survival of Argopecten irradians and Anomia simplex within eelgrass meadows. *Exp. Marine Bio. Eco.* **106** (1987) 165–191.
9. J. D. Ackerman, Diffusivity in a marine macrophyte canopy: Implications for submarine pollination and dispersal, *Am. J. Botany* **89** (2002) 1119–1127.
10. T. Asaeda, T. Fujino and J. Manatunge, Morphological adaptations of emergent plants to water flow, *Freshwater Biol.* **50** (2005) 1991–2001.
11. J. G. Duan, B. Barkdoll and R. French, Lodging velocity for an emergent aquatic plant in open channels, *J. Hydraulic Eng.* **132**(10) (2006) 1015–1020.
12. S. Patil, V. P. Singh and A. K. Rastogi, Monami wave in submerged vegetation, *6th International Symposium on Stratified Fluid*, Perth, Western Australia, Australia (2006), pp. 348–353.
13. E. R. Vivoni, Turbulence structure of a model seagrass meadow, Master's Thesis, Massachusetts Institute of Technology, Cambridge, Massachusetts, (1998), 128 pp.
14. R. G. Dean and R. A. Dalrymple, *Water Wave Mechanics for Engineers and Scientists* (Prentice Hall, World Scientific, Singapore, 1991), 353p.
15. R. E. Grizzle, F. T. Short, C. R. Newell, H. Hoven and L. Kindblom, Hydrodynamically induced synchronous waving of seagrasses: 'monami' and its possible effects on larval mussel settlement, *J. Exp. Marine Biol. Ecol.* **206** (1996) 165–177.
16. J. J. Finnigan, Turbulence in plant canopies, *Ann. Rev. Fluid Mech.* **32** (2000) 519–571.

Advances in Geosciences
Vol. 11: Hydrological Science (2007)
Eds. Namsik Park *et al.*

APPLICATION OF THE GENERALIZED EXTREME VALUE DISTRIBUTION TO AUSTRALIAN PEAK DISCHARGE DATA

R. SRIKANTHAN[*,‡], MURRAY C. PEEL[†]
THOMAS A. MCMAHON[†] and DOERTE JAKOB[*]
[*] *Water Division, Bureau of Meteorology, GPO Box 1289*
Melbourne 3001, Australia
[‡] *r.srikanthan@bom.gov.au*

[†] *Department of Civil and Environmental Engineering*
The University of Melbourne, Victoria 3010, Australia

Flood frequency analysis has been a controversial topic among engineers, and a general agreement as to methods to adopt is hard to come by. The Log Pearson type III distribution (LPIII) has been widely used in the United States and Australia with parameters estimated by the method of moments in the log domain. Recently, there is a trend to move toward the use of the Generalized Extreme Value (GEV) distribution with the parameters estimated by using probability weighted moments or L moments. Furthermore, in Australia some analysts are estimating the parameters of the GEV distribution using higher order L (LH) moments. The order of the LH moment is not fixed and needs to be estimated from the data. The first part of the paper determines the order of LH moment by a simulation experiment, where the GEV distribution is assumed to be the population distribution. The second part deals with the flood quantiles obtained from GEV and LPIII parent populations. Finally, the GEV distribution was fitted to 102 annual maximum flood series from Australian stations.

1. Introduction

Statistical flood frequency analysis has been a controversial topic among engineers, and a general agreement as to the methods to adopt is hard to come by. The Log Pearson type III distribution (LPIII) has been widely used in the United States[5] and Australia[4] with parameters estimated by the method of moments in the log domain. Recently, there is a trend to move toward the use of the Generalized Extreme Value (GEV) distribution with the parameters estimated by using probability weighted moments[3] or L-moments.[2] Wang[7] proposed a higher order L moments called the

LH-moments, to give more weight to the larger events. The order of the LH-moments is not fixed and needs to be determined from the data by using an approximate goodness of fit test.[8] In Australia some analysts are estimating the parameters of the GEV distribution using LH-moments. Hewa *et al.*[1] compared the L- and LH-moments for low-flow frequency analysis and recommended LH-moments with a shift of 2. Jakob *et al.*[6] investigated the use of LH-moments to estimate the parameters of the GEV distribution for deriving design rainfall for Australia. Since the shift parameter varied considerably in a non-systematic way and the use of LH-moments resulted in lower quantiles, they decided to use the L-moments for the estimation of the GEV parameters.

The objectives of this paper are to (i) determine the order of the LH-moments which recovers the parameters of the GEV distribution for synthetic data with known population parameters, (ii) determine the bias and standard error of flood quantiles for several average recurrence intervals (ARI), and (iii) compare the flood quantiles obtained from the GEV with those from the LPIII distribution. Since the underlying distribution is not known, two populations, GEV and LPIII, are used to assess the robustness of the distribution. Finally, the GEV is fitted to 102 annual maximum flood series from Australian stations, and the results are presented.

2. Parameter Estimation of GEV Distribution

The probability density function of the generalized extreme value (GEV) distribution, in terms of the shape (κ), scale (α), and location (ξ) parameters, is given by

$$f(x) = \alpha^{-1} e^{-(1-k)y - e^{-y}}, \tag{1}$$

where

$$y = \begin{cases} -k^{-1}\log\{1 - k(x-\xi)/\alpha\}, & \kappa \neq 0 \\ (x-\xi)/\alpha, & \kappa = 0. \end{cases}$$

The probability distribution function is

$$F(x) = e^{-e^{-y}}. \tag{2}$$

The quantiles for the GEV distribution can be obtained from

$$x(F) = \begin{cases} \xi + \alpha\left\{1 - (-\log F)^{\kappa}\right\}/\kappa, & \kappa \neq 0 \\ \xi - \alpha \log(-\log F) & \kappa = 0. \end{cases} \tag{3}$$

L-moments for the GEV distribution are defined for $\kappa > -1$ as

$$\lambda_1 = \xi + \alpha\{1 - \Gamma(1+\kappa)\}/\kappa, \tag{4}$$

$$\lambda_2 = \alpha(1 - 2^{-\kappa})\Gamma(1+\kappa)/\kappa, \tag{5}$$

$$\tau_3 = 2(1 - 3^{-\kappa})/(1 - 2^{-\kappa}) - 3, \tag{6}$$

$$\tau_4 = \{5(1 - 4^{-\kappa}) - 10(1 - 3^{-\kappa}) + 6(1 - 2^{-\kappa})\}/(1 - 2^{-\kappa}), \tag{7}$$

where $\kappa \neq 0$.

Wang[7] introduced the LH-moments, and the first four LH-moments for GEV distribution with a shift of η are given below:

$$\lambda_1^\eta = \xi + \frac{\alpha}{\kappa}[1 - \Gamma(1+\kappa)(\eta+1)^{-\kappa}], \tag{8}$$

$$\lambda_2^\eta = \frac{(\eta+2)\alpha\Gamma(1+\kappa)}{2!\kappa}[-(\eta+2)^{-\kappa} + (\eta+1)^{-\kappa}], \tag{9}$$

$$\lambda_3^\eta = \frac{(\eta+3)\alpha\Gamma(1+\kappa)}{3!\kappa} [-(\eta+4)(\eta+3)^{-\kappa} + (\eta+3)(\eta+2)^{-\kappa} - (\eta+2)(\eta+1)^{-\kappa}], \tag{10}$$

$$\lambda_4^\eta = \frac{(\eta+4)\alpha\Gamma(1+\kappa)}{4!\kappa}[-(\eta+6)(\eta+5)(\eta+4)^{-\kappa} + 3(\eta+5)(\eta+4)(\eta+3)^{-\kappa} - 3(\eta+4)(\eta+3)(\eta+2)^{-\kappa} + (\eta+3)(\eta+2)(\eta+1)^{-\kappa}], \tag{11}$$

where $\kappa \neq 0$.

The selection of the shift parameter is carried out by a goodness of fit test, and the following approximate test-statistic is used[8]:

$$z = \frac{\hat{\tau}_4^\eta - \tau_4^\eta}{\sigma(\hat{\tau}_4^\eta \mid \hat{\tau}_3^\eta = \tau_3^\eta)}. \tag{12}$$

The test-statistic z is approximately normally distributed with zero mean and unit variance. If the test-statistic for a given value of the shift parameter exceeds the critical value, then the selected shift value is considered as inappropriate. Wang[8] derived the following approximation for the variance

Table 1. Values of coefficients in Eq. (14).

η	b_0	b_1	b_2	b_3	b_4
0	0.0745	0.0555	0.0067	−0.3090	0.2240
1	0.0579	−0.0328	0.1524	−0.4102	0.2672
2	0.0488	−0.0527	0.1620	−0.3856	0.2566
3	0.0380	−0.0309	0.0354	−0.1233	0.0878
4	0.0241	0.0024	−0.0813	0.0733	−0.0210

Table 2. Values of coefficients in Eq. (14).

η	c_0	c_1	c_2	c_3	c_4
0	1.0100	−0.0282	−2.9336	4.0801	−1.0874
1	1.3403	−0.8291	−3.8777	9.5371	5.7866
2	1.8800	−2.2233	−2.5825	10.435	−7.3887
3	2.6784	−4.8418	3.5255	2.3736	−3.2076
4	3.7793	−8.3485	11.517	−7.9095	1.9459

of the kurtosis.

$$\sigma^2(\hat{\tau}_4^\eta \mid \hat{\tau}_3^\eta = \tau_3^\eta) = \frac{b}{n} + \frac{c}{n^2}, \tag{13}$$

where

$$b = b_0 + b_1[\tau_3^\eta] + b_2[\tau_3^\eta]^2 + b_3[\tau_3^\eta]^3 + b_4[\tau_3^\eta]^4 \tag{14}$$

$$c = c_0 + c_1[\tau_3^\eta] + c_2[\tau_3^\eta]^2 + c_3[\tau_3^\eta]^3 + c_4[\tau_3^\eta]^4. \tag{15}$$

The coefficients b_0, b_1 , b_2 , b_3, and b_4 are given in Table 1 and the coefficients c_0, c_1 , c_2, c_3, and c_4 are given in Table 2.[8] For a group of stations (N), the test statistic $\sum z^2$ is formed and it can be approximated by a Chi-squared distribution with N degrees of freedom. If the test-statistic is smaller than the critical value, then the null hypothesis that the shift η and the GEV distribution are appropriate for a group of stations is accepted.

3. Simulation Experiments

Two simulation experiments were carried out. In the first experiment, the GEV distribution was assumed to be the population distribution, and the order of the LH-moment that recovers the parameters of the GEV distribution satisfactorily was determined. For a given value of scale ($\alpha = 10$) and location ($\xi = 50$) parameters, 10,000 values were generated for different values of the shape parameter (κ) varying from -0.6 to 0.3. These values were subdivided into samples of size 50, 80, and 100, and the GEV distribution parameters estimated from the subsets using LH-moments for orders equal to 0, 1, 2, and 4 were compared with the population values. The shift zero case corresponds to the L-moments. The estimates from different replicates of the same size are averaged. The differences between the averaged estimates and the population values are presented in Fig. 1. As expected, the figure shows that the largest sample size always gives the best parameter estimate. There is a general tendency to overestimate the parameters α and κ. This tendency is more pronounced for higher shift and negative shape (κ) parameters. For negative shape parameters and higher shifts, there is a negative tendency to underestimate ξ. The L moments recovered α and κ parameters satisfactorily in almost all cases. The L moments and LH2 recovered the parameter ξ satisfactorily for κ in the range -0.5–0.3, while LH1 recovered ξ satisfactorily for the whole range of κ. Both the LH3 and LH4 did not recover the parameter ξ for κ less than -0.2. However, the differences are small (of the order of 0.5%) and L moments can be seen to recover the parameters satisfactorily overall.

In the second simulation experiment, two populations were assumed: one GEV and the other LPIII distribution. In this experiment, the effectiveness of the two distributions in estimating the flood quantiles is assessed for the two assumed populations. For the GEV population, using the theoretical values of the GEV distribution parameters, quantiles corresponding to the average recurrence intervals of 10, 20, 50, and 100 years were estimated. These are the true estimates of the quantiles. Also, for each sample from the long simulated series, these quantiles were obtained using the GEV parameters estimated from the samples and averaged. In addition, the flood quantiles were also estimated from the samples using LPIII distribution and averaged. The bias and the relative root-mean-square error (RMSE) of the estimated values for the two distributions are shown in Fig. 2.

It can be seen from Fig. 2 that the bias is within $\pm 5\%$ for the GEV up to the shape parameter $\kappa = -0.5$ and has a large negative bias for $\kappa = -0.6$. The LPIII has a bias within $\pm 5\%$ for all the values of κ except $\kappa = -0.5$. There is very little difference in the bias for the GEV and LPIII distributions. With regard to the relative RMSE, both distributions resulted in small values for κ down to -0.2. The LPIII distribution has large RMSE for $\kappa \leq -0.3$. The estimates from the LH-moments appear to result in larger bias than the L-moment estimates. The RMSE is similar.

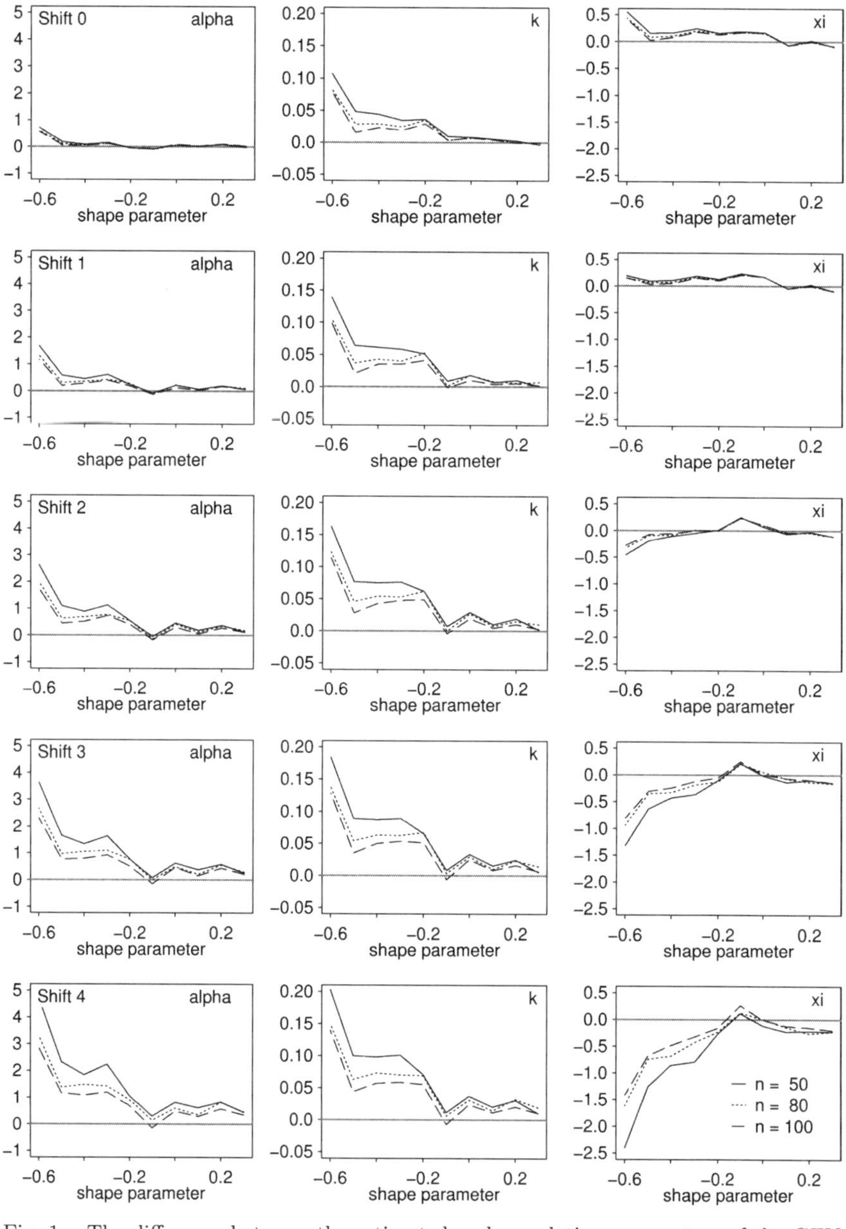

Fig. 1. The difference between the estimated and population parameters of the GEV distribution for different sample sizes (50, 80, and 100) from simulated sequences with $\alpha = 10$ and $\xi = 50$.

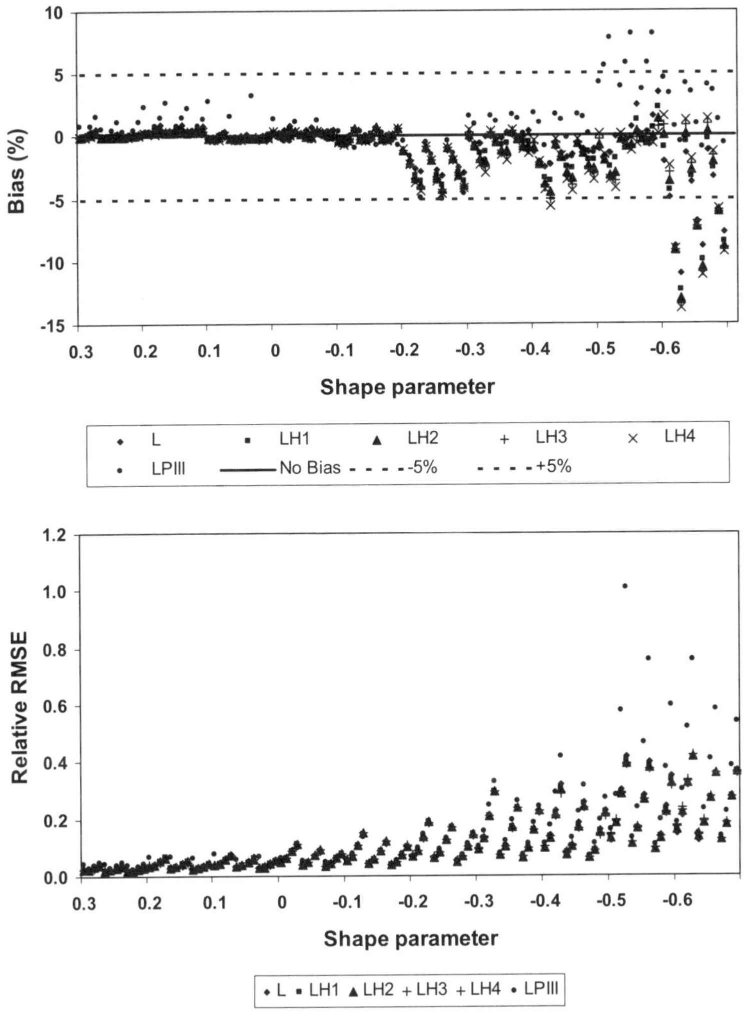

Fig. 2. The bias and the relative RMSE of the flood quantiles estimated using GEV with different shifts and LPIII distributions for GEV population.

The above comparison was repeated using LPIII as the population distribution. Again 10,000 values were generated using the LPIII distribution for skewness in the log domain varied from −1 to 1. The mean in the log domain was kept constant at 1 and the standard deviation varying from 0.1 to 0.3 in steps of 0.1. The true values of the flood quantiles for the same ARI as above were calculated using the population values of the

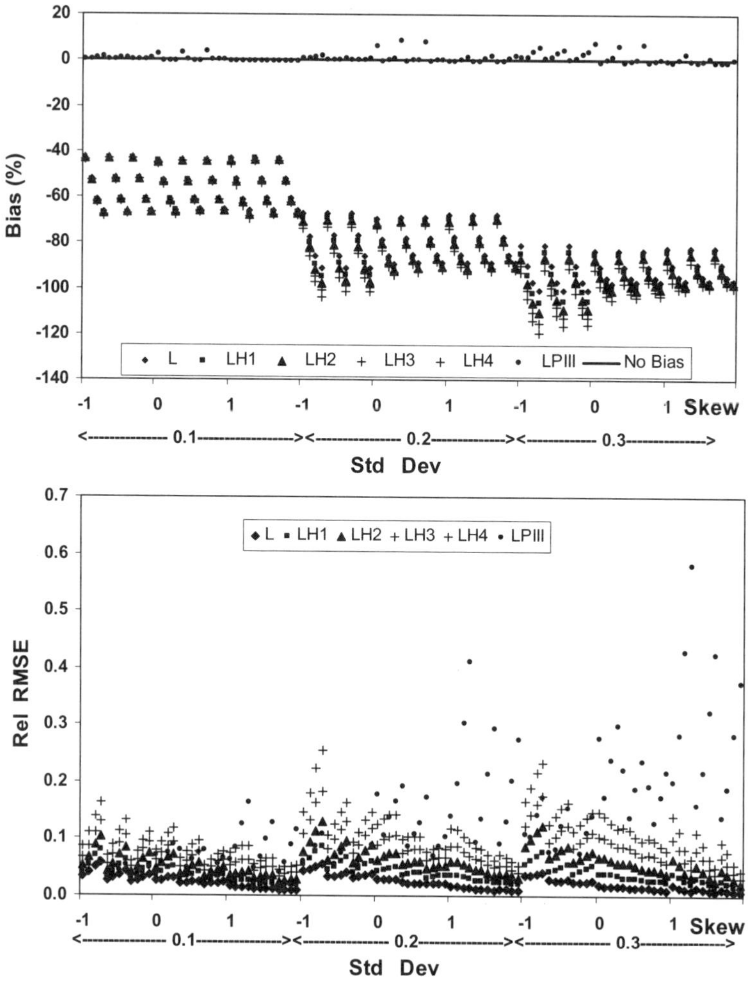

Fig. 3. The bias and relative RMSE of the flood quantiles estimated using GEV with different shifts and LPIII distributions for the LPIII population.

parameters. As before, the generated values were broken into three samples of size 50, 80, and 100, and the flood quantiles were estimated using both the LPIII and GEV distributions and averaged. The bias and the relative RMSE for the two distributions are shown in Fig. 3. It is surprising to see that all the cases of GEV resulted in large bias while, as expected, LPIII has very small bias. The GEV resulted in smaller relative RMSE than the LPIII.

Table 3. The z-statistic for all the sites. The lowest shift for which the z-statistic was smaller than the critical value is shown in bold. Stations in bold (6004, 6018, 6055, 6078, 6136) indicate that GEV was not acceptable for any shift.

No	LH0	LH1	LH2	LH3	LH4	No	LH0	LH1	LH2	LH3	LH4	No	LH0	LH1	LH2	LH3	LH4
6001	**0.27**	0.42	0.66	0.78	0.85	6052	**0.50**	0.54	0.98	1.27	1.45	6100	2.99	2.16	*1.55*	1.04	0.71
6003	4.28	3.69	2.94	2.15	**1.43**	6053	**0.07**	0.49	0.58	0.37	0.12	6102	**1.92**	0.47	−0.67	−1.27	−1.34
6004	4.27	3.86	3.41	2.94	2.45	**6055**	2.98	2.43	2.66	2.70	2.72	6103	**0.37**	−0.23	−0.18	0.19	0.64
6005	**0.49**	−0.67	−1.10	−0.99	−0.62	6057	2.59	**1.98**	1.73	1.51	0.00	6104	**0.00**	−0.14	0.11	0.53	0.90
6007	**1.86**	1.59	1.38	1.25	1.23	6058	**−0.27**	−1.07	−1.31	−1.03	−0.71	6109	**0.27**	−0.15	−0.31	−0.24	−0.17
6008	**1.43**	0.69	0.53	0.70	0.96	6059	3.21	**1.68**	0.86	0.48	0.27	6110	2.21	**1.77**	0.00	1.00	0.62
6009	**−1.49**	−0.60	0.01	0.36	0.60	6061	**0.96**	0.00	−0.87	−0.92	−0.78	6114	**0.71**	0.97	1.56	1.97	2.06
6010	**1.01**	−0.08	−0.74	−1.12	−1.28	6062	**−0.39**	−1.39	−1.42	−0.92	−0.42	6115	**0.42**	0.49	0.61	0.55	0.43
6011	**−0.08**	−0.38	−0.66	−0.74	−0.72	6064	**−0.35**	0.09	0.56	0.90	1.14	6116	**−0.08**	−0.91	−1.18	−1.32	−1.45
6012	**−0.65**	−0.57	−0.71	−0.65	−0.59	6065	**1.63**	1.46	1.01	0.67	0.57	6117	2.49	2.24	*1.91*	1.49	1.07
6013	**0.85**	0.45	0.16	0.27	0.56	6069	**−0.49**	0.16	0.29	0.14	−0.08	6120	**1.70**	2.07	2.48	0.00	2.82
6014	**1.93**	2.03	2.36	2.46	2.43	6070	**0.86**	0.10	−0.17	−0.12	0.08	6121	**1.58**	0.66	0.35	0.21	0.19
6016	**1.59**	0.30	−0.41	−0.52	−0.27	6071	**1.95**	1.77	2.18	2.22	2.24	6122	**0.74**	0.89	1.28	1.41	1.45
6017	**0.56**	−0.18	−0.87	−1.23	−1.29	6072	**1.36**	0.60	0.36	0.45	0.61	6123	**1.10**	0.05	−0.22	−0.14	−0.06
6018	2.83	2.85	3.01	2.83	2.52	6073	1.51	1.23	0.00	1.08	0.98	6124	**1.35**	0.00	0.00	−1.26	−1.52
6019	2.26	**0.68**	−0.32	−0.85	−1.10	6075	0.16	1.04	1.16	1.00	0.79	6126	**−0.02**	−0.34	0.24	0.78	1.10
6021	**0.00**	0.30	−0.19	−0.21	0.08	6076	**1.63**	1.65	1.48	1.51	1.67	6129	2.27	**0.00**	0.53	0.01	−0.32
6025	**0.64**	0.34	0.34	0.37	0.38	*6078*	3.01	2.97	2.83	2.54	2.22	6130	3.50	2.39	*1.53*	0.85	0.36
6026	**1.57**	1.83	1.98	1.70	1.21	6079	**0.28**	0.25	0.64	0.80	0.74	6131	**0.92**	0.61	0.27	−0.32	−0.89

(*Continued*)

Table 3. (*Continued*)

No	LH0	LH1	LH2	LH3	LH4	No	LH0	LH1	LH2	LH3	LH4	No	LH0	LH1	LH2	LH3	LH4
6028	**1.45**	1.31	1.29	1.11	0.87	6082	**−0.32**	−1.25	−1.30	−1.10	−0.92	6132	**−0.90**	−0.41	−0.45	−0.70	−0.85
6029	**1.11**	1.07	1.16	1.07	0.92	6083	**1.47**	1.28	1.47	1.53	1.38	6134	2.34	2.22	*1.91*	1.58	1.14
6031	**1.97**	−0.01	−1.37	−2.09	−2.26	6084	**−0.19**	−0.35	0.00	0.57	0.96	**6136**	3.42	3.20	3.23	3.10	2.93
6034	**0.77**	0.05	0.10	0.33	0.59	6085	2.13	2.08	1.63	1.07	0.61	6138	2.79	**1.61**	0.92	0.64	0.51
6036	**1.08**	0.49	0.55	0.69	0.76	6087	**1.51**	0.74	0.45	0.41	0.49	6140	−0.35	−0.86	−0.89	−0.83	−0.82
6038	**−1.09**	−0.50	0.23	0.72	0.96	6088	**1.91**	0.88	0.39	−0.01	−0.27	6143	**0.44**	1.45	1.61	1.40	1.10
6041	2.27	**1.27**	0.48	−0.09	−0.37	6090	2.39	**1.79**	1.59	1.50	1.36	6145	1.39	0.06	−0.89	−0.93	−0.34
6042	**1.73**	1.38	2.00	2.16	2.26	6091	2.08	1.83	2.17	2.59	2.88	6147	−0.29	0.16	0.62	0.88	1.08
6043	**0.81**	0.71	0.69	0.52	0.00	6092	2.59	1.53	0.81	0.44	0.36	6148	0.60	0.00	1.03	1.09	0.91
6044	**−0.09**	0.04	0.18	0.23	0.23	6093	**1.45**	1.37	0.96	0.58	0.36	6150	2.58	1.93	1.57	1.34	1.22
6046	**0.89**	−0.98	−2.04	0.00	−1.46	6094	**1.49**	2.07	2.38	2.51	2.55	6151	**0.30**	0.00	0.96	1.26	1.28
6048	3.25	2.71	2.16	**1.76**	1.54	6095	2.19	**0.79**	−0.44	−1.38	−1.96	6152	**0.11**	−0.46	−0.45	−0.15	0.25
6049	**1.35**	1.79	2.13	2.11	1.97	6096	**1.57**	0.58	0.20	−0.05	−0.19	6153	**1.17**	0.00	−0.35	−0.27	−0.10
6050	**1.62**	1.23	0.61	−0.21	−0.98	6098	**1.81**	1.74	1.82	1.78	1.68	6154	**0.40**	−0.13	−0.78	−1.13	−1.28
6051	**0.50**	−0.08	−0.19	−0.24	−0.28	6099	**1.66**	0.90	0.32	0.02	−0.06	6156	**1.51**	1.28	1.30	1.26	1.23

4. Application to Australian Peak Discharge Data

The annual maximum peak discharge data were obtained for 102 Australian rivers and the parameters of the GEV distribution were estimated using L- and LH-moments of order 1–4. Less than 5% of the stations have the shape parameter κ less than -0.5. The z-statistics was then calculated for the 102 annual maximum discharge series for shifts of 0, 1, 2, 3, and 4 using Eqs. (12)–(15) and Tables 1 and 2. A shift of zero corresponds to the GEV distribution with L moments. Table 3 presents the z-statistic for all the sites. The lowest order for which the z-statistic was smaller than the critical value (1.96) at 5% significant level is shown in bold in Table 3. In these cases the assumed shift is appropriate as the null hypothesis is accepted. The number of these cases is given in Table 4.

Table 5 gives the statistic $\sum z^2$ for the L- and LH-moments. The value of $\sum z^2$ is larger than the critical value (126.6) for all shifts. A similar result was obtained by Wang[8] based on the analysis of 107 Australian peak discharge data.

The empirical distribution of the peak discharge series was plotted using the plotting position formula[4]

$$P = \frac{m - 0.4}{n + 0.2}, \tag{16}$$

where P is the probability of non-exceedance, m is the rank, and n is the number of years of data. The average recurrence interval is the reciprocal of the probability of non-exceedance. Along with this, the theoretical curves for GEV with estimation procedures based on different shifts are shown in Fig. 2 for two sites. It can be seen from Fig. 4 that GEV (shift 0 or L-moments) in general appear to fit the data satisfactorily over the whole range. It is noticeable that as the shift increases, the theoretical curves tend to bend down and away from the observed points. Even though the lower

Table 4. The number of cases for which the z-statistic is less than the critical value for the L-moments and the additional stations for the LH-moments.

L	LH1	LH2	LH3	LH4	None
78	11	6	1	1	5

Table 5. $\sum z^2$-statistic for the L and LH moments.

L	LH1	LH2	LH3	LH4
293	188	176	159	158

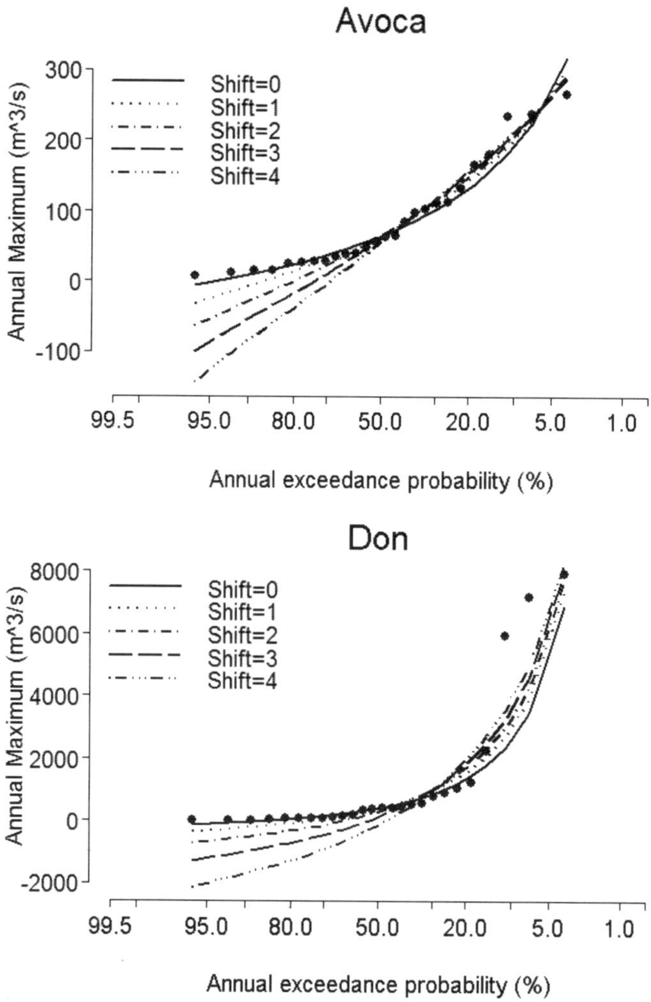

Fig. 4. Theoretical probability distribution of GEV using different estimation methods (lines). Annual maxima are denoted by points.

end of the curve is not relevant in flood estimation, the distribution only fits the higher end of the curve.

5. Conclusions

The parameter estimation for the GEV distribution was first investigated using synthetic data with known population values. L-moments and

LH-moments with shifts from 1 to 4 were used to estimate the parameters of the GEV distribution for three sample sizes 50, 80, and 100. The results showed that the L-moments were able to estimate the parameters better than the LH-moments. The flood quantiles of ARI 10, 20, 50, and 100 years were then compared for both the GEV and LPIII parent populations. For the GEV parent population, both the GEV and LPIII distributions gave similar bias and RMSE for values for the shape parameter greater than -0.3. When the shape parameter is less than -0.3, the LPIII tended to give larger RMSE than that of GEV. However, for the LPIII parent population, estimates of quantiles from the GEV resulted in larger bias but smaller RMSE values. From the simulation experiments, the GEV distribution with L-moments was found to give better results than the other alternatives. The GEV distribution was fitted to 102 annual maximum peak discharge series from Australia, and it was found that the L-moments fitted most of the data satisfactorily. From this study, the use of L-moments is recommended for the estimation of GEV parameters for flood estimation.

References

1. G. A. Hewa, Q. J. Wang, T. A. McMahon, R. J. Nathan and M. C. Peel, Generalized extreme value distribution fitted by LH moments for low-flow frequency analysis, *Water Resour. Res.* **43** W06301 (2007), doi: 10.1029/2006WR004913.
2. J. R. M. Hosking and J. R. Wallis, *Regional Frequency Analysis: An Approach Based on L Moments* (Cambridge University Press, Cambridge, UK, 1997).
3. J. A. Greenwood, J. M. Landwehr, N. C. Matalas and J. R. Wallis, Probability weighted moments: Definition and relation to parameters of several distributions expressible in inverse form, *Water Resour. Res.* **15** (1979) 1043–1054.
4. Institution of Engineers, Australia, *Australian Rainfall and Runoff — A Guide to Flood Estimation* (Institution of Engineers, Australia, Barton, ACT, 1987).
5. Interagency Committee on Water Data (IACWD), *Guidelines for Determining Flood Flow Frequency* (Hydrol. Subcommittee, Washington, DC, 1982), Bulletin 17B.
6. D. Jakob, B. Taylor and K. Xuereb, A pilot study to explore methods for design rainfalls for Australia, in *29th Hydrology and Water Res. Symp.* (Engineers Australia, Canberra, 2005).
7. Q. J. Wang, LH moments for statistical analysis of extreme events, *Water Resour. Res.* **33**(9) (1997) 2841–2848.
8. Q. J. Wang, Approximate goodness-of-fit tests of fitted generalized extreme value distributions using LH moments, *Water Resour. Res.* **34**(10) (1998) 3497–3502.

Advances in Geosciences
Vol. 11: Hydrological Science (2007)
Eds. Namsik Park *et al.*

NUMERICAL SIMULATION OF WATER LEVEL REDUCTION DUE TO FLUSHING OF SANDBARS IN AGANO RIVER MOUTH

MASASHIGE TAYASU
Department of Civil Engineering, Fukui National College of Technology
Geshi, Sabae, Fukui 916-8507, Japan
tayasu@fukui-nct.ac.jp

TOKUZO HOSOYAMADA and PUJIRAHARJO ALWAFI
Department of Civil and Environmental Engineering
Nagaoka University of Technology
Nagaoka, Niigata 940-2188, Japan

Flushing process of river mouth sandbars in flood time is considerably important for the urban safety against flood disaster. In this study, numerical simulation of interaction between the flood waves and the bed level change were carried out for Agano river mouth located in Niigata City in Japan. Governing equations of flood waves were two-dimensional wave equations based on the shallow water approximation. Equations of suspended load and bed load in flood were added to the wave equations to estimate total flux of sediment on sandbars and river bed. The deformation of the sandbars is calculated with net sediment flux on the sandbars. The numerical simulation is applied to flood in Agano River in 2002. The numerical results show that flushing of sandbars in Agano River reduced maximum water level by 2 m for the flood in 2002.

1. Introduction

River mouth sandbar is easy to be formed by onshore sediment transport due to wave motion.[1] The river mouth sandbar obstructs streamflow at the time of a flood. Rise of the water level caused by the river mouth increases the possibility of flood disaster in the downriver city area. For many rivers which have river mouth in the Japan Sea, river mouth sandbar develops by the sever wave condition in winter due to seasonal winds, and possibility of flood disaster increases during the flood time in the next summer. However, even if flood occurs, flushing of sandbars reduces rise of water level, eventually defending urban security. Understanding of these processes is considerably important for flood control management. However, the fundamental physical process is not elucidated sufficiently because of lack of understanding of the interaction of river flows and sediment transport flux on the river mouth.[2] Governing equation of sediment flux caused by river flow has several empirical constants. Applicability of each constant has each limit, which makes things more complicated.

In the recent studies for deformation of river mouth, Sato *et al.*[3] and Mano[4] modeled sediment transport by wave motion, tidal currents, and a river stream. They proposed equation of the opening width of the river mouth in the equilibrium condition. Authors carried out numerical experiments on generation of river mouth induced by incidence of onshore waves and coastal currents. In the study, multi-directional irregular wave signal is used as incident wave conditions.[1]

A river which runs into the Japan Sea has annual cycle of sediment transports in the river mouth. In winter, seasonal wind enhances wave height and wave induced coastal currents develop sandbars in its river mouth. River mouth in Agano River does not have many coastal structures and sandbars changes mainly depending on wave and currents.

In this study, two-dimensional numerical wave simulation with sediment transport and bathymetry change is carried out to simulate temporal deformation of sandbars by the Agano River flood in 2002. Reduction of water level caused by flushing of sandbars is estimated quantitatively. Expansion of cross section of river mouth, calculated by the simulation is compared with real field measurements.

2. Flood Wave, Sediment Transport, and the Bed Level Change Model

Governing equations of flood wave are wave equations based on shallow water approximation. Suspended load and bed load equations are added to the wave equations to calculate deformation and flushing of sandbars. Each process is included in the same time integration step to evaluate interaction between flood wave and sandbars. A general curvilinear coordinate system was employed for the treatment of arbitrarily meandering shape of the river.

The models describe free surface flow with vertically averaged variables and temporal bathymetry change in Cartesian coordinates:

$$\frac{\partial h}{\partial t} + \frac{\partial M}{\partial x} + \frac{\partial N}{\partial y} = 0, \tag{1}$$

$$\frac{\partial M}{\partial t}+\frac{\partial uM}{\partial x}+\frac{\partial vM}{\partial y}$$
$$=-gh\frac{\partial(z_{\mathrm{b}}+h)}{\partial x}-\frac{\tau_{\mathrm{bx}}}{\rho}+\frac{\partial}{\partial x}(-\overline{u'^2}h)+\frac{\partial}{\partial y}(-\overline{u'v'}h), \tag{2}$$

$$\frac{\partial N}{\partial t}+\frac{\partial uN}{\partial x}+\frac{\partial vN}{\partial y}$$
$$=-gh\frac{\partial(z_{\mathrm{b}}+h)}{\partial x}-\frac{\tau_{\mathrm{by}}}{\rho}+\frac{\partial}{\partial x}(-\overline{u'v'}h)+\frac{\partial}{\partial y}(-\overline{v'^2}h), \tag{3}$$

$$(1-\lambda)\frac{\partial z_b}{\partial t}=-\left(\frac{\partial q_{\mathrm{B}x}}{\partial x}+\frac{\partial q_{\mathrm{B}y}}{\partial y}\right)+w_0(C_a-E_s). \tag{4}$$

Equations (1)–(4) are vertically integrated continuity equation, momentum equations of vertically integrated velocity, and bathymetry change equation, where x and y are the horizontal coordinates, t is the time, h is the water depth, M and N are the vertically averaged flux, u and v are vertically averaged velocities in x and y directions, ρ is the water density, g is the gravitational acceleration, z_b is the bed elevation, $\overline{u'^2}$, $-\overline{u'v}$ and $-\overline{v'^2}$ are Reynolds stresses, λ is the sediment porosity, $q_{\mathrm{B}x}$ and $q_{\mathrm{B}y}$ are the sediment transport of bed load in x and y directions, w_0 is the fall velocity of sand, C_a is reference concentration, E_s is the non-dimension raise velocity, $\tau_{\mathrm{b}x}$ and $\tau_{\mathrm{b}y}$ are the bottom shear stress in x and y directions:

$$\tau_{\mathrm{bx}}=\frac{\rho g n^2 u\sqrt{u^2+v^2}}{h^{1/3}},\quad \tau_{\mathrm{by}}=\frac{\rho g n^2 v\sqrt{u^2+v^2}}{h^{1/3}} \tag{5}$$

and n is the Manning roughness coefficient.

As mentioned before, these equations are converted to general curvilinear coordinate. Chain rule is applied to wave equations in Cartesian coordinates to convert independent variables. Unknown variables are velocity components which are contravariant velocity components obtained by scalar product of velocity components in Cartesian coordinate and normal vector of line of curvilinear coordinate. By using contravariant velocity components as unknown variables, implementation of conservation law is easy because net flux of flow rate, sediment concentration etc. are expressed by product of line element length, contravariant velocity and each quantity.

2.1. *Sediment transport model*

Sediment transport equations consisted of bed load and suspended load equation. The bed load transport rate is calculated by equation of Meyer-Peter and Mullere[10] as follows:

$$q_{\mathrm{B}}^{*} = 8.0 \times (\tau^{*})^{3/2} \left(1 - \frac{\tau_{\mathrm{c}}^{*}}{\tau^{*}}\right)^{3/2}, \tag{6}$$

where q_{B}^{*} is the non-dimensional bed load, τ^{*} is Shields number, and τ_{c}^{*} is the non-dimensional critical tractive force. The non-dimensional critical tractive force is calculated with the formula proposed by Iwagaki.[7]

The suspended load generation rate is added to the sediment continuity equation as forcing term in the right-hand side of the equation. The term involves reference concentration of suspended load and non-dimension rise velocity on the bed. The reference concentration is estimated as the vertically averaged concentration which is found by advective diffusion equation of suspended sediment concentration:

$$\begin{aligned} h\left(\frac{\partial C}{\partial t} + u\frac{\partial C}{\partial x} + v\frac{\partial C}{\partial y}\right) \\ = -\frac{hw_0^2}{\varepsilon_{sz}}C + w_0 E_s + \frac{\partial}{\partial x}\left(\varepsilon_{sx}\frac{\partial Ch}{\partial x}\right) + \frac{\partial}{\partial y}\left(\varepsilon_{sy}\frac{\partial Ch}{\partial y}\right), \end{aligned} \tag{7}$$

$$C = C_{\mathrm{a}}\frac{\varepsilon_{sz}}{hw_0}\left[1 - \exp\left(-\frac{hw_0}{\varepsilon_{sz}}\right)\right] \approx C_{\mathrm{a}}\frac{\varepsilon_{sz}}{hw_0}, \tag{8}$$

where C is the vertically averaged concentration, and ε_{sz} is the diffusion coefficient.

The non-dimension raise velocity is estimated by Einstein as a function of u^{*}, d, q_{B} as is shown in Eq. (9).

$$E_s = C_{ae} = \frac{q_{\mathrm{B}}}{(11.6u^{*})(2d)}, \tag{9}$$

where C_{ae} is the bed load concentration, q_{B} is the bed load transport rate per unit width, u^{*} is the friction velocity, d is the average diameter of sediment particle. Einstein defined two times of the diameter of sediment as thickness of the layer of bed load and calculated a mean concentration of the layer of bed load for bed load concentration. Non-dimension raise velocity is equal with bed load concentration when it is presupposed that sedimentation flux matches with flux rolled up.

3. Calculation Condition and Boundary Condition

The objective of this numerical simulation was flood in Agano river in 2002. In this flood, maximum water level did not go beyond crown height of the levee fortunately. The parts of sandbars in Agano river mouth were flushed out. It is supposed that the flushing reduced backwater level rise in the river. Figure 1 shows the aerial photographs of the Agano river mouth, taken before and after the flood. Both left and right tip divisions of the sandbars were flushed though the sandbars near both riverbanks remained.

Figure 2 shows the map of the Agano River mouth and the calculation grids of the general curvilinear coordinate system. The computation domain is about 5 km in length from river mouth to upstream direction. Each grid position in the computational domain is generated by solving the Laplace equation with coordinate of the levee as boundary condition. The domain area is divided into 15 grids in cross direction and 72 grids in mainstream direction, which is reefed as structured grid. Height data of the riverbed was measured by the Agano river office of Hokuriku Regional Development Bureau data in February 2000. The data is shown as contour lines. Height data on numerical grids are obtained by converting data on contour lines to grid data.

Upstream boundary condition for inflow rate was 72 h discharge data of Typhoon 6 in 2002. For downstream boundary condition, the tidal level was for the same period (Fig. 3). The time series data of the tidal level were modified by the observed value of the nearest port.

In the calculation, Manning roughness coefficient was 0.032, particle diameter of the sediment was 0.25 mm, and the time step was 0.1 s.

4. Results

Numerical simulation cases were mainly classified into two types. One is sediment transport and bathymetry change (that is to say flush of sandbars may occur) is enabled and the other is that for no bathymetry change. These cases compare difference of water level rise between flush and no flush of sandbars in the river mouth.

4.1. *Velocity vector distribution*

Figure 4 shows the velocity vector distribution of the river mouth in the time of peak discharge. The upper figure is case of no flushing sandbar

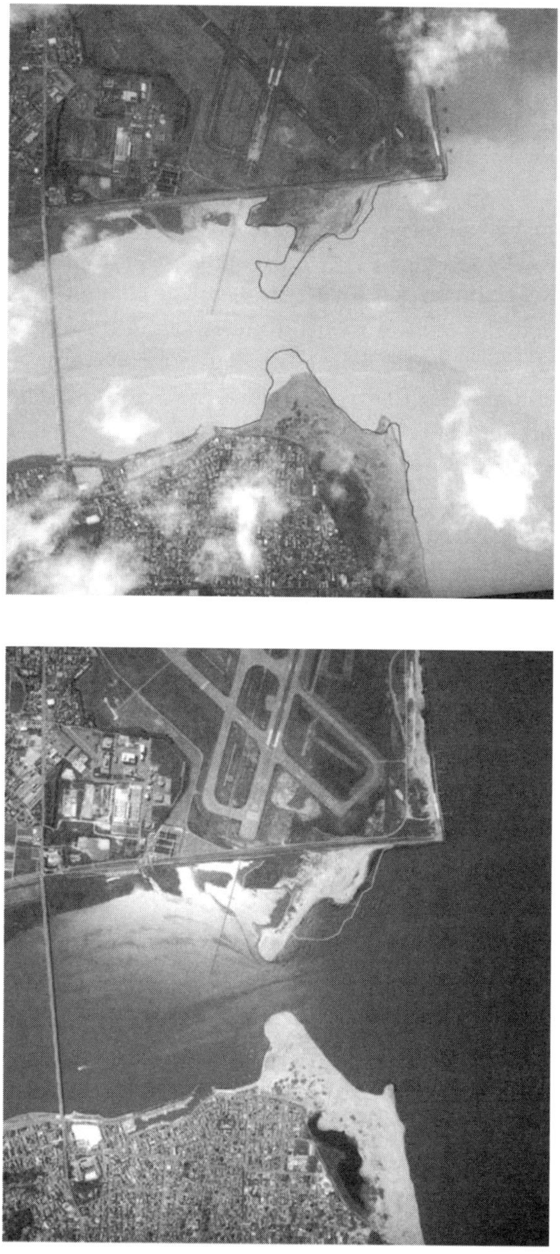

(a) (b)

Fig. 1. Aerial photos of Agano River mouth in 2002. (a) Before the flush. (b) After the flush.

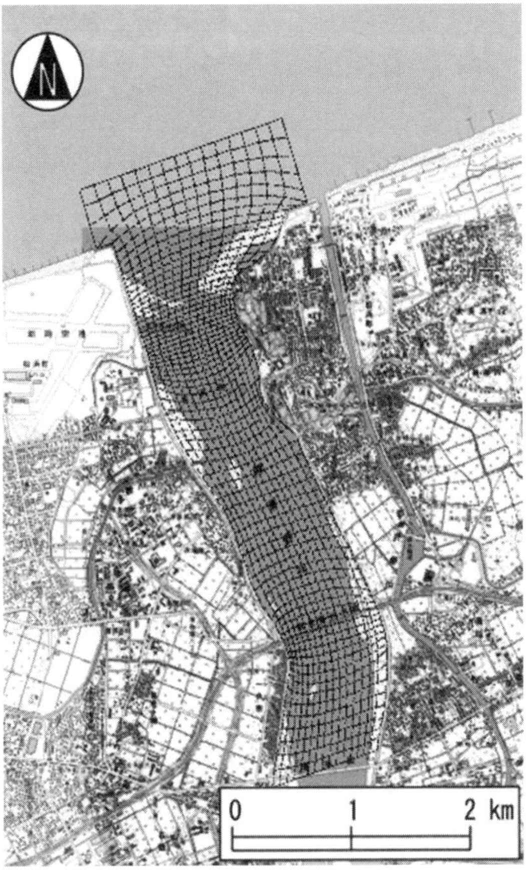

Fig. 2. The map of Agano River mouth and calculation grids.

(fixed bed). The lower figure is case of flushing sandbar (movable bed). In the fixed bed case, water level raised more in the upper streams side of the sandbar than the movable bed case. Therefore, overland flow just above the sandbar was enabled in this numerical scheme, velocity vectors are shown on the sandbars. The velocity of movable bed case at the river mouth is slower than fixed bed case. This is because for the case of movable bed, the sand bars were flushed out and the river mouth cross section was larger than the fixed bed case. On the other hand, as for the velocity in the river channel far from the sandbars, the movable bed case is faster.

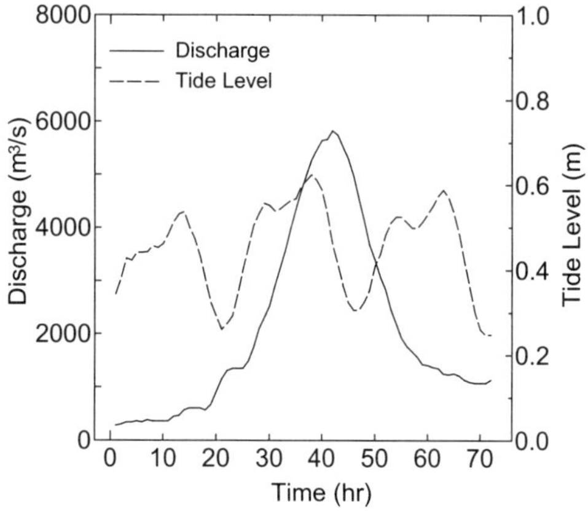

Fig. 3. The boundary conditions. Upstream boundary condition is water discharge (solid line). Downstream boundary condition is tide level (dashed line).

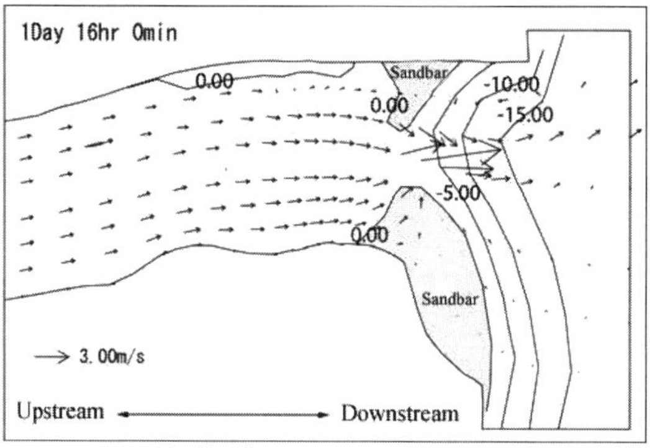

No flushing sandbar (fixed bed)

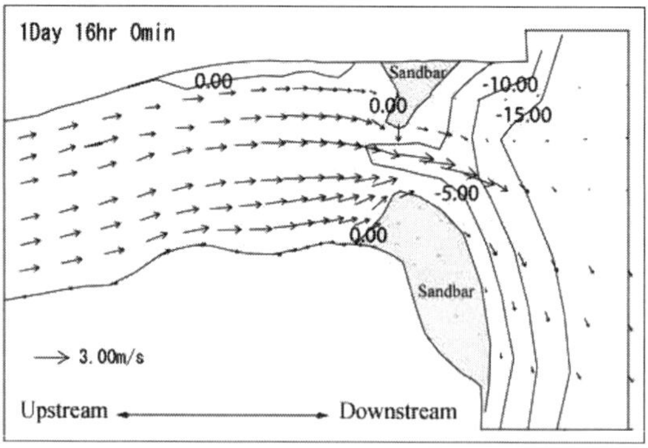

Flushing sandbar (movable bed)

Fig. 4. Velocity vector distribution of the river mouth in the peak discharge.

4.2. *The reduction of water level rise due to change of bed level*

Figure 5 shows comparison between flush and no flush cases on the water level distribution along the stream direction just on the center of the river. The left side of each figure is the upstream direction. The dotted line, the solid line, and dashed line stand for case of flushing, no flushing, and right levee elevation, respectively. In the alignment of Fig. 5, temporal transition

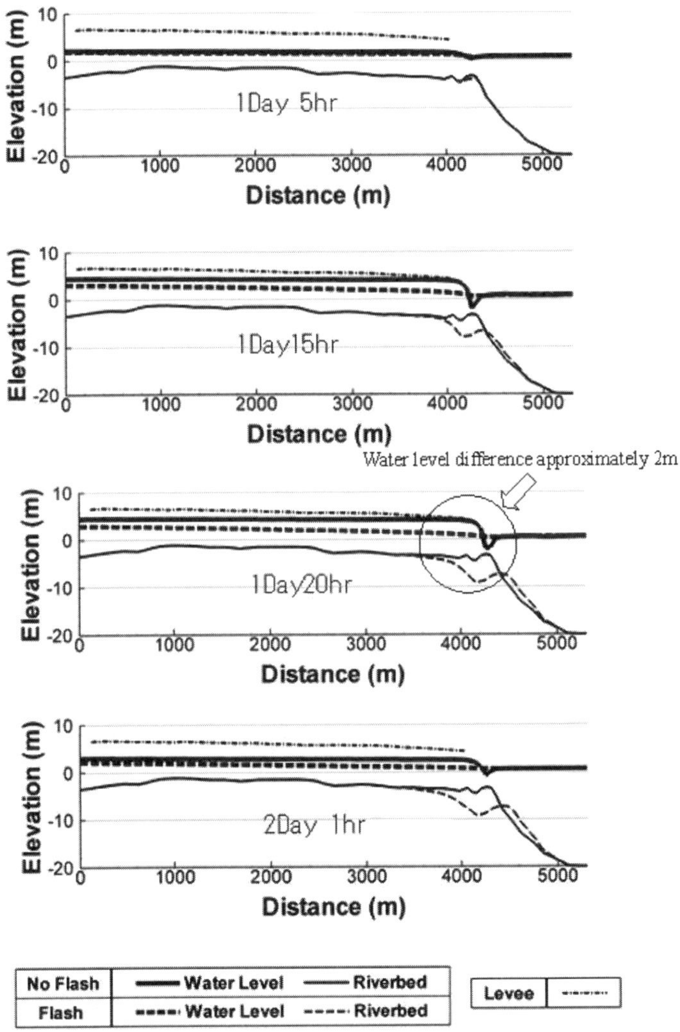

Fig. 5. Water level and bed level of simulation.

is from up to downward. In the figure of 1 day 5 h, the difference of the water level between flushing and no flushing cases is not visible because flushing of sandbars does not occur even for movable bed cases. With progress of time, with increase of discharge, bathymetry change in the river mouth comes to be remarkable for the case of flushing at the river mouth. Strong river stream undermined the sandbars in the peak of discharge and sediment accumulates in the downstream side. On the other hand, as for the case of no flushing, sandbars in the river mouth remain, and a water level rise occurs more than flushing. The water level rise reaches as high as crest elevation of the levee. The difference of maximum water level at river mouth between flushing and no flushing condition is approximately 2 m.

4.3. *The comparison of landform change quantity*

Figure 6 shows the cross-sectional shape of the Agano river mouth for numerical simulation and field survey. Cross-sectional area of the river mouth after the flood in 2002 is 2145 m^2 for field survey and 2185 m^2 for numerical simulation of this study. Initial conditions for both ways are supposed to be the same. As for the amount of flushed cross-sectional area at the river mouth, numerical study agrees well with the field measurements. However, there may be some difference on shape of the cross section. Cause of the difference is that numerical study does not have side erosion which may occur even if the land is not submerged. In this numerical study, erosion occurs only when the land is submerged. The side erosion should be involved to make a more precise numerical model.

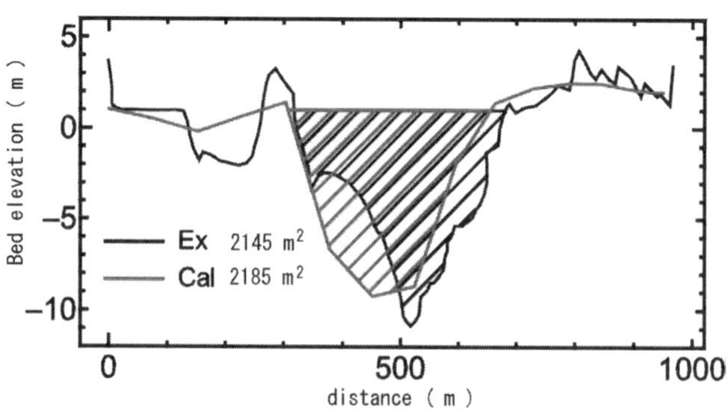

Fig. 6. Water level and bed level of simulation.

5. Conclusions

Numerical simulation of interaction of flood wave and the bed level change with general curvilinear coordinate system was applied to Agano River mouth.

In case of flood in 2002 with no flushing sandbars, the water level rises up to the crest of levee. On the other hand, in case of bed level change of river mouth, that is to say case of flushing phenomena, rise of water level is reduced by approximately 2 m. As the result, the effect of reduction of water level due to the flushing of the sandbar is decisively important against flood. Comparison of numerical results with real field measurement for water level and change of cross-sectional area of Agano river mouth shows good agreement. As for the shape of cross-section, numerical result shows difference from real filed measurements. Lateral erosion of the river cross-section should be considered to further agreement between numerical and field results.

Acknowledgments

The author thanks Prof. Shoji Fukuoka (Chuo University), Prof. Tetsuro Tsujimoto (Nagoya University), for their valuable advices. Provision of real filed data for Agano River mouth and temporal history of water level by Ministry of Land, Infrastructure and Transport Hokuriku Regional Development Bureau Agano River office support is highly appreciated.

References

1. T. Hosoyamada, Z. Tyou and T. Hoshino, Fundamental numerical experiment of generation of river mouth sandbar about the influence of shore wave motion and river water, *Ann. J. Coastal Eng.*, *JACE* **52** (2005) 561–565.
2. T. Hosoyamada, T. Sato, T. Noda, Y. Sakai and H. Sakamukai, A numerical study for flushing of sandbars by flood in the Agano River mouth, *Adv. River Eng.*, *JACE* (2006) 73–78.
3. T. Sato, T. Suproijo and A. Mano, The most narrow opening equilibrium condition of the River Having River mouth sandbar, *Ann. J. Coastal Eng.*, *JACE* **51** (2004) 526–530.
4. A. Mano, Study on the morphological equilibrium condition on river mouths and tidal inlets, Grant-in-Aid for Scientific Research (C), Research Accomplishment Report, 2005.
5. N. Nagata, Numerical analysis of two-dimensional unsteady flow by general curvilinear coordinate system, The Class Lecture Document of the Numerical

Computation of the Hydraulic Engineering with the Computer, *JACE* (1999) 61–76.

6. N. Hosoda, N. Nagata and Y. Muramoto, Numerical analysis of unsteady open channel flows by means of moving boundary fitted coordinate system, *J. Hydraulic, Coastal Environ. Eng., JACE* **533**(II-34) (1993) 53–56.
7. Y. Iwagaki, Hydrodynamic study on critical tractive force, *Journal of the Japan Soc. Civil Engrs., JACE* **41** (1956) 1–21.
8. M. Sekine, *Hydraulics of the Movable Bed Stream* (Kyoritsu Publishing Co., Ltd, 2005) 211 pp.
9. H. A. Einstein, The bed-load function for sediment transportation in open channel flows, *Tech, Bull.*, No. 1026, USDA, Soil Conservation Service, 1950, pp. 1–70.
10. E. Meyer-Peter and R. Muller, Formulas for bed-load transport, in *Proc. 2nd Cong. ITAH*, Stockholm, Sweden, 1948, pp. 39–64.

Advances in Geosciences
Vol. 11: Hydrological Science (2007)
Eds. Namsik Park *et al.*

DEVELOPMENT OF DISCHARGE-STAGE RELATION USING ARTIFICIAL NEURAL NETWORK

TAPESH K. AJMERA
Department of Earth and Environmental Sciences
University of Waterloo, Waterloo, Ontario, Canada-N2L 3G1
tkajmera@sciborg.uwaterloo.ca

A. K. RASTOGI
Department of Civil Engineering
Indian Institute of Technology, Bombay, India 400076
akr@civil.iitb.ac.in

The present study focuses on artificial neural network (ANN) in the development of rating curve, for generating relation between stream flow and stage, using four different training algorithms and linear regression (LR) for Peachtree creek watershed, Atlanta. The aim of this work is to show the applicability of feed forward back propagation (FFBP) neural network for discharge prediction and evaluate their performance using different training algorithms. The results indicate that Levenberg–Marquardt (LM) algorithm has better efficiency in estimating the discharge compared to the three different training algorithms and linear regression.

1. Introduction

Stream flow is a very important phenomenon in surface hydrology. For many years, hydrologists have attempted to understand the relation between stream flow and stage for purposes such as water supply, flood control, and power generation. A traditional rating curve for a stream at a gauging station refers to a relationship between the height of water level above a datum and the flow passing at a corresponding cross section. Current meter is normally used to measure the flow velocity at several points across the stream section to calculate the flow rate for that particular stage. By taking such measurements for a number of different stages and the corresponding discharge over a period of time, a number of points can be plotted on a stage–discharge curve to produce a rating curve. Establishing a rating curve relationship requires nonlinear mapping. The regression and curve fitting techniques are not adequate in view of the complexity of the problem. The present study is concerned with the application of the more powerful tool involving artificial neural networks (ANNs).

Artificial neural networks have been successfully used to directly map nonlinear complex relations and applied to different problems in civil engineering since the beginning of the last decade.[2,5,6] Maier and Dandy[9] have reviewed several papers dealing with the prediction and forecasting of water resource variables using neural network models. Very few works have been done in the past on the rating curve development using ANNs.[7,8,12,13] The present study differs from the previous ones exemplified in that it develops rating curve applying different methods of ANN.

2. Description of Site Study and Data Collection

Peachtree Creek is a major tributary of the Chattahoochee River in Atlanta, Georgia. The Peachtree Creek watershed is located at a longitude of 33° 49′ 10″ N, latitude 84° 24′ 28″ W, at an altitude of 233 m, above the mean sea level. The watershed encompasses an area of 224.29 km^2. Data has been collected by the US Geological Survey (USGS), at Northside Drive gauge of Peachtree Creek at every 15 min interval from 1 Oct 2006 at 2.15 am to 23 Dec 2006, till 8.00 am.

3. Artificial Neural Network

An artificial neural network consists of a number of interconnected processing elements, known as nodes, which are logically arranged in two or more layers and interact with each other through weighted connections. The number of nodes in each layer and the number of layers in the network depend on the nature of the problem. There is no unique guiding theory for the proper selection of the number of nodes and the number of layers. The scalar weights determine the nature and strength of the influence between the interconnected nodes. Each neuron is connected to all the nodes in the next layer.

The neural network consists of: (1) an input layer, where data are presented to the network, and (2) an output layer that holds the response of the network to the input. However, the intermediate layers, also termed as hidden layers, enable these networks to represent and compute complicated associations between inputs and outputs. In correspondence with a computational node, each one of the entering values is multiplied by a connection weight. Such products are then all summed with a neuron-specific parameter, called bias, used to scale the sum of products into a useful range. The computational node finally applies an activation function to the above sum producing the node output. Weights and biases are determined by means of a nonlinear optimization procedure (training) that aims at minimizing a learning function expressing closeness between observations and estimated outputs. A set of observed input and data pairs, the training data set, is processed repeatedly, changing the parameters until they converge to values such that each input vector produces outputs as close as possible to the desired target vectors. The structure of a typical neural network is shown in Fig. 1.

The training process of a feed-forward neural network with the back-propagation algorithm consists of two distinct passes through the different layers of the network: a forward pass and a backward pass. In the forward pass, inputs are applied to the node in the input layer of the network, and their effects propagate through the network, layer by layer. Finally, a set of outputs is produced as the actual response of the network. During the forward pass, the weights of the network are all fixed. During the backward pass, on the other hand, the weights are all adjusted in accordance with the error-correction rule. The actual response of the network is subtracted

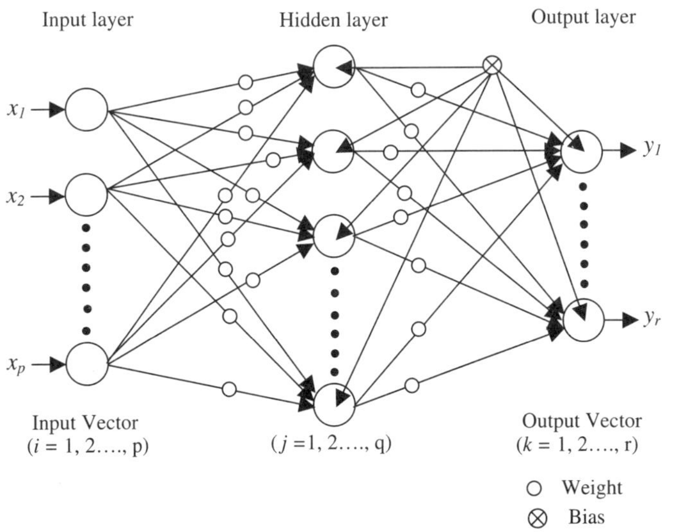

Fig. 1. A typical three-layer feed-forward neural network.

from a desired (target) response to produce an error. This error is then propagated backward through the network, from the output layer to the input layer via each hidden layer. The weights are adjusted so as to make the actual response of the network move closer to the desired response.

4. Training Algorithms

It is very difficult to know which training algorithm will be the fastest for a given problem. It depends on many factors, including the complexity of the problem, the number of data points in the training set, the number of weights and biases in the network, the error goal, and whether the network is being used for function approximation (regression). These algorithms are explained briefly in the following, the details of which can be found in Demuth *et al.*[4] Conjugate Gradient Polak–Ribiere update (CGP) algorithm is another version of Conjugate gradient algorithm in which the search direction for iteration is determined with the help of the inner product of previous change in the gradient with the current gradient divided by the norm-squared of the previous gradient. One-step secant (OSS) method requires less storage and computation per epoch. The Levenberg–Marquardt (LM) algorithm is a variation of Newton methods,[4] and this algorithm is designed to approach the second-order training speed without having to compute the Hessian matrix.

5. Performance Evaluation Criteria

The closeness between the network prediction and the actual observation is judged by three performance evaluation criteria, namely, mean square error (MSE), mean absolute error (MAE), and correlation coefficient (R^2).

Mean Square Error (MSE)

$$\mathrm{MSE} = \frac{1}{N}\sum_{i=1}^{N}(Q_{oi} - Q_{ei})^2, \tag{1}$$

Mean Absolute Error (MAE)

$$\mathrm{MAE} = \frac{\sum_{i=1}^{N}|Q_{oi} - Q_{ei}|}{N}, \tag{2}$$

Coefficient of Correlation (R^2)

$$R^2 = \sqrt{\frac{\sum_{i=1}^{N}(Q_{oi} - \bar{Q}_o)^2 - \sum_{i=1}^{N}(Q_{oi} - Q_{ei})^2}{\sum_{i=1}^{N}(Q_{oi} - \bar{Q}_o)^2}}, \tag{3}$$

where $\bar{Q}_o = \frac{1}{N}\sum_{i=1}^{N} Q_o$, Q_{oi} is the observed discharge (m^3/s), Q_{ei} is the estimated discharge (by ANN) (m^3/s), and $\bar{Q}_o$ is the mean of discharge (m^3/s). The performance of the models applied in the present study is evaluated based on the various numerical criteria listed above, which are indicative of the model performance.

6. Results and Discussion

Determining the ANN architecture is the most important part in ANN modeling and is normally attempted by a trial and error process. The process entails finding the number of input nodes, the number of output nodes, the network's transfer functions, the error functions which monitor the network's performance, the number of hidden layers, and the number of nodes in the hidden layers. The numbers of input and output nodes are decided by the nature of the problem. In the present problem, input variable is the stage of the creek, and output variable is the rate of stream flow of Peachtree creek. Consequently, there is only one input node (stage) and one output node (stream flow). The number of hidden nodes in FFBP is decided by trials, till the most acceptable testing performance is reached. For the present study, feed-forward back-propagation network architecture (FFBP, $1 \times 4 \times 1$) is selected in the model. The logsig transfer function is applied in the hidden layer, and the purelin transfer function is applied in the output layer.

About 60% of the available data is used for training the network and the remaining ones are employed to test it. The testing period is involved from 22 Nov 2006, 2.15 am to 23 Dec 2006, 8.00 am. It is found that sometimes a few lower-value predictions at a stretch had a tendency to systematically over-predict the actual observations, which could be due to the gaps in the measurements. In such a case an empirical correction is applied in that if there is a systematic over-prediction for two previous time steps then the future predictions would be reduced by the corresponding ratio between the actual and the predicted values. This has been found to considerably improve the prediction of the lower stream flow.

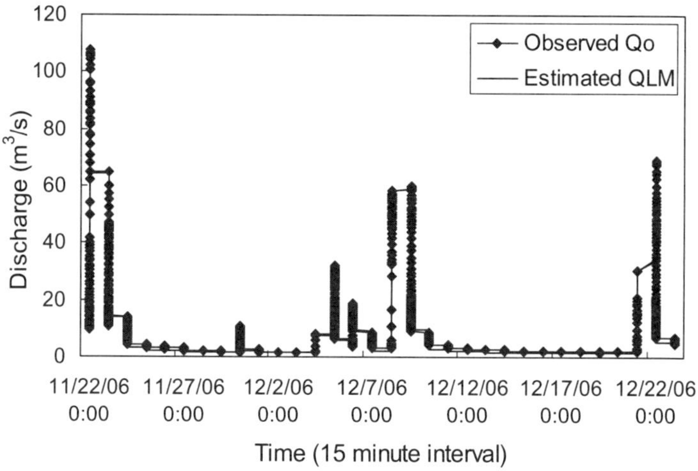

Fig. 2. Time series graph.

Table 1. Comparison of different models.

	Performance evaluation criteria		
Models	MSE $((\mathrm{m}^3/\mathrm{s}))^2$	MAE $(\mathrm{m}^3/\mathrm{s})$	R^2
FFBP-LM	0.4729	0.5577	0.9995
FFBP-CGF	0.9724	0.6929	0.9984
FFBP-CGP	0.9834	0.6982	0.9984
FFBP-OSS	3.5081	1.0647	0.9941
LR	6.1264	1.5530	0.9714

Figure 2 shows the time history graph based comparison between the observed discharge and their predictions made by the FFBP-LM model. Generally, satisfactory match between the two may be noted. This is confirmed by the values of the error measures, namely, the mean square error (MSE), the mean average error (MAE), and the correlation coefficient (R^2).

Table 1 shows the comparison of the results from various training algorithms. The magnitude of R^2 is high, ranging from 0.9995 to 0.9984, while the MSE and MAE are very low, ranging from 0.4729 to 3.5081 $(\mathrm{m}^3/\mathrm{s})^2$ and 0.5577–1.0647 m^3/s, respectively, indicating satisfactory performance of this scheme. LM training algorithm gives relatively better results compared to other algorithms.

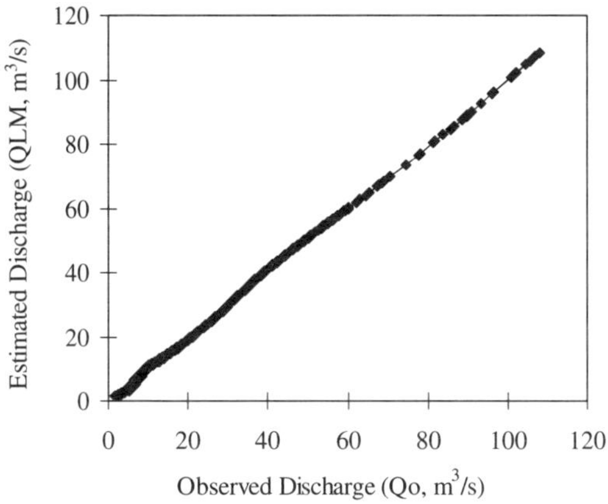

Fig. 3. Scatter plot between the observed and the estimated discharge.

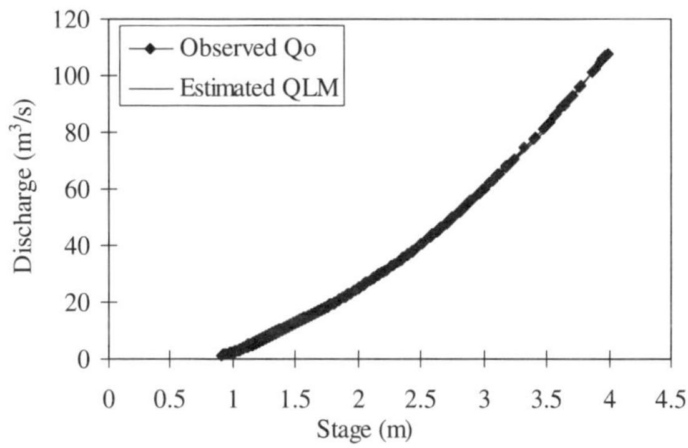

Fig. 4. Rating curve computed from FFBP-LM model along with the observed curve.

Figure 3 shows the scatter point plot-based comparison between the observed and estimated discharge made by the LM algorithm. As seen, all the data lie on the exact fit line in all training algorithms, and the correlation coefficient (R^2) approaches 0.9995.

A comparison of LM algorithm with other algorithms is shown in Table 1 and the former clearly stands apart considering the R^2 values. In Fig. 4, a plot of estimated discharge (using LM) against different stage values is compared with the observed values (Q_0). Even though all the algorithms in Table 1 are good in predicting discharge, it can be seen that observed and estimated discharge values exactly coincide in LM's case

Table 2. Statistical comparison.

	Input	Observed output	Estimated output	
Parameter	Stage (m)	Q_O (m^3/s)	Q_{LM} (m^3/s)	Q_{LRA} (m^3/s)
Mean	1.2231	8.1764	7.7387	8.1764
Standard error	0.0094	0.2676	0.2711	0.2637
Median	1.0149	2.7371	1.9699	2.3560
Mode	0.9540	1.8821	1.5555	0.6514
Standard deviation	0.5166	14.6569	14.8477	14.4461
Sample variance	0.2668	214.8249	220.4533	208.6903
Kurtosis	7.7392	14.6879	14.1685	7.7392
Skewness	2.7615	3.6032	3.5484	2.7616
Minimum	0.9144	1.4459	1.3355	−0.4566
Maximum	3.9898	107.9277	108.5351	85.5417

(Fig. 4). This indicates that LM algorithm is a desirable tool to predict discharge values.

Analyzing both models, regression model uses linear combinations of variables and, therefore, is not adept at modeling grossly nonlinear complex interactions as has been demonstrated in complex hydrological systems. ANN models are flexible systems that show robust performance in dealing with noisy or incomplete data, and have the ability to generalize from the input data. They may be better suited than other modeling systems to predict outcomes when the relationships between the variables are complex, multidimensional, and nonlinear, as found in complex hydrological systems. It can be used in both linear and nonlinear systems, and allows us to study the covariate interactions. By adjusting weights, multivariate outputs can also be evaluated. The error obtained from ANN models is less, compared to the regression model that shows the higher accuracy in prediction (Table 1). Table 2 shows the statistical analysis that indicates that the LM algorithm gives better results compared to LR.

7. Summary and Conclusion

In the present study, the feed-forward back-propagation network model developed with different training algorithms is applied to a Peachtree creek in Atlanta, Georgia for discharge prediction. Performance evaluation criteria were applied between the observed and estimated discharge values, and it was found that the values are very close, as shown in Table 1. Hence, it can be said that the developed model works satisfactorily for real-world problems. FFBP network trained with LM gives relatively better results as compared to other training algorithms.

The current study is in line with previous research efforts in recommending LM algorithm to be a desirable tool for hydrological, water resources modeling and applications.[1,3,9–11,14]

Acknowledgment

The data used in this study were provided by Brian E. McCallum, Assistant Director and Anthony J. Gotvald, Hydrologist USGS Georgia Water Science Center, Atlanta, Georgia. The authors wish to thank him and the staff of the USGS who are associated with data observation and processing. The authors are also thankful to Sunand Sreeramachandran for his efforts in preparing the manuscript.

References

1. A. J. Adeloye and A. D. Munari, Artificial neural network based generalized storage–yield–reliability models using the Levenberg–Marquardt algorithm, *J. Hydrol.* **362** (2006) 215–230.
2. ASCE Task Committee, Artificial neural networks in Hydrology I: Preliminary concepts, *J. Hydrologic Eng.* **5**(2) (2000) 115–123.
3. P. Coulibaly, F. Anctil, R. Aravena and B. Bobee, Artificial neural network modeling of water table depth fluctuations, *Water Res. Res.* **37**(4) (2001) 885–896.
4. D. Howard and B. Mark, *Neural Network Toolbox User's Guide* (The Math Works, Inc, 2000).
5. I. Flood and N. Kartam, Neural networks in civil engineering I: Principles and understanding, *J. Comp. Civil Eng.* **8**(2) (1994) 131–148.
6. I. Flood and N. Kartam, Neural networks in civil engineering II: Systems and applications, *J. Comp. Civil Eng.* **8**(2) (1994) 149–162.
7. S. K. Jain, Development of integrated sediment rating curves using ANNs, *J. Hydraulic Eng.* **127**(1) (2001) 30–37.
8. S. K. Jain and D. Chalisgaonkar, Setting up stage discharge relations using ANN, *J. Hydrologic Eng.* **5**(4) (2000) 428–433.
9. H. R. Maier and G. C. Dandy, Neural network for the prediction and forecasting of water resource variable: A review of modeling issues and application, *Environ. Modeling Software* **15** (2000) 101–124.
10. H. R. Maier and G. C. Dandy, Empirical comparison of various methods for training feed-forward neural networks for salinity forecasting, *Water Res. Res.* **32**(8) (1999) 2591–2596.

11. N. Samani, G. M. Moghadam and A. A. Safavi, A simple neural network model for the determination of aquifer parameters, *J. Hydrol.* **340** (2007) 1–11.
12. K. P. Sudheer and S. K. Jain, Radial basis function neural network for modeling rating curves, *J. Hydrologic Eng.* **8**(3) (2003) 161–164.
13. M. Tawfik, A. Ibrahim and H. Fahmy, Hysteresis sensitive neural network for modeling rating curves, *J. Comp. Civil Eng.* **11**(3) (1997) 206–211.
14. E. Toth, A. Brath and A. Montanari, Comparison of short-term rainfall prediction models for real-time flood forecasting, *J. Hydrol.* **239** (2000) 132–147.

Advances in Geosciences
Vol. 11: Hydrological Science (2007)
Eds. Namsik Park *et al.*

SEDIMENT DEPOSITION RATES DOWNSTREAM OF HARVESTED AND NATURAL PEAT BOGS IN NEW BRUNSWICK, CANADA

BRONWYN PAVEY[*,†,§], SIMON COURTENAY[†,‡], ANDRÉ ST-HILAIRE[*] and TAHA OURDA[*,¶]

[*]*Institut National de la Recherche Scientifique*
Centre Eau, Terre et Environnement 490 rue de la Couronne
Québec, Québec G1K 9A9, Canada

[†]*Department of Fisheries and Oceans*
Gulf Fisheries Centre, Oceans and Science Branch PO Box 5030
Moncton, New Brunswick E1C 9B6

[‡]*Canadian Rivers Institute, University of New Brunswick*
Fredericton, New Brunswick E3B 6E1

[§]*Current address: Jacques Whitford Ltd.*
115 Harrisville Blvd., Moncton, New Brunswick E1H 3T3
[¶]*taha_ouarda@ete.inrs.ca*

Peat harvesting is an important industry in many countries, including Canada. To harvest peat, bogs are drained and drainage water is evacuated toward neighboring rivers, estuaries, or coastal waters. Previous research in New Brunswick (Canada) has demonstrated that suspended solids in this drainage water can compromise fish habitat where it settles out and smothers the benthos. The objectives of this study were to quantify the particle size distribution and organic content of deposited sediments in addition to determining deposition rates. Results show that most of the material settling immediately downstream of harvested bogs is inorganic. Grain sizes were different between the reference and harvested sites, with the former showing a greater percentage of sand than the latter. Deposition rates are significantly higher at harvested sites than at the reference, unperturbed bog.

1. Introduction

Peatlands cover approximately 113 million hectares of Canada's land mass (over 11% of the total surface area) and over 75% of Canada's wetland area. Primary ombrotrophic (rain-fed) peat bog vegetation is sphagnum moss, a product that is commercially extracted and sold for its water retention properties, a quality desirable in the horticultural industry. Canada is the top producer of horticultural peat in the world with the province of New Brunswick producing 30% of Canada's total exports. In 2003 and 2004, peat shipments from New Brunswick were valued at \$CDN 95 million and \$CDN 103 million, respectively.[1]

One of the primary functions of natural wetland systems is sediment retention.[2] Because harvested peat bogs are drained and vegetation is removed, major physical effects of peat harvesting include:

1. Compaction of soils and changes to the structure of peat leading to decreased water storage capacity.[3,4]
2. Increased surface runoff and hence accelerated canal/ditch erosion.
3. Exposition of underlying friable peat layers and clay soils.

A major consequence of these physical changes is the disruption of the natural sediment regime in receiving waters, in particular the volume and timing of sediment delivery.[5]

A number of studies[5–11] have shown that very high suspended sediment concentrations frequently occur downstream of the harvested bogs. Ouellette *et al.*[9] studied one harvested bog on the Richibucto River system in Southeastern New Brunswick and found that a significant volume of peat was settling in the estuarine portion of the system, where the channel slope was minimal.

To our knowledge, there have been no studies performed in North America on sediment deposition rates downstream of peat bog harvesting operations. Deposition of sediments in freshwater streams or rivers can decrease the habitat value by reducing photosynthesis, smothering benthic organisms and spawning areas, and reducing the survival rates for fish eggs.[12–15]

As part of a larger project evaluating the effects of peat bog harvesting on sediment characteristics, this study describes the downstream settling rates at three harvested peat bogs and one unharvested reference site in New Brunswick, Canada. In addition, we quantify the grain size distribution and organic content of the deposited sediments speculating that the organic component of the deposited sediments would primarily be peat particles released from the upstream harvesting operations. The objectives of the study were to:

1. quantify deposition rates including the mass of sediment settling per season, per year and the average sedimentation rate per unit area;
2. determine the organic content of settled fine sediment (as an estimate of peat content); and

3. quantify particle size distribution and determine differences between the harvested and reference sites.

2. Materials and Methods

2.1. *Study area*

Sediment traps were deployed at five sites downstream of ombrotrophic bogs located in the central and eastern part in the province of New Brunswick, Canada (Fig. 1). The climate of New Brunswick (NB) is classified as mid-latitude continental with a yearly temperature range of 31°C.[16] Mean total annual precipitation in the study area is 817.5 mm, with nearly a third falling as snow.

Site 1, located in Kouchibouguac National Park, served as the unharvested reference site. Sites 2 and 3 were located downstream of a harvested bog in St Charles Plain located near Rexton, NB, and corresponding to the study sites used by St Hilaire *et al.*[19] Sites 4 and 5 were located approximately 70 km west of Sites 2 and 3, near Rogersville, NB on two separate harvested bogs.

At Site 1, the sediment traps were placed in a narrow stream exiting Rankin bog toward Black River. Sites 2 and 3 were located downstream of the harvested bog on the St Charles Plain, a 19.3 km^2 wetland. Peat has

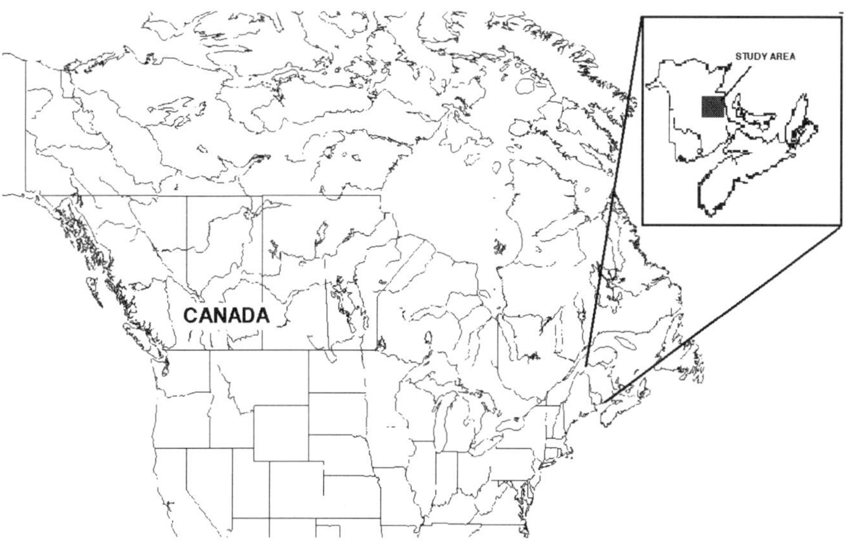

Fig. 1. Map of study area.

been extracted at this site since 1985, the longest operating site in our study. More specifically, the Site 2 sediment traps were located in a drainage ditch, 50 m downstream of the sedimentation pond outflow. Site 3 was located 1000 m downstream of Site 2 in Malpec Brook. This was the only location that permitted the installation of an upstream and downstream station.

The sediment traps at Site 4 were located 50 m downstream of the sedimentation pond. This bog was initially ditched in 2002, and harvesting began in 2003. The sediment traps at Site 5 were located on a second bog near Rogersville, 50 m downstream of the sedimentation pond. This sedimentation pond was enlarged during Summer 2003 from 2160 m^3 to its current size of 6048 m^3. The bog was ditched in 1992 and harvesting began in 1994. Table 1 summarizes the characteristics at each monitoring site.

2.2. *Field methods*

In order to evaluate the sedimentation rate, four to eight sediment traps were installed 50 m downstream of the outflow of peat bog sedimentation ponds. The traps, called Sedibacs^{TM} (Bio Innov Inc.), consisted of two identical cylinders (12 cm in diameter, 13.5 cm in height, and a total volume of 1 l) placed one inside the other. Each cylinder has 1.3 cm diameter holes throughout its surface providing a total open side area of 39.8 cm^2. For installation and removal, the cylinders can be rotated so that all holes are closed preventing unwanted collection or spilling of the sediment. Finally, a mesh bag is filled with clean gravel (2–4 cm) and placed in the inner cylinder prior to installation.

The Sedibacs^{TM} were buried in the streambed with the top of the trap flush with the surrounding streambed. In order to calculate net deposition, the top was removed and the cylinders were rotated to open all holes thus permitting fine particles to enter and exit or accumulate in the interstices of the large gravel pieces. In some cases, the Sedibacs^{TM} were installed in clay beds and little intra-bed flow of sediments was expected.

In 2003, traps were deployed during two separate periods: Spring (May to late June or early July) and Fall (October to November). During Spring, deployment periods varied between stations, from 59 days (station 5) to 74 days (stations 2 and 3). During Fall, traps were deployed for 43 days at station 4, 46 days at station 5, and 56 days at the other stations. In 2004, traps were deployed for most of the ice-free season, from May to November, with deployment periods varying from 174 days (stations 1 and 2) to 188 days (stations 4 and 5). No Sedibacs^{TM} were installed at Site 1 in Spring 2003 or at Site 2 in 2004.

Table 1. Summary of monitoring site characteristics.

Site	Site 1	Site 2	Site 3[a]	Site 4	Site 5
Location (GPS)	46.8139N 64.9391W	46.6388N 64.9188W	46.6371N 64.9189W	46.7121N 65.5890W	46.7045N 65.4169W
Elevation above sea level (m)	16	28	13	16	88
Total bog Area (ha)	204.1	366.6		202	360
Harvested/ cleared area (ha)	0	232		82	198.7
Total area being drained into studied sedimentation ponds (ha)	100	175		52	105
Sedimentation pond volume (m^3)	na	5478[b]		2280	2160[c] 6048[d]
Ratio: Drainage area to pond volume (m^2/m^3)	na	319		228	174
Number of harvest years (2003)	0	23		2	10

[a]Site 3 is downstream from Site 2;
[b]Combined volume of three ponds all draining past Sites 2 and 3;
[c]Size of pond in Spring 2003;
[d]Size of pond for the remainder of the study period, Fall 2003 and 2004; na = not applicable.

In addition to deploying traps at all sites, a more detailed experiment was conducted during the 2004 season in which 28 sediment traps were deployed at Site 3. Two traps were removed every two weeks for 18 weeks to determine whether the sediment traps reached a saturation point above which no new sediments were retained.

Instantaneous discharge measurements were taken weekly except when prohibited by shallow water conditions (less than 3 cm) using a flow meter (Flo-Mate model 2000). Continuous daily flow series were estimated using data from nearby hydrometric stations according to the method used by St Hilaire *et al.*[19]

2.3. *Laboratory methods*

After collection, the contents of each Sedibac™ were air-dried at room temperature for 1 week, and then oven-dried at 65°C for 24 h. The contents were passed through a series of Canadian standard sieves (64 μm to 4 mm) and agitated in a portable sieve shaker for 10 min. The top sieve containing large gravel pieces originally placed in the Sedibac™, was discarded and the dry weights of the remaining size classes were recorded. Sediments were divided into six size classes according to the Wentworth Scale[17]: very fine gravel (2–4 mm); very coarse sand (1–2 mm); coarse sand (0.5–1 mm); medium sand (0.25–0.5 mm); fine sand (0.063–0.25 mm); and clay and silt (<0.063 mm).

Once all of the samples were sieved and size-classed, the organic content of each size class at each site was determined. Each sample was placed in labeled aluminum dishes and weighed to the nearest 0.001 g. Percent organic matter was calculated as weight loss following 3 h in a Muffle oven heated to 475–500°C.

2.4. *Statistical methods*

The grain size distribution (i.e. percent mass of each sample in each size fraction) was compared between sites using the Pearson Chi square analysis. Data were non-normally distributed (Shapiro Wilks test, $p < 0.05$); thus, the organic content and net deposition rate were compared between sites using the Kruskal–Wallis test followed by Mann–Whitney pairwise tests. Medians, rather than means, are presented in this study, except where indicated.

Net annual sediment deposition rates (kg/m^2/y) were calculated from the mass of sediment retained by the trap, taking into consideration the number of days of deployment and the surface area of the trap (0.0113 m^2). Net deposition rate rather than total deposition was used, because sediment traps were deployed for varying lengths of time and sediments were allowed to enter and exit. Net deposition rates were compared between sites for each grain size and between sites (all grain sizes combined).

3. Results

A total of 84 Sedibacs™ were installed over all seasons of which 62 were recovered. Between 3 and 7 Sedibacs™ were collected per site.

3.1. *Accumulation differences*

Pairs of Sedibacs™ removed biweekly from Site 3 showed no evidence of saturation (i.e. no plateau in mass over time) (Fig. 2), but, rather, appeared to maintain a state of flux. Possible reasons for this variability are discussed further in the final section. Overall, the variable results suggest that saturation was not reached.

3.2. *Grain size distribution*

Grain size distributions were compared using percentage of the total weight of deposited sediments in each grain size category. The grain size distributions at the harvested sites (combined) and control site were found to be significantly different (Pearson $\chi^2 = 17.148$; $p = 0.004$).

The dominant grain size category at the harvested sites was medium sand (0.25–0.5 mm) (Fig. 3a), which constituted 42% of the total sediment weight. This was followed by fine sand (22%) and coarse sand (18%). The grain size of deposited sediment was much more uniformly distributed at the unharvested, reference site (1) varying between 12% for clay and silt to 24% for fine sand (Fig. 3b).

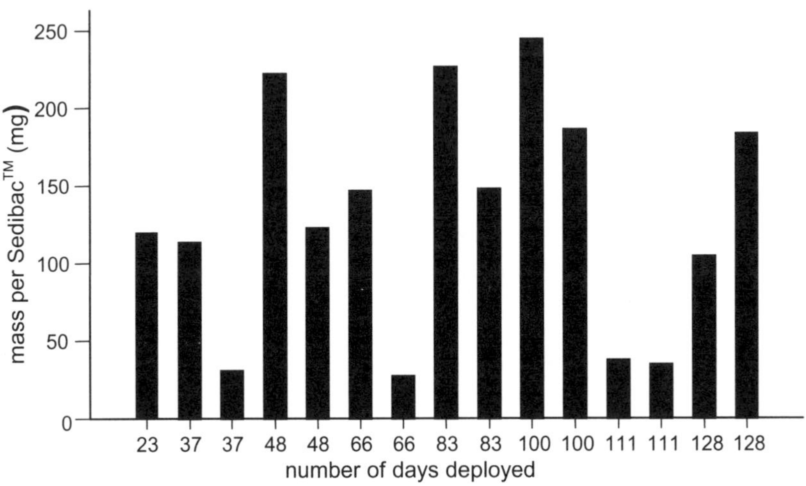

Fig. 2. Sedibac™ total mass after x days of deployment at Site 3. Note that two traps were removed on every site visit except on day 23 when one trap was removed.

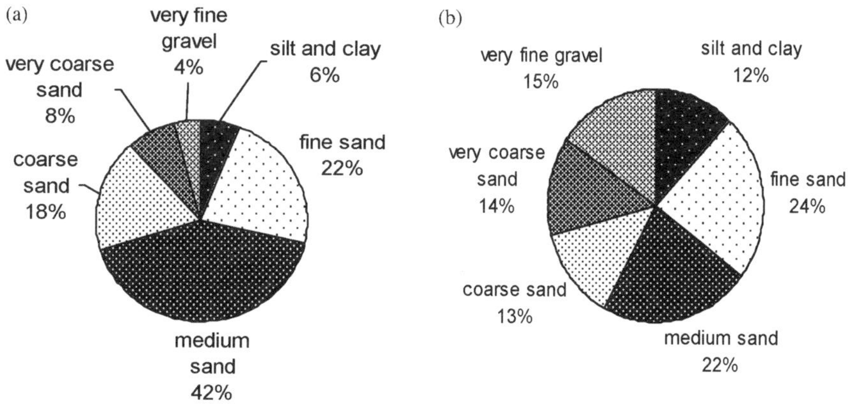

Fig. 3. Sediment grain size class distribution (median percent of total mass) in traps placed downstream of harvested and unharvested peat bogs. (a) Harvested sites, $n = 38$; (b) Control site, $n = 9$.

3.3. *Deposition rates*

At the harvested sites, the net deposition rates varied between 4.18 and 210.69 kg/m^2/y (Table 2, column (i)). The pooled median for all harvested sites over all sampling periods was 60.97 kg/m^2/y compared to 12.76 kg/m^2/y at the reference site. Data were extrapolated from seasonal to annual deposition rates to enable comparison. Considerable seasonal variability between Spring and Fall 2003 indicates that annual deposition rates based on one season may require correction. The 2003 data do show that deposition rates were higher in Fall than in Spring at all sites.

In 2004, all sites had lower net deposition rates than in spring 2003 and fall 2003. Mean flow values for all stations varied between 24 m^3/s and 66 m^3/s during the spring season in 2003. During the Fall of the same year, mean flow varied between 18 m^3/s and 39 m^3/s. In contrast, mean flows calculated in 2004 varied between 9 m^3/s and 18 m^3/s. The lower deposition rates in 2004 may result from lower sediment loads associated with lower discharge.

3.4. *Percentage organic content*

Percentage of organic content of deposited sediments varied between 2.88% and 7.88% at the harvested sites with an overall pooled median of 4.39% for all sampling seasons (Table 3). The reference site had an organic content of 11.99% in Fall 2003 and 33.29% in 2004 resulting in a pooled median of 22.64%. Significant differences between sites were found in Fall 2003 and in

Table 2. Median deposition rate (kg/m^2/y) and Kruskal–Wallis results are reported for each grain size per site.

(a) Site	(b) n	(c) Gravel	(d) Very coarse sand	(e) Coarse sand	(f) Medium sand	(g) Fine sand	(h) Clay and silt	(i) Pooled median Deposition rate per site
SPRING 2003								
Site 1	0	*	*	*	*	*	*	
Site 2	3	1.35	3.83	7.15	18.81	10.05	8.89	50.37^a
Site 3	3	2.25	4.52	11.08	13.97	4.80	11.99	53.07^a
Site 4	0	*	*	*	*	*	*	*
Site 5	4	3.12	5.63	13.25	22.95	10.39	11.46	77.40^b
X^2		2.78	0.69	1.46	0.69	5	3.36	6.71
p-value		0.249	0.706	0.480	0.706	0.082	0.186	0.035
FALL 2003								
Site 1	5	1.22	1.44	1.69	3.26	3.30	2.04	13.37^a
Site 2	4	1.26	1.69	4.94	38.18	14.92	5.65	68.86^b
Site 3	6	8.92	10.42	17.80	50.30	22.92	6.66	108.23^c
Site 4	4	3.67	5.09	16.14	57.99	57.36	7.13	147.64^c
Site 5	7	8.98	14.79	40.63	79.94	37.99	6.28	210.69^d
X^2		16.25	19.59	20.27	19.32	18.53	13.04	18.72
p-value		0.002	0.000	0.000	0.000	0.001	0.011	0.001
2004								
Site 1	4	2.83	2.04	1.73	2.29	2.39	0.60	12.15^a
Site 2	0	*	*	*	*	*	*	*
Site 3	2	2.83	2.01	5.28	24.59	1.82	1.69	36.20^a
Site 4	4	1.63	1.92	3.42	7.07	4.87	0.75	19.81^a
Site 5	3	0.66	0.53	0.66	1.19	1.04	0.25	4.18^a
X^2		4.75	5.14	4.75	6.35	3.41	4.30	7.75
p-value		0.093	0.076	0.093	0.096	0.182	0.117	0.051

Notes: [a,b,c,d]Sites with similar medians based on the Mann–Whitney U test. Grain size descriptions are given in the text.
*No data.

2004, with the reference site (Site 1) depositing significantly more organic material than the harvested sites except Site 4 in Fall 2003 and Site 3 in 2004 (Table 3). No sediment traps were deployed at the reference site in Spring 2003; thus, a comparison could not be made.

At both the reference and harvested sites, percent organic content within each grain size followed a bimodal pattern with higher organic content in gravel, very coarse sand, and then in clay and silt. The highest percent organic content occurred at Site 1 in very coarse sand in Fall 2003 and in 2004 (25.10% and 67.64%, respectively) (Table 3).

Table 3. Median percent organic content per site and season. Kruskal–Wallis results are reported for each grain size and the total % organic content per site.

(a) Site	(b) n	(c) Gravel	(d) Very coarse sand	(e) Coarse sand	(f) Medium sand	(g) Fine sand	(h) Clay and silt	(i) Pooled median deposition rate per site
SPRING 2003								
Site 1	0	*	*	*	*	*	*	*
Site 2	3	11.36	12.84	4.87	1.26	1.59	6.72	3.71[a]
Site 3	3	9.24	13.45	4.72	1.53	2.36	6.78	4.45[a]
Site 4	0	*	*	*	*	*	*	*
Site 5	4	3.69	5.79	3.02	1.36	3.35	6.12	4.33[a]
X^2		1.94	5.00	1.50	1.07	5.11	0.75	1.86
p-value		0.378	0.082	0.472	0.587	0.078	0.689	0.394
FALL 2003								
Site 1	5	21.88	25.10	20.55	12.38	1.99	5.29	11.99[a]
Site 2	4	14.70	14.71	8.03	4.03	3.38	13.19	6.23[b]
Site 3	6	4.49	4.93	3.60	3.96	1.99	8.12	3.87[c]
Site 4	4	5.96	6.08	3.81	4.84	4.69	7.06	7.88[a,b,c,d]
Site 5	7	2.32	2.18	3.90	2.78	4.04	4.06	3.70[b,c,d]
X^2		15.66	20.71	13.26	9.78	5.77	12.47	13.14
p-value		0.0035	0.0004	0.0101	0.0444	0.217	0.0142	0.01
2004								
Site 1	4	50.36	67.64	30.75	13.69	4.70	14.40	33.29[a]
Site 2	0	*	*	*	*	*	*	*
Site 3	2	13.10	13.30	7.17	6.60	6.87	7.91	7.38[a,b]
Site 4	4	6.12	5.33	3.08	2.65	1.41	4.97	2.88[b]
Site 5	3	11.95	10.79	5.52	3.22	2.10	5.83	5.96[b]
X^2		7.48	8.33	7.85	7.21	6.55	6.63	8.519
p-value		0.0238	0.0156	0.0198	0.0272	0.038	0.0364	0.036

Notes: *No data;
[a,b,c,d]Similar medians based on the Mann–Whitney U test.

Within grain sizes, there were no significant organic content differences between sites in Spring 2003; yet in Fall 2003 and in 2004 there were differences between sites in all cases except "fine sand" in Fall 2003.

4. Discussion and Conclusions

In this first Canadian study of sedimentation downstream of harvested peat bogs, it was important to demonstrate at the onset that the use of sedimentation traps was appropriate. The detailed experiment undertaken at Site 3 showed that the traps did not become saturated. As shown in

Fig. 2, there was considerable variability between Sedibacs™ removed on the same day, at the same site despite the proximity of the traps (within 25 m of each other). This is most likely a result of placement in runs, riffles or pools which experience varying hydraulic characteristics. The differences in net deposition calculated from traps collected at 2-week intervals may be caused by rain events or changing flow conditions leading to increased deposition or resuspension. Differences in the amount and size of sediment grains collected may also be reflecting local and immediate conditions at the time of Sedibac™ removal rather than long-term net deposition. Regardless of the actual weight accumulated in the sediment traps, the median percentage by weight in each size class followed a similar unimodal distribution for each monitoring period.

Grain size distributions were significantly different between the harvested sites and the reference site. At the harvested sites, the dominant grain size found in the Sedibacs™ was medium sand (0.25–0.5 mm) followed by fine sand (0.063–0.25 mm) and coarse sand (0.5–1 mm) totaling 82% of the total mass of deposited sediments. At the reference site, these same categories made up 59% of the total mass. Wood and Armitage[5] found that medium sand particles of size 0.2–0.5 mm are more easily entrained into the water column and transported downstream than either larger or smaller particles. Similarly, Walling and Amos[18] found that most sediment less than 2 mm in diameter are deposited 5–7 km from the source. High deposition rates can affect aquatic biota, including benthic invertebrates and early life stages of local fish.

Overall, Site 1 (unharvested bog) had a lower deposition rate than the other sites (harvested sites). Pavey *et al.*[11] also found that Site 1 had significantly lower suspended sediment concentrations (SSC) than the harvested sites. In 2003, net deposition rates at all sites were higher in Fall compared to Spring. Interestingly, Pavey *et al.*[11] found that SSC levels were also greater in Fall. Our results support the findings of Barton[18] who reported that deposition of sediment in traps varies directly with suspended solid levels.

A study by St Hilaire *et al.*[19] found a net deposition rate, referred to as a storage rate, of $40\,kg/m^2/y$, in a relatively pristine river in Northern New Brunswick. Harvested Sites 3, 4, and 5 all recorded deposition levels an order of magnitude greater than the value reported by St Hilaire *et al.*[19] The high deposition rates at those harvested bogs are likely evidences of a disturbed hydrological and sedimentological regime. Site 3 in Malpec Brook is the only site within an area of potential fish habitat. Site 3 recorded a median deposition rate of 53 kg/m^2/y over all sampling seasons and had the highest deposition rate of all sites in 2004.

Sallantaus[20] found that the annual discharge of suspended particulate organic matter out of harvested bogs was not greatly different from the discharge of organic matter leaving natural peat bogs. When considering relative values (i.e. percent grain size), our results show that the organic content in deposited matter was higher at the reference site.

At the harvested sites, fine, medium, and coarse sand represented the highest percent of total mass. There are two potential reasons why the organic content found in the harvested sediment traps was low: there is little organic content exiting harvested peat bogs; or organic particles are settling upstream or downstream of the sediment traps. Baumgaertner[21] found that the settling velocities for peat, thought to be the primary source of organic matter in this study, were of the same order as fine silt. Madej[22] confirmed this finding noting that organic particles have less bulk-density and a higher surface-to-volume ratio than inorganic particles causing them to remain in suspension longer. This suggests that at our study sites, peat may be settling further downstream than the Sedibacs™, which were deployed only 50–500 m from settling ponds. Ouellette *et al.*[9] confirmed this hypothesis downstream of Sites 2 and 3 where a large peat delta was found at the confluence of Malpec Brook and Mill Creek in the Richibucto Estuary.

Ditch erosion and the easily-dislodged top peat layer are suspected to be the key sources of sediment found downstream at Site 3 and just after the sedimentation ponds at Sites 2, 4, and 5. The field ditches are characterized by clays, low-plasticity organic soil, and sand which are among the most easily erodible soils.[23] The top peat layer is very friable, caused in part, by desiccation. Dry peat falls from the soil surface into the ditches when the large industrial vacuums collect peat and move between fields by crossing the ditches. Ditches were observed to extend into the mineral soil layer. Peat and sediment are also transported into the field ditches via gullies and rills caused by rain events.[3]

In summary, this study demonstrated that

1. sedimentation ponds may be trapping sediment but both fine and coarse sediments are escaping to the receiving waters, either through resuspension or incomplete trapping;
2. fines and peat are traveling more than 1 km from the sedimentation pond outflow;

3. sand is depositing in the area 50–500 m downstream, where our sediment traps were located;
4. more sand is leaving harvested sites than the reference site;
5. the absolute amount of organic content leaving the harvested and reference sites are in the same order of magnitude;
6. more sediment is depositing in Fall than in Spring;
7. overall, the reference site had a lower deposition rate than the harvested sites which corresponds to previous results demonstrating lower suspended sediment concentrations at the reference site when compared to the harvested sites.

These first results indicate the need for further study to fully understand the sedimentation dynamics in and downstream of peat drainage systems. Future studies should include numerous reference sites to develop a complete picture of deposition rates. A greater number of replicates (sediment traps) at each site within a given monitoring period may help to further explain intra- and inter-site variability. Further work could also be done to explain interseasonal variability within a site. The use of sediment traps should also be combined with the collection of resident matrix material (scoop samples) for comparison. Finally, the differences between total and net deposition could be examined.

Acknowledgments

The following individuals provided technical support for this study: Monica Boudreau, Stéphan Leblanc, and Isabel Moreau. Financial support was graciously provided by the National Science and Engineering Council (NSERC-IPS1), Premier Horticulture Ltd., and Nirom Peat Moss. Special thanks are due to Andrew Boghen, Claudie Beaulieu, and Chantale Dubé for logistical support. The logistical support of Éric Tremblay at Kouchibouguac National Park is also acknowledged.

References

1. J. Thibault, *Peat Industry Review 2005*, NB DNRE (Bathurst, NB, 2006).
2. K. G. Boto and W. H. Patrick Jr., Role of wetlands in the removal of suspended sediments, *Proc. National Symp. Wetlands*, 1979, Minneapolis, USA, Vol. 479, pp. 479–489.
3. B. Kløve and L. Bengtsson, Runoff generation in a plough-drained cutover fen in Central Finland, *J. Hydrol.* **218** (1999) 157–168.

4. S. M. Schlotzhauer and J. S. Price, Soil water flow dynamics in a managed cutover peat field, Quebec, *Water Res. Res.* **35**(12) (1999) 3675–3683.
5. P. J. Wood and P. D. Armitage, Biological effects of sediment in the lotic environment, *Environ. Manage.* **21**(2) (1997) 203–217.
6. J. D. Gregory, R. W. Skaggs, R. G. Broadhead, R. H. Culbreath, J. R. Bailey and T. L. Foutz, Hydrologic and water quality impacts of peat mining in North Carolina, Report No. 214 (1984).
7. B. Kløve, Erosion and sediment delivery from peat mines, *Soil Tillage Res.* **45** (1998) 199–216.
8. S. Joensuu, E. Ahti and M. Vuollekoski, The effects of peatland forest ditch maintenance on suspended solids in runoff, *Boreal Environ. Res.* **4** (1999) 343–355.
9. C. Ouellette, S. C. Courtenay, A. St-Hilaire and A. D. Boghen, Impact of peat moss released by a commercial harvesting operation into an estuarine environment on sand shrimp *Crangon septemspinosa*, *J. Appl. Ichthyol.* **22**(1) (2006) 15–24.
10. A. St Hilaire, S. C. Courtenay, C. Diaz-Delgado, B. Pavey, T. B. M. J. Ouarda, A. Boghen and B. Bobée, Suspended sediment concentrations downstream of a harvested peat bog: Analysis and preliminary modeling of exceedances using the logistic regression, *Canad. Water Resour. J.* **31**(3) (2006) 139–156.
11. B. Pavey, A. St-Hilaire, S. C. Courtenay, T.B.M.J. Ouarda and B. Bobée, Exploratory study of suspended sediment concentrations downstream of harvested peat bogs, *Environ. Monitor. Assess.* **135** (2006) 369–382.
12. E. M. Stern and W. B. Stickle, Effects of turbidity and suspended material in aquatic environments, US EPA Dredged Material Research Program, Final report, Vicksburg, MS (1978).
13. R. A. Sweeting, *The Rivers Handbook: Hydrological and Ecological Principles*, Vol. 2 (Oxford University Press, 1994), pp. 23–32.
14. Trow Consulting Engineers Ltd., Instream sediment control techniques field implementation manual, OMNR, Northeast Science and Technology. FG-007, (1996).
15. A. R. Curry and S. W. MacNeill, Population-level responses to sediment during early life in brook trout. *J. North Am. Benthol. Soc.* **23**(1) (2004) 140–150.
16. D. Phillips, *The Climates of Canada* (Canadian Government Publishing, 1990).
17. C. K. Wentworth, A scale of grade and class terms for clastic sediments, *J. Geol.* **30** (1922) 377–392.
18. B. A. Barton, Short-term effects of highway construction on the limnology of a small stream in southern Ontario, *Freshwater Biol.* **7** (1977) 99–108.
19. A. St Hilaire, D. Caissie, R. A. Cunjak and G. Bourgeois, Streambed sediment composition and deposition in a forested stream: Spatial and temporal analysis, *River Res. Appl.* **21** (2005) 883–898.
20. T. Sallantaus, Quality of runoff from Finnish fuel peat mining areas, *Aqua Fennica* **14** (1984) 223–233.

21. I. V. Baumgaertner, Settling velocity, in *Encyclopedia of Sedimentology* (Dowden, Hutchinson and Ross, 1978), p. 735.
22. M. A. Madej, The role of organic matter in sediment budgets in forested terrain, *Proc. Symp. S* (IAHS Publication, 2005), p. 292.
23. J. S. Fifield, Effective sediment and erosion control for construction sites, *Stormwater* **2**(1) (2001) 52–56.

Advances in Geosciences
Vol. 11: Hydrological Science (2007)
Eds. Namsik Park *et al.*

FINE SEDIMENT DEPOSITION IN A CURVED CHANNEL WITH AN INTAKE WEIR*

YOSHIYUKI SHIROTORI*, KOZO WATANABE, KEIKO UDO and AKIRA MANO
Graduate School of Engineering, Tohoku University, 6-6-11
Aoba, Sendai 980-8579, Japan
**shirotori@potential1.civil.tohoku.ac.jp*

On the upstream of Atago weir, Hirose River, Japan, the sedimentation phenomenon is complex due to the coexistence of the weir and channel bend. Moreover, fine sediment accumulates on the river bed notably. From the observed results, it is found that silt is the major component in the flood water and accumulates significantly in the discharge increasing stage by the backwater of the weir. From the theoretical computation considering the settling and resuspension of the suspended sediment, the cohesion of fine sediment is an important factor of the net deposition process.

1. Introduction

Deposition of fine sediment forms the medium of vegetation, and can bring thick growth of vegetation.[1] Thick growth of such vegetation brings the trouble of the flood passage. Moreover, thick sedimentation of fine sediment under calm water promotes oxygen consumption, and produces anoxic condition. Since deposition of fine sediment brings problems in both a flood control and an environmental preservation, the control of fine sediment deposition is one of the most important subjects in the river channel management.

Intake weirs crossing river channels are common examples of such issues.[2] The weirs produce stagnant water body at the upstream and promote sedimentation. Moreover, river bend makes the phenomena more complex. The secondary flow by the bend leads sedimentation near the inner bank and erosion near the outer bank. This makes the channel unstable. This study targets the flow and sediment transport at the coexistence of

*This work was supported by the contact with Miyagi Prefecture.

weir and bend. The Atago weir site, in the Hirose River, Japan, was selected as the study area, which is located near our university.

This study purposes to get good understanding on the complex deposition phenomenon; we conducted field observation and theoretical computation. For the field observation, we researched the suspended sediment at the flood and the bottom deposits. With the aim of clarifying the deposition factor, we conducted theoretical computation of one-dimensional unsteady flow and sediment transport. In one of existing studies, Zhou and Lin[3] have conducted a calculation of the suspended sediment transport for the Yangtze River. In these studies, however, several adjusting parameters are required, and we could not apply the model for our study area. So, we attempt an analysis which is able to deal with settling and resuspension separately without that kind of parameters.

2. Study Area and Secular Change of Point Bar Area

The Hirose River with the basin area of 311 km^2 flows through the central part of Sendai city, and the citizens and the river administrator are highly interested in the safety and environment of the river. The Atago weir with the weir height of 2.2 m constructed in 1954 at the lower reach of the river is used to distribute water for agriculture. The river has bends at the upstream as shown in Fig. 1 and has been promoting sedimentation near the right bank. As a result, a point bar shown by dashed line in Fig. 1 has been formed on the right bank and the bar emerges above the water. The study area was set as the area bounded by two bridges, Otamaya bridge and Atago bridge, which were points for the water sampling.

Figure 2 shows secular change of the area of the right point bar. The area was measured from the aerial photographs. Before a large dam has been constructed on the upstream in 1961, the area was stable and the bar surface was covered by gravels. This would be interpreted as meaning that frequent floods passage has washed out the surface material of the bar and has suppressed the development of the bar. The area rapidly decreases in the 1970s due to the new channel formation in the bar by a large flood passage. The vegetation was observed to progress on the bar after 1975, and was thick in 1984.

After 1980, the area gradually increased up to 1997. The growth of vegetation has a function to fix the bar and expand. In 1999, a part of the bar was artificially removed for river management purposes.

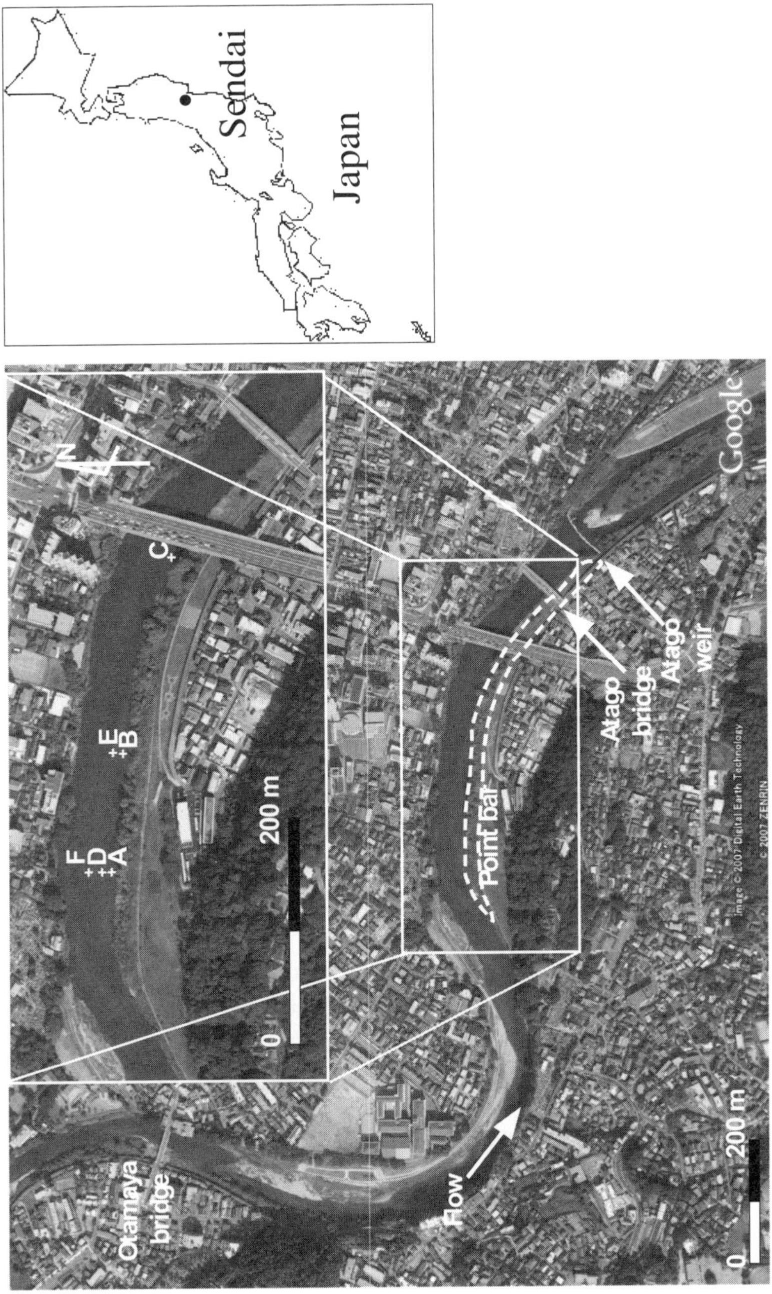

Fig. 1. Study area and deposits sampling points edited from Google Earth.

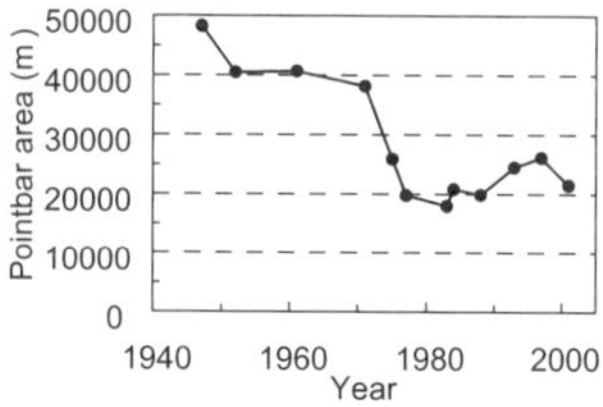

(a)

(b)

(c)

(d)

Fig. 2. (a) Secular change of point bar area. (b) Bed surface was covered by gravels in 1947. (c) The new channel was formed and the vegetation was progressing in 1983. (d) The vegetation was growing thickly in 2001.

3. Methodology

3.1. *Field observation and laboratory analysis*

In order to know the sediment transport characteristics in the study area, it is important to comprehend the suspended sediment flux and the bed deposits quantitatively. Then, we measured sediment concentration and grain distribution from the suspended sediment during the flood event. We sampled the suspended sediment at Otamaya bridge and Atago bridge. We also took core samples from the bed and measured the grain size distribution.

The rainfall event of the precipitation 20 mm/h (Nikkawa, which is about 40 km on the upstream from the study area) in the maximum occurred in 27th September 2006, and the peak discharge was 190 m^3/s (Fig. 3) with the occurrence frequency of several times per year.

For the flood observation, we cooperated with Miyagi Prefecture. The flood water near the center of channel was sampled at the upper bridge by us and at the lower bridge by Miyagi Prefecture. At the upper bridge, we obtained the flood water using a bucket with rope at intervals of 3 h. At the lower bridge, the cooperator obtained the flood water using a metal sampler at intervals of 1 h until 6 h after the discharge peak, and then sampled the flood water in response to the behavior of discharge.

Sediment concentration and grain distribution of the samples were analyzed. 50–200 ml water was filtered by the glass fiber filter of nominal pore size, 1.0–1.2 μm to obtain suspended solid concentration (SS).

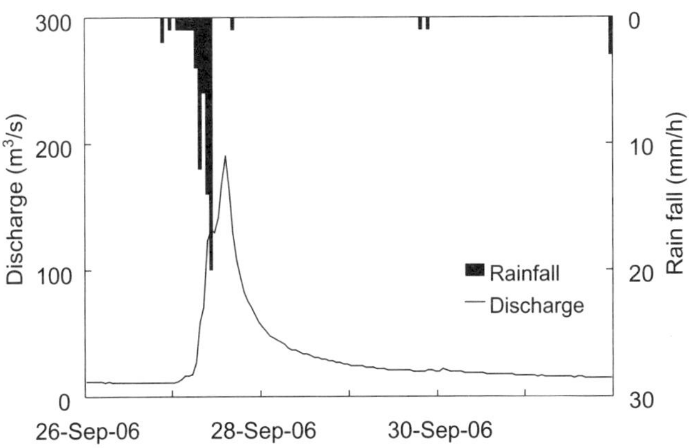

Fig. 3. Observed flood in September 2006.

We regarded SS as suspended sediment concentration, referring to Togashi and Mano.[4]

Grain distribution was analyzed by laser, Nikkiso MicroTrack with measurable range of 0.1–700 μm and Shimadzu SALD-300S with measurable range of 0.08–3000 μm.

Surface deposits at the riverbed were sampled in several meters point from the waterside by using the core sampler with the depth of 20–30 cm. The sampling points, A, B, C, D, E, and F are shown in Fig. 1. The sampled deposits were analyzed with sieves and the laser grain distribution meter.

3.2. *Numerical computation*

3.2.1. *Open channel flow and sediment transport*

To clarify the deposition factor of suspended sediment, we have developed a numerical method to compute the flow and sediment transport. One-dimensional unsteady flow is governed by the continuity equation (1) and momentum equation (2).

$$\frac{\partial \bar{\eta}}{\partial t} + \frac{1}{B}\frac{\partial Q}{\partial x} = 0, \tag{1}$$

$$\frac{\partial Q}{\partial t} + \frac{\partial}{\partial x}\frac{Q^2}{A} = -gDB\frac{\partial \bar{\eta}}{\partial x} - \frac{\bar{n}^2 g}{D^{7/3}}\frac{|Q|}{B}Q, \tag{2}$$

where $\bar{\eta}$ is the water level, B is the river width, Q is the discharge, A is the cross-section area, g is the gravity acceleration, D is the water depth, and $\bar{n}$ is the manning roughness coefficient. The overbar denotes the average over cross-section.

The suspended sediment transport averaged in the cross-section is governed by the mass conservation (3). Here, we take the effect of vertical distribution of the concentration into an account on the settling flux at the bottom. Mano *et al.*[5] measured the suspended sediment distribution in flood water and found that Rouse's theoretical distribution shows good agreement for the vertical distribution of the sediment concentration on various grain sizes. Therefore, we assume Rouse's distribution on the vertical sediment profile and find the relationship between the averaged and bottom concentrations.

$$\frac{\partial \bar{C}}{\partial t} + \frac{1}{BD}\frac{\partial(\bar{C}Q)}{\partial x} = \frac{\bar{C}}{BD}\left(\frac{\partial Q}{\partial x}\right) + \frac{\partial}{\partial x}\left(\bar{\varepsilon}_{\mathrm{s}}\frac{\partial \bar{C}}{\partial x}\right) + \frac{w_0}{D}C_{\mathrm{a}} - \frac{w_0}{D}C_{\mathrm{bottom}}, \tag{3}$$

where $\bar{C}$ is the suspended sediment concentration averaged in vertical direction, w_0 is the settling velocity calculated by Rubey formula, $\bar{\varepsilon}_s$ is the diffusion coefficient of suspended sediment in vertical direction calculated by Eq. (4).

$$\bar{\varepsilon}_s = \frac{u_* D}{15\sigma_s}, \tag{4}$$

where σ_s is the turbulent Schmidt number (It was assumed 1.0), u_* is the friction velocity, the third term in the right-hand side of Eq. (3) signifies resuspension, and C_a is the concentration at the reference level near the bottom calculated by Eq. (5), which is an expanded equation of the van Rijn formula[6] by considering the portion of the target grain particle p_d.

$$C_a = 0.015 p_d \frac{d}{a} \frac{T_*^{1.5}}{d_*^{0.3}}, \tag{5}$$

$$d_* = d \cdot \sqrt[3]{\frac{(\rho_s - \rho)g}{\nu^2}}, \tag{6}$$

$$T_* = (\theta - \theta_c)/\theta_c, \tag{7}$$

where d is the grain diameter; d_*, the non-dimensional grain diameter defined by Eq. (6); ρ_s, the sediment density; ρ, the water density; ν, the kinematic viscosity; a, the reference level and it is defined as the roughness height or a half of sand wave height. In this study, the latter was adopted. θ is the Shields number, θ_c is the critical Shields number.

Moreover, the fourth term in the right-hand side of Eq. (3) signifies settling, and C_{bottom} is calculated by Rouse's formula (8).

$$C_{\text{bottom}} = \bar{C} \left[\frac{1}{1 - z_a^*} \int_{z_a^*}^{1} \left[\left(\frac{1 - z^*}{z^*} \right) \left(\frac{z_a^*}{1 - z_a^*} \right) \right]^{R_o} dz^* \right]^{-1}, \tag{8}$$

$$z^* = z/D, \tag{9}$$

$$z_a^* = a/D, \tag{10}$$

where R_o is the Rouse number.

These equations were computed by leap-frog difference scheme and upwind scheme for the advection term. The grid interval was 5 m and the time step was set as 0.5 s from Courant–Friedrichs–Lewy stability condition. Linearly interpolated values of the observation results were used as inflow discharge and sediment concentration from the upper boundary. The lower boundary condition is the critical depth to consider effect of the weir.

3.2.2. *Shields number*

For fine sediment such as silt, it is important to consider the cohesion. So we adopted θ_c derived logically by Righetti and Lucarelli.[7]

$$\theta_c = \theta_{c0} + \frac{\theta_{c0}}{\alpha_3} \frac{c}{(\rho_s - \rho) g d^2} + \frac{\theta_{c0}}{\alpha_3} \frac{A}{(\rho_s - \rho) g d}, \tag{11}$$

where θ_{c0} is the classical Shields curve; $\alpha_3 = \pi/6$; c, the cohesion coefficient that arises from van der Waals forces, and A is the adhesion coefficient that arises from biological activity.

4. Results and Discussion

4.1. *Observed results*

4.1.1. *Time series behavior of discharge and sediment concentration*

Figure 4 shows time series of observed discharge and sediment concentration at the upper and lower bridges. It is found that the sediment concentration peaks at both bridges lag 2–3 h behind the discharge peak. Sediment concentration at the lower bridge shows remarkable decrease at the water rising stage.

Figure 5 shows the relationship between the discharge and sediment concentration. The arrows in this figure indicate the way of time elapse. Both curves at the two bridges show anticlockwise hysteresis curves. The curve at the lower bridge shows remarkable decrease of sediment

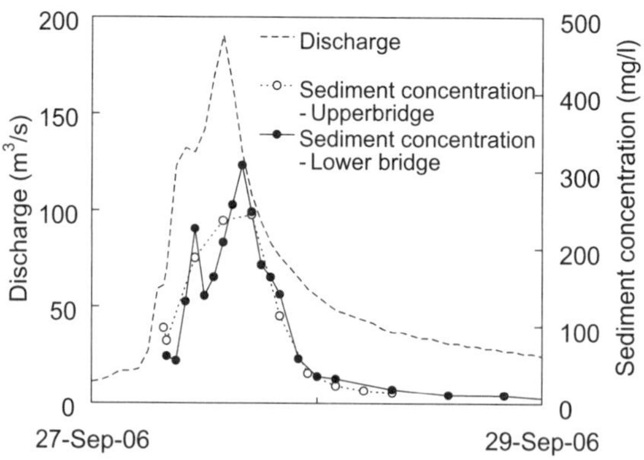

Fig. 4. Time series of observed discharge and sediment concentration.

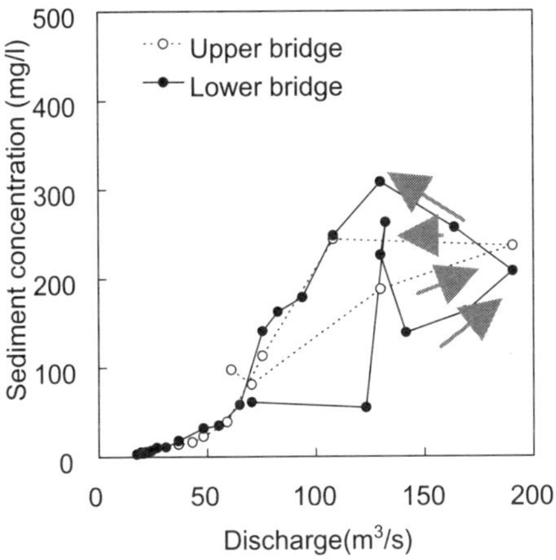

Fig. 5. Hysteresis curves in discharge and sediment concentration.

concentration at the increasing stage of the discharge and represents significant sedimentation at the study area. It could be the reason that the sedimentation proceeded by the relatively calm water body due to the backwater to the weir. On the hysteresis, both clockwise and anticlockwise curves are reported at the other places.[8,9] Several mechanisms have been proposed so far. As one of the reasons, the delay of the sediment concentration peak could be attributed to the slower transport characteristics of the sediment than the propagation of floods. The propagation speed is theoretically 5/3 times faster than the flood flow velocity when we consider the kinematic wave approximation and Manning's roughness formula.

4.1.2. *Grain size distribution of suspended sediment and deposits*

Grain size distributions of the suspended sediment are shown in Fig. 6. The grain distributions did not have a notable temporal variety. The portion of sediment smaller than 0.062 mm occupies approximately 60%–80% at the upper bridge, and 50%–70% at the lower bridge.

The riverbed deposits show the portions of sediment which is smaller than 0.062 mm are approximately 55% at the points A, B and C, 35% at points C, E, and 15% at point F, respectively, see Fig. 7. These distributions indicate that the deposition of finer sediment is dominant

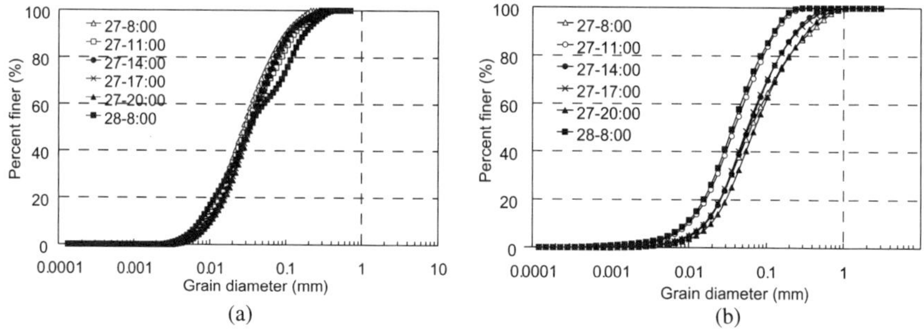

Fig. 6. Grain distribution of suspended sediment at (a) the upper bridge and (b) the lower bridge.

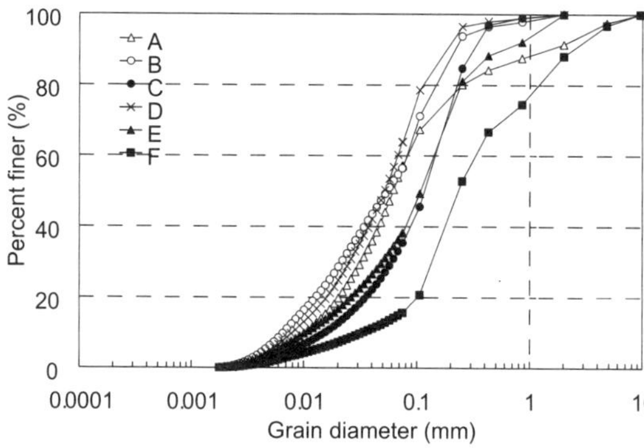

Fig. 7. Grain distribution of riverbed deposits in December 2006.

near the inner bank. From this fact, it is considered that the secondary flow which is the flow from the inner bank from the outer bank near the bed affected the deposition process, and resulted in the sediment sorting in cross direction. In comparison with the suspended sediment, the soil particles of approximately 001–0.1 mm are dominated for the deposits as well as the suspended sediment. This makes sure that the suspended sediment at the flood has a major effect on the deposition phenomenon of the study area.

4.2. *Computational result*

4.2.1. *Computation neglecting cohesion and adhesion*

Focusing our attention to silt (0.004–0.062 mm), the computation is conducted about the soil particles of 0.016 mm which is the representative diameter of silt. The portion of silt p_d was uniformly set at 15%, from the observed result at point F. The computational period is a period from 8:00 27th to 8:00 28th September 2006, corresponding to the observed period.

In this case, we used the classical Shields curve without considering the cohesion of fine sediment (Fig. 8). Figure 9 shows the comparison of

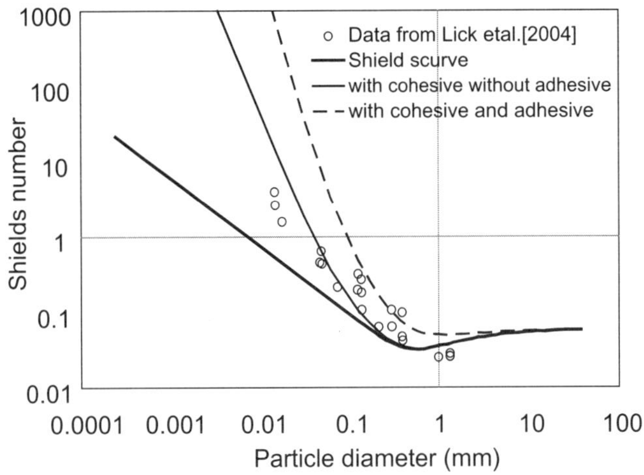

Fig. 8. Comparison of the critical Shields number for a single particle.[7]

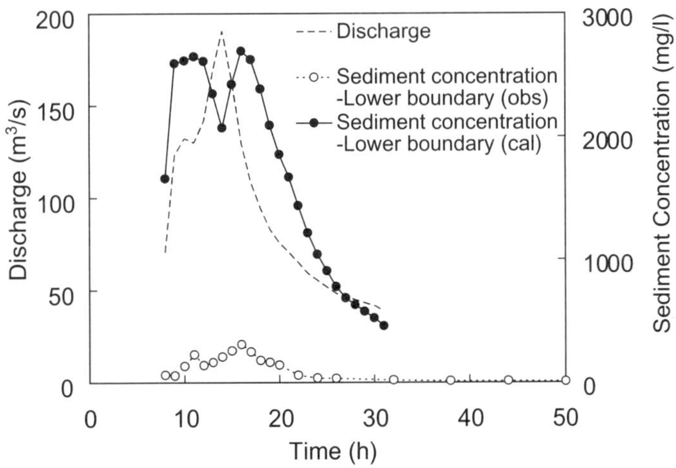

Fig. 9. Time series of observed concentration and calculated concentration at the lower boundary (neglecting cohesion and adhesion).

the concentration at the lower bridge. The computed concentration was about 10 times higher than the observed one. This would be mainly due to the overestimation of the resuspension. The initiation motion of particles is subject to the critical Shields number. So we validated the importance of cohesion by calculating with the Shields curve considering the cohesion.

4.2.2. *Computation considering cohesion*

Since there was a discrepancy in the computation neglecting the cohesion and adhesion, we considered the cohesion of fine sediment ($c = 5 \times 10^{-5}$ N/m). We did not consider the adhesion, which is caused by biological activity. It is because the adhesion varies by the organic matter condition and so on, as Righetti and Lucarelli[7] reports. Therefore, we regarded the adhesive coefficient as zero. Figure 10 shows the comparison of the concentration at the lower bridge. The computation was drastically improved by adding a tinge of the cohesion, and reproduced the characteristic that the concentration series was bimodal. This objective indicates that the cohesion is an important factor for the deposition of fine sediment.

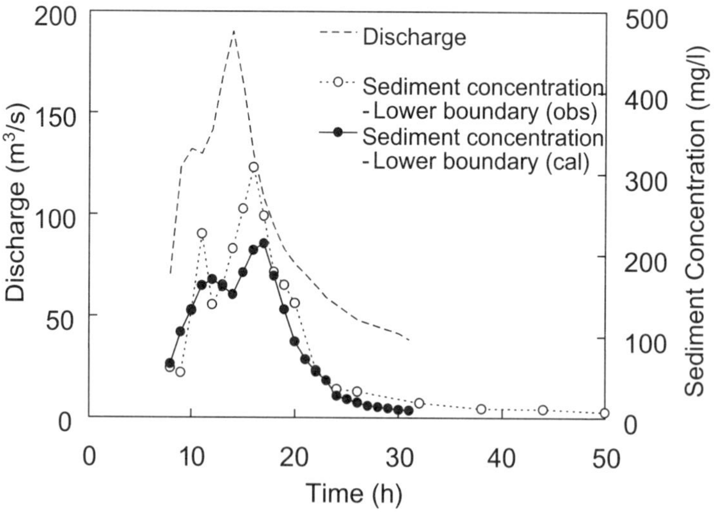

Fig. 10. Time series of observed concentration and calculated concentration at the lower boundary (considering cohesion).

5. Conclusion

From the observed results, we found that silt is the major component in the flood water and accumulates significantly in the discharge increasing stage by the backwater of the weir. We have shown long-term deposition characteristics near the inner bank of the bend, from the aerial photographs. Decrease of the large-scale flood passage by the dam construction on the upstream could explain the bed surface change from gravels to silt deposition. Once the silt accumulates on the river bed, it is difficult to be resuspended due to the cohesion. As a result, net deposition of the fine sediment is considered to proceed. Also by the comparison of the observation result and the computation result, we clarified effect of the cohesion could not be neglected to evaluate the resuspension of fine sediment. As mentioned above, it has been clarified that the cohesion is an important factor of the deposition of fine sediment. Through the one-dimensional model, we have shown the sedimentation mechanism upstream of the weir. However, the model cannot explain the lateral sedimentation distribution at the river bend. Therefore, 2D or quasi-3D model should be developed to get complete understanding in the future.

References

1. S. Lee, K. Fujita, T. Tsukahara, S. Watanabe, K. Yamamoto and T. Mochizuki, *Annu. J. Hydraulic Eng. JSCE* **42** (1998) 433 (in Japanese).
2. H. Miwa and E. Kurashima, *Paddy and Water Environment* **1** (2003) 207.
3. J. Zhou and B. Lin, *J. Hydraulic Eng.* **124** (7) (1998) 712.
4. N. Togashi and A. Mano, *Annu. J. Hydraulic Eng., JSCE* **48** (2004) 949 (in Japanese).
5. A. Mano, M. Sugiki and Y. Sato, *J. Global Environm. Eng.* **3** (1997) 53.
6. L. C. van Rijn, *Principles of Sediment Transport in Rivers, Estuaries, and Coastal Seas* (Aqua Publications, 1993).
7. M. Righetti and C. Lucarelli, *J. Geophysical Res.* **112** (2007) C05039.
8. K. Sakai, A. Yoshinaga and K. Ohsawa, *J. Japan Soc. Hydrol. & Water Resour.* **16**(6) (2003) 595 (in Japanese).
9. G. P. Williams, *J. Hydrol.* **111** (1989) 89.

Advances in Geosciences
Vol. 11: Hydrological Science (2007)
Eds. Namsik Park *et al.*

VARIATION OF WATER DEPTH AND VELOCITY BY ACCUMULATED DEBRIS AT PIER

GYE-WOON CHOI*, SANG-WOOK CHO and MYUNG-SOO HAM
Civil and Environment Engineering, University of Incheon
177 Dohwa-Dong Nam-Gu, Incheon 402-748, Korea
**gyewoon@incheon.ac.kr*

YOUNG-KYU KIM
Technological Development Department, Dohwa Consulting Eng.
736-6 Yeoksam-Dong Gangnam-Gu, Seoul 135-080, Korea

Accumulated debris is influenced by increasing water depth and velocity, so this phenomenon could cause a major flood disaster. This study shows how debris accumulation is influenced by water depth and velocity. From this experiment, water depth and velocity were a little influenced at upstream, but near the pier they were abruptly influenced by accumulated debris. If the accumulated debris area was the same, velocity was more influenced that debris was accumulated rather than spread at the pier. But water depth was more influenced by discharge than accumulated size. The accumulated debris was submerged, the water depth was more increased and velocity was faster than the accumulated water surface one.

1. Introduction

Floating debris can accumulate at various locations and at obstructions within the river such as bridge piers and abutments, mid-channel bars, point bars, island heads, the streambed, or in pools along the base on the outside bank of bends.

Debris accumulations initially form at the water surface, grow toward and eventually become part of the streambed. As the water surface increases during the flood, floating debris already existing on the bridge usually remains in place as additional floating debris accumulates at the water surface. When the flood subsides, the new accumulated debris usually slides downward until it rests on the bed or on the previous debris accumulation to form a solid mass with irregular protrusions around the base of the pier.

(a) (b)

Fig. 1. Span blockage accumulation bridge failure. (a) Louisiana, USA, (b) Gangwon province, Korea.

The blockage of flow area from debris accumulations can cause a significant increase in the flow velocities through the bridge structure. This increase in flow velocities and boundary shear stresses may cause an increase in the contraction scour through the bridge if the entire bridge opening is affected.

Debris accumulations can cause an increase in forces due to the increase in upstream water surface elevations, increase in the flow velocities through the bridge, and increase in projected area of these forces on the structure. Increases of these forces may cause the bridge structure to collapse either by the buckling of the bridge substructures, shearing of roadway deck supports, or overturning of the structures as shown in Fig. 1.

The subject of debris accumulation at bridge pier has been investigated by Melville and Sutherland[4] and Melville and Dongol.[3]

Melville and Sutherland[4] presented a design method for estimating local scour depth at bridge pier by laboratory experiment. After a few years, Melville and Dongol[3] have modified the design equation which conservatively predicts by using effective pier diameter. But these studies only focus on local scour, not flow characteristics.

The point that we have to think about flow characteristics is that the accumulated debris is caused by raising water levels and then the water will overflow the bank. And also the higher velocity by narrowed conveyance will attack the bank. Those kinds of situations would frequently occur in some countries which are developed or surrounded by many trees or mountains during the flood. Predicting the rising water level with accumulated debris is important in preventing the flood.

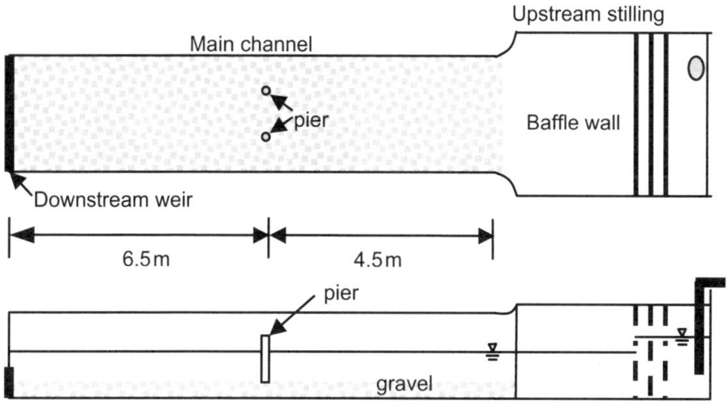

Fig. 2. General view of experiment.

2. Experimental Investigation

2.1. *Experimental setup*

The arrangement of the experiment is shown in Fig. 2. The experiment was conducted in 1.0 m wide and 0.8 m deep horizontal flume of approximately 11 m length with glass sides. Inflow water was changed to a steady condition through three baffle walls.

Gravel was spread in 50 mm thickness with a single size (20 mm) on the channel bed and the gravel's specific gravity is 2.69.

Two cylindrical piers which were made from clear PVC pipes of 50 mm diameter were used. Piers were installed at 4.5 m downstream and each pier was installed 30 cm from the side as shown in Fig. 3.

Debris rafts were idealized as smooth, impermeable, and varied size. The upper surface of debris rafts were placed at the water surface. The used debris raft size was between 200 mm and 500 mm of width and 50 mm to 125 mm of thickness.

2.2. *Experiment procedure*

A constant discharge was supplied from an overhead constant head tank, ranging from 0.073 CMS to 0.116 CMS. The upstream end of the channel has a stilling tank provided with a triangle sharp-crested weir.

At the downstream end of the channel, a weir has been set up to adjust the water depth which was 162 mm at every experimental case.

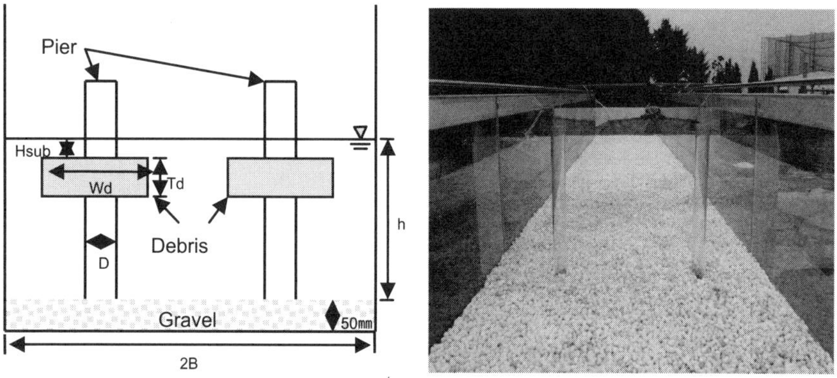

Fig. 3. Cross section of the flume.

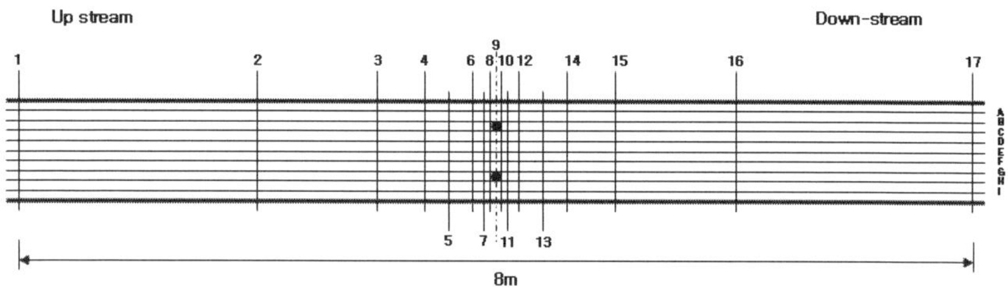

Fig. 4. Water depth and velocity measure point.

The measured point was shown 2 m up and downstream from piers installed position as in Fig. 4. The water depth measured point was 144 which were 16 sections and the sections were measured 9 points every 10 cm to across direction, and the number of the velocity measured points were measured 45 which was nine sections and every section was measured every 5 cm to across direction. The measured sections are represented in Table 1.

The water depth of flow was measured using a point gauge with a reading accuracy of ± 0.1 mm. The velocity was measured using Acoustic Doppler Velocimeter (ADV-11000) with 1 Hz.

In this paper, $V_{\max}$ was the maximum velocity which was measured for 30 s after installed debris raft. V_{mean} was the average velocity with measured values before installed debris raft. Velocity fluctuate coefficient (V_{fc}) was defined such that $V_{\max}$ was divided by V_{mean} like Eq. (1). V_{fc} is represented

Table 1. Water depth and velocity measured sections.

Measuring section	Upstream			Measuring section	Downstream		
	L/B	Water depth	Velocity		L/B	Water depth	Velocity
No.1	4.0	O	O	No. 9	0	O	O
No.2	2.0	O		No. 10	0.05	O	
No.3	1.0	O		No. 11	0.1	O	O
No.4	0.6	O		No. 12	0.2	O	O
No.5	0.4	O	O	No. 13	0.4	O	O
No.6	0.2	O	O	No. 14	0.5	O	
No.7	0.1	O	O	No. 15	2.0	O	
No.8	0.05	O		No. 16	4.0	O	O

B: Half of channel width.
L: The distance from the piers.

as the degree of variation of flow as

$$V_{\mathrm{fc}} = \frac{V_{\mathrm{max}}}{V_{\mathrm{mean}}}. \tag{1}$$

The experiments were processed by changing discharge, debris size, and submerged depth, so that it was performed in 14 cases, and the list of experiment cases is shown in Table 2.

3. Experiment Results

3.1. *The trend of flow condition through accumulated debris at pier*

As debris accumulates at a pier, the variation of flow is very complex. In this paper, a trend of water depth and velocity between debris and without debris at pier was analyzed.

Figure 5 shows the figuration of water depth and velocity at 0.096CMS.

When it was only installed piers, velocity did not change between upstream and downstream, and water depth is represented as increasing and decreasing near the piers.

If the debris accumulated at piers, velocity decreased slowly, and water depth slightly increased at upstream. Also the velocity was faster between the piers than other places. But velocity rapidly increased near the piers

Table 2. Experiment cases.

Type	Debris raft size ($W_d \times T_d$)			Discharge (CMS)	Submerged depth
	Left	Right	Area		
Experiment 1	500 × 50	500 × 50	25,000	0.073	
• Effect by increment of debris width				0.091	
				0.116	
• Effect by increment of discharge	400 × 62.5	400 × 62.5	25,000	0.073	0 cm
				0.091	
				0.116	
	300 × 83.3	300 × 83.3	25,000	0.073	
				0.091	
				0.116	
	200×125	200×125	25,000	0.073	
				0.091	
				0.116	
Experiment 2	500 × 50	500 × 50	25,000	0.073	0 cm
• Effect by increment of submerged depth					2 cm
					4 cm
				0.136	0 cm
					2 cm
					4 cm

and the fastest velocity area was represented at 0–40 cm downstream from the pier.

Water depth was slightly increased at upstream, but just after the pier, water depth rapidly decreased. It recovered to original water depth but recovered length increased by discharge increment.

3.2. *The variation of velocity fluctuation coefficient by accumulated debris at pier*

The variation of velocity fluctuation coefficient (V_{fc}) is shown in Fig. 6. and through this experiment, the area of used debris was the same as 25,000 and it changed the width and thickness.

V_{fc} is gradually decreasing as it gets closer to the pier in all cases of discharge, and the minimum value occurs just in front of the piers

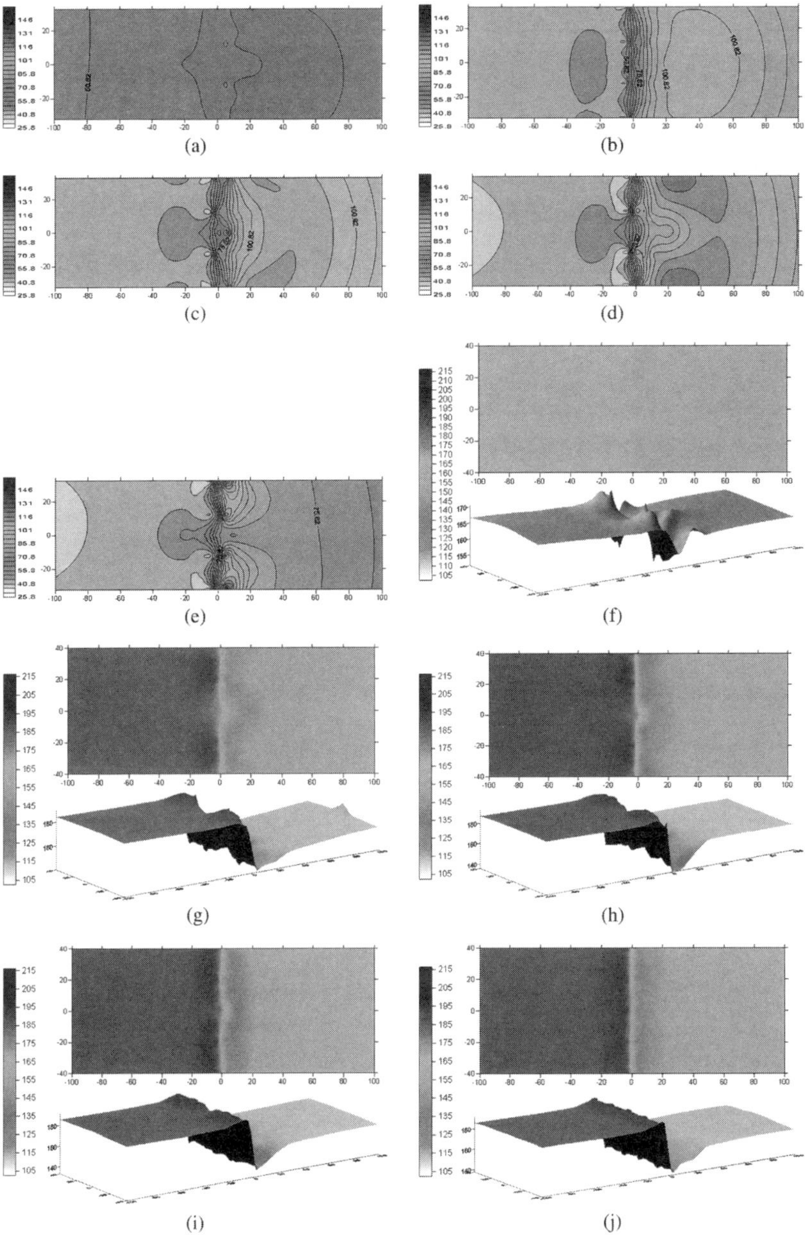

Fig. 5. The variation of flow condition by debris accumulation at 0.097 CMS. (a) velocity contour without debris; (b) velocity contour with 200*125 debris; (c) velocity contour with 300*83.3 debris; (d) velocity contour with 400*63.3 debris; (e) velocity contour with 500*50 debris; (f) water depth contour without debris; (g) water depth contour with 200*125 debris; (h) water depth contour with 300*83.3 debris; (i) water depth contour with 400*63.3 debris and (j) water depth contour with 500*50 debris.

Table 3. Maximum and minimum of V_{fc}.

Debris size	Discharge (CMS)	Maximum V_{fc}	Minimum V_{fc}	Max. − Min.
200*125	0.073	1.79	0.64	1.15
	0.091	1.66	0.72	0.95
	0.116	2.05	0.69	1.36
300*83.3	0.073	2.31	0.71	1.60
	0.091	2.01	0.62	1.36
	0.116	2.55	0.61	1.94
400*63.3	0.073	2.41	0.76	1.64
	0.091	1.82	0.69	1.13
	0.116	2.71	0.60	2.10
500*50	0.073	1.71	0.78	0.93
	0.091	1.79	0.75	1.05
	0.116	2.44	0.70	1.75

(see Figs. 5a–5e). The values are represented as 0.64–0.85 at 0.073CMS, 0.62–0.75 at 0.091CMS, and 0.60–0.72CMS at 0.116CMS. Therefore, as the discharge increases, the V_{fc} decreases until 40%.

As it passes the piers, V_{fc} increases sharply and the maximum V_{fc} is represented as 2.41, 2.01, 2.71 at 0.073 CMS, 0.091 CMS and 0.116 CMS, respectively (see Figs. 5b–5f).

Finally, this various flow is represented similarly to the flow which had no accumulated debris case within the length of 4B far from the piers.

Maximum V_{fc}, minimum V_{fc} and the difference size of accumulated debris, are presented in Table 3.

As difference accumulated debris, maximum V_{fc} is represented as 2.05, 2.55, 2.71, 2.44 at 200 * 125, 300 * 83.3, 400 * 63.3, and 500 * 50, respectively. Maximum V_{fc} is increasing and decreasing as increased width at different size of accumulated debris.

Also the minimum V_{fc} is represented as 0.64, 0.61, 0.60, 0.78 at 200 * 125, 300 * 83.3, 400 * 63.3, and 500 * 50, respectively. Minimum V_{fc} is decreasing and increasing as increased width at different size of accumulated debris.

The gap of maximum and minimum V_{fc} is represented between 0.93 and 2.10, and the maximum gap is represented by 400 * 63.3, 300 * 83.3, and 400 * 63.3 at discharge of 0.073 CMS, 0.091 CMS, and 0.116 CMS, respectively. Therefore, if the debris accumulated on the central part at pier, it represents bigger velocity variation than debris spread toward across or vertical.

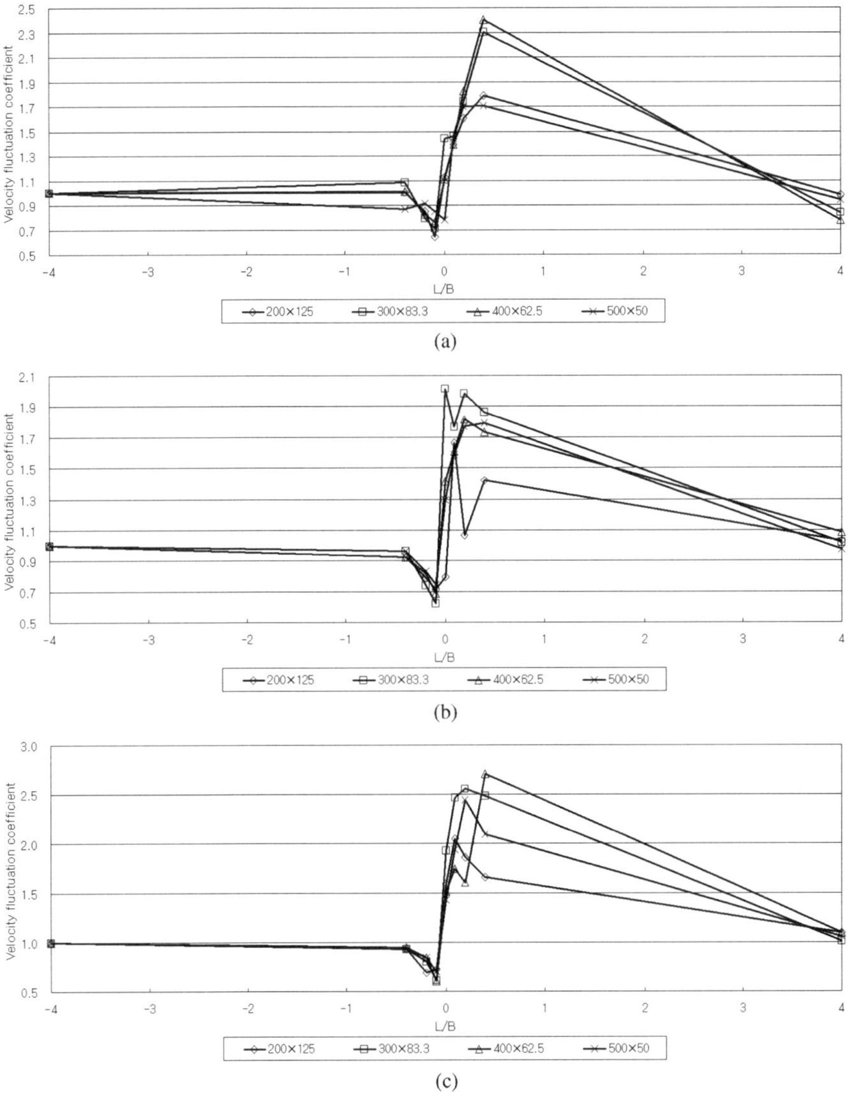

Fig. 6. The variation of velocity fluctuation coefficient against L/B (a) 0.073 CMS; (b) 0.073 CMS; and (c) 0.116 CMS.

3.3. *Variation of water depth by accumulated debris at pier*

The variation of water depth is represented by the ratio of water depth with pier and accumulated debris against without water depth of any debris accumulation and pier, and Fig. 7 shows the variation at each discharge.

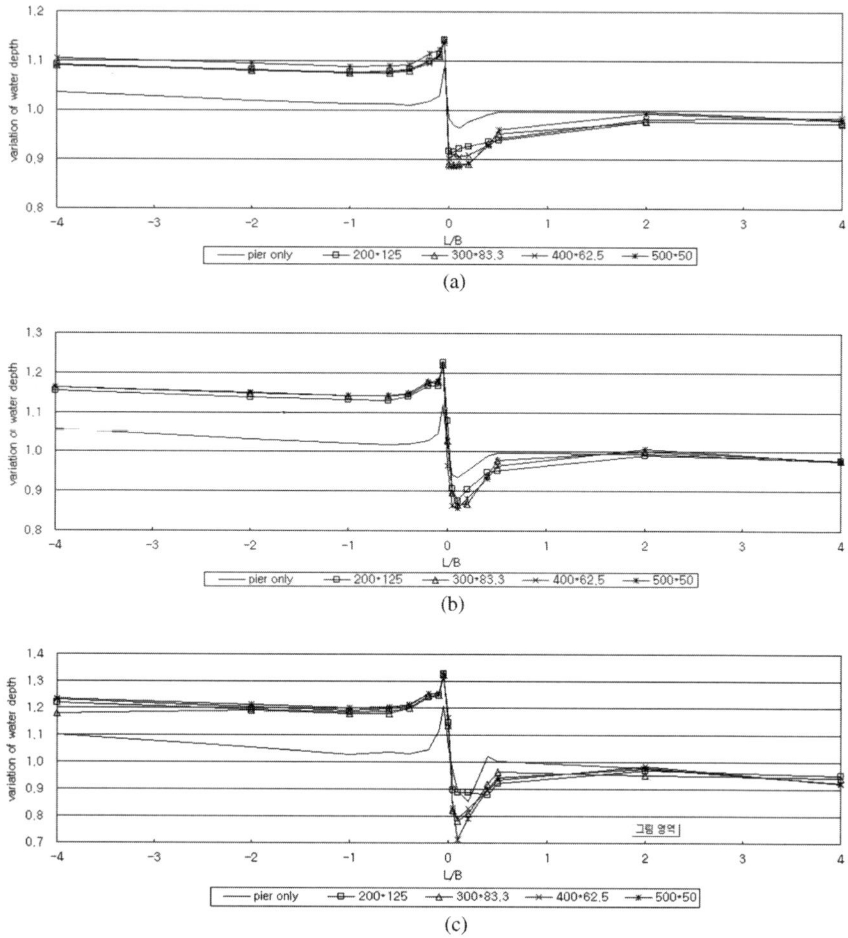

Fig. 7. The variation of water depth against the distance from piers (a) 0.073 CMS; (b) 0.073 CMS; and (c) 0.116 CMS.

When Piers or piers with accumulated debris were installed, the depth decreased as the closer to pier, but the water rapidly increased between 1.12 and 1.26 near piers. The length depends on the discharge and the length extent increases as discharge increases.

Increased water depth dropped until 0.71 where the flow just passed piers. The water depth which raises and falls is stabilized from 2B far from the pier and accumulated debris.

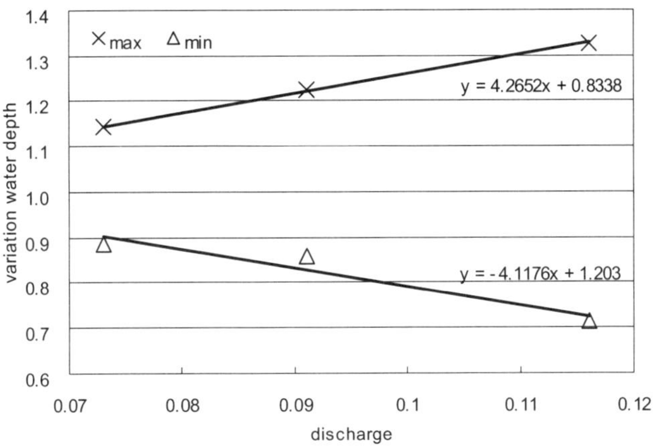

Fig. 8. The variation of water depth vs discharge.

The variation of water depth is represented as little as debris size, but a big variation of discharge and the variation is shown in Fig. 8. As discharge increases, the maximum variation of water depth increases to 1.33 and the minimum variation of water depth decreases to 0.71, so that the gap of maximum variation and minimum variation increases or starts to increase.

3.4. *Variation of flow condition through debris submerged*

For the analysis, the variation of flow characteristics as debris was submerged, it was carried out experiencing all across section blocked case $(500 * 50)$.

Figure 9 shows the variation of velocity and water depth as submerged depth.

As submerged depth of accumulated debris was increasing, shown in Fig. 9 (a), V_{fc} was also increasing at discharge of 0.073 CMS or 0.136 CMS. This kind of present state is that flow discharge is influenced by accumulated debris raft and the discharge is apt to flow under the accumulated debris raft than flow over it. Because the flow is under the gravity and as submerged the raft, the space to flow decreases then V_{fc} was represented as bigger. But as increased discharge, V_{fc} slope decreases from 5.35 to 1.79.

In case of water depth, as accumulated debris was submerged every 2 cm, water depth increased 36% and 47% than 0 cm submerged at discharge was 0.073 CMS, and it increased 19% and 81% than 0 cm submerged

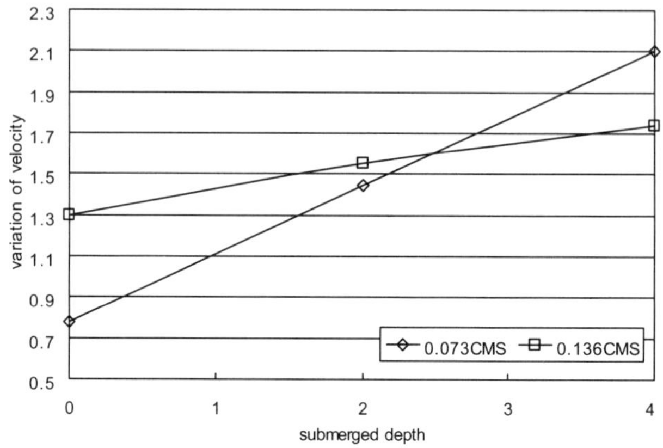

(a)

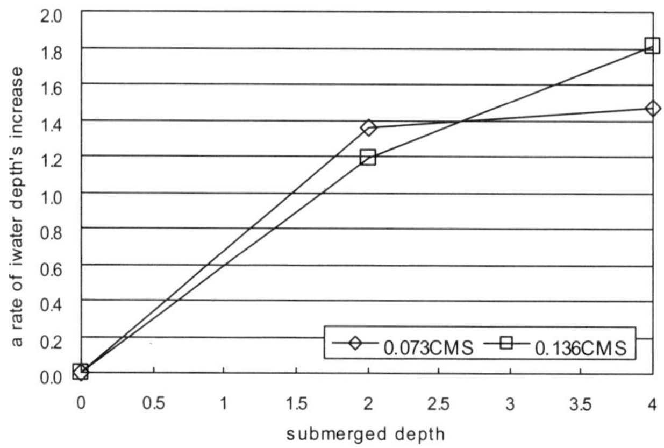

(b)

Fig. 9. The variation of flow through debris submerged. (a) Variation of velocity; (b) variation of water depth.

at discharge was 0.136 CMS, as shown in Fig. 9(b). Lower discharge, 0.073 CMS, the rate of increase rapidly decreased at 2 cm submerged, but at higher discharge, 0.136 CMS, it slightly decreased.

4. Conclusion

In this study, for the sake of analysis of flow through changing debris accumulation, the experiments were processed with various debris size and discharge. So that it has drawn the followed results.

1. Water depth and velocity were not a big variation at upstream by accumulated debris at piers. But near the piers, the velocity rapidly increased and water depth increased and suddenly dropped at just past the pier.
2. Varied water depth returned to original water depth at 2B and velocity returned to original velocity at 4B from the piers.
3. Velocity fluctuation coefficient (V_{fc}) represented the maximum value of 2.71 and the minimum value of 0.60 at 400*63.3 debris accumulated size.
4. In additional, V_{fc} was more influenced when debris was accumulated between vertical or horizontal direction than vertical or horizontal direction at pier if accumulated debris area was same.
5. The water depth was more influenced by discharge than accumulated debris size. As discharge increased, the rate of maximum water depth increment increased and the rate of minimum water depth increment decreased.
6. As submerged accumulated debris, water depth and velocity increased. Water depth represented a big range of variation at big discharge, but velocity represented a big range of variation at small discharge.

Acknowledgments

This study was supported by the 2003 Innovation Project for Construction Technology (03-SANHAKYOUN-C01-01) through the Urban Flood Disaster Management Research Center in KICTTEP of MOCT, Korea.

References

1. G. W. Choi, G. H. Kim and Y. S. Par, Changes in water depth and velocity by debris around piers, *Journal of Korea Water Resources Association* **36**(2) (2003) 273–284 (in Korean).
2. G. W. Choi, Y. K. Kim and S. W. Cho, The analysis of water depth through accumulated debris, *Journal of University of Incheon* **22**(2) (2006) 129–135 (in Korean).
3. B. W. Melville and D. M. Dongol, Bridge scour with debris accumulation, *Journal of Hydraulic Engineering* **118**(9) (1992) 1306–1310.
4. B. W. Melville and A. J. Sutherland, Design method for local scour at bridge piers, *Journal of Hydraulic Engineering* **114**(10) (1988) 1210–1226.

Advances in Geosciences
Vol. 11: Hydrological Science (2007)
Eds. Namsik Park *et al.*

ASSESSMENT OF INSTITUTIONAL CAPACITIES OF FLOOD MANAGEMENT INSTITUTION IN PAKISTAN

NOOR M. KHAN
Civil Engineering Department
University of Engineering and Technology
Lahore, 54890, Punjab, Pakistan
muhammad@gmail.com

Pakistan is frequently devastated by floods. The flood impacts can be reduced if the flood management institutional capacities are improved. This paper reviews and assesses the capacities of flood management institution[a] in Pakistan. Citing a number of case studies about the flood management practices in Pakistan, the study estimates the weaknesses and strengths of the institution with respect to various phases of flood management, namely, mitigation, preparedness, response, and rehabilitation and also with respect to various characteristics of institutions, namely, deliberation, coordination, implementation, and evaluation, using an improved capacity assessment framework. It has been found that the performance of the mitigation and rehabilitation phases is not satisfactory and that of preparedness and response is satisfactory. It is concluded that the functions of deliberation need to be improved while the other three characteristics of institution namely, coordination, implementation, and evaluation are performing well. The study will help the policy makers to concentrate on the identified weak capacities.

1. Introduction

After droughts, floods are the largest culprits in terms of the number of people affected in Asia,[1] and Pakistan is no exception in this regard. Pakistan is one of the top ten worst disaster-hit countries of the world in 2006. Flood disasters have persistently played havoc in Pakistan and affected an estimated 47 million people.[2] International disaster databases[1,2] show that 8 out of top 10 disasters (in terms of number of people affected)

[a]Pattern of human relationships and/or behavior which has continued to exist for a significant period of time and has thus demonstrated its usefulness in satisfying some human needs or desires.

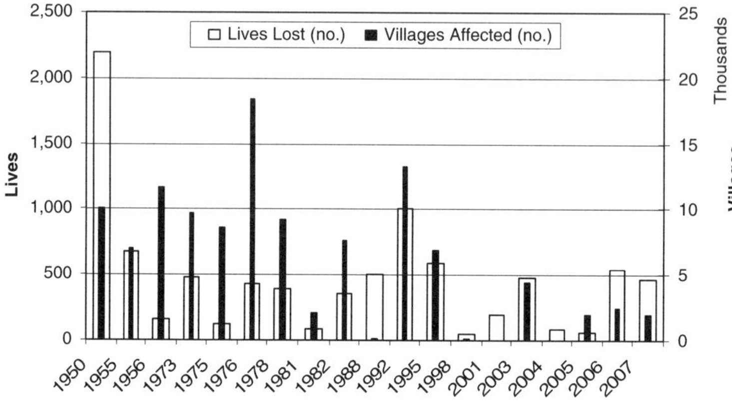

Fig. 1. Major floods and the resulting damages in Pakistan.[3,4]

in Pakistan are the flood disasters. Out of these top ten disasters, only the event of 1992 affected well over 12 million people. Human errors have played its role in adding to the miseries of the people. The data of the last 60 years shows that Pakistan have lost more than 8000 lives due to flooding, which affected 115 thousand villages across the country (Fig. 1) and flooded more than half a million square kilometers.

Since the unprecedented flood of 1992, billions of dollars have been spent on enhancing the protection against the flooding in the country through structural interventions and improvements of flood forecasting and warning capacities. More recently (late 2006), institutional aspect is being invigorated under the umbrella of newly created National Disaster Management Authority (NDMA).

Due to the persistent nature of flooding and the resulting socioeconomic losses (Fig. 1), the assessment and then improvement of individual flood management organizations (governmental and non-governmental, formal and informal) are not bearing fruits as were expected. The need is felt to evaluate and assess the capacities of flood management as an institution in Pakistan, using a scientific framework of capacities assessment. The current study will fill this gap of assessment of Pakistan's flood management institution for figuring the weak and strong links in the institutional setup.

2. Literature Review

Integrated flood risk management is a comprehensive approach where equal emphasis is placed on mitigation, preparedness, relief, and recovery through the involvement of all relevant sectors and stakeholders with the overall goal to reduce flood risks.[5] Not many studies are found assessing the flood management as an institution or providing a framework to assess it. A few studies have assessed the institutional capacities of integrated water resources management[6] and integrated flood management[7] in country perspectives. Lebel *et al.*[8] have devised a framework for assessing institutional capacities for flood disaster risk reduction. They propose that institutional capacity to flood management should be checked against four functions, namely, deliberation, coordination, implementation, and evaluation during the four phases of integrated flood disaster management cycle (mitigation, preparedness, emergency response/relief, and rehabilitation/recovery). This framework, with certain improvements will be employed in this study.

Review of flood management organizations in Pakistan is done more frequently by a number of national and international organizations. The Ministry of Water and Power[9] has reviewed Pakistan's overall water-related performance and proposed a 10-year medium-term investment plan (MTIP), including the flood protection sector. The plan outlines the structural and non-structural measures to reduce the flood damages, including comprehensive allied measures like the creation of Water Resources Council (headed by the Prime Minister), institutional streamlining, and application of IWRM concepts for water management.

WCDR[10] has pointed out that flood management in Pakistan is mainly restricted to rescue and relief efforts. After each disaster episode, the government incurs considerable expenditure directed at rescue and relief. Also, it is pointed out that the response and relief efforts carry strategic biases, favoring those who are economically, politically, and strategically significant at the cost of people and areas with less importance and significance. After the earthquake of October 2005, OCHA[11] reviewed the overall disaster response agencies of Pakistan. They have pointed out that "*it is only for the flood hazards, that frequent central and southern plains, that the country has a comprehensive disaster management system in place.*" This comment of OCHA is true if considered in comparison to the management of other disasters (earthquake, oil-spillage, epidemics etc.) in Pakistan. While describing the achievements of flood year-2006, FFC[3] has outlined the current flood management infrastructure available in the country and also those which are planned to be executed.

Apart from the review of organization, a few authors have pointed out the problems and gaps in the flood management institutions. Mustafa[12,13] has identified the structural causes of the vulnerability in Pakistan and addressed the problems of those who are already suffering due to poverty in addition to the flooding.

3. Organizational Setup for Flood Management in Pakistan

Federal Flood Commission (FFC) created in 1977 is the principal institution for flood protection planning and control in Pakistan. Its mandate includes the preparation of the national flood protection plan, approval of flood control schemes, review of flood damages, plans for reconstruction works, improvements in flood forecasting and warning system, monitoring, and evaluation, etc.

Other organizations that play a major role in flood management are Provincial Irrigation and Drainage Authorities (PIDAs) (formally known as Provincial Irrigation Departments, PIDs), Pakistan Water and Power Development Authority (WAPDA), Emergency Relief Cell, Provincial Relief Commissioners, Pakistan Army, Pakistan Navy, Police, Rangers, and Pakistan Commissioner for Indus Waters, in addition to a number of local and international NGOs (Fig. 2).

The detailed objectives and performance of each of these organizations are covered in many review reports,[3,7,10,11] and are being skipped to concentrate more on capacity assessment of the institution as a whole. A recent change in the institutional setup is the creation of NDMA in 2006, which will bring all the disasters-related agencies in the country under one authority. As such, FFC will now effectively follow the guidance of NDMA which seems to be concentrating more on vulnerability reduction, but the impact and improvement due to this change is still needed to be seen after a few disaster cycles.

4. Institutionalized Capacities of Flood Management in Pakistan

Following the concept of strength, weakness, advantages/opportunities, and threats analysis (SWAT or SWOT), Lebel *et al.*[8] proposed an initial framework to assess the institutional capacity of flood management institution. They suggested that the capacity should be checked against four functions of the institute, namely, deliberation, coordination, implementation and evaluation during the four phases of integrated flood disaster management cycle (mitigation, preparedness, response,

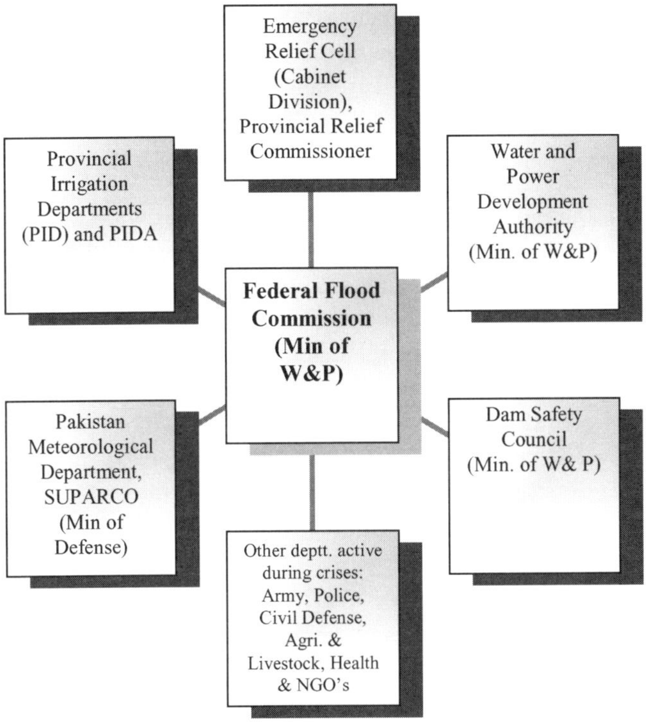

Fig. 2. Organizations responsible for flood management in Pakistan.

and rehabilitation). The framework poses various questions, which when answered help in assessing the capacities of the flood management institution. Some shortcomings found in the framework confuse various functions together. These shortcomings along with the revised questions in the framework are listed as under:

- Under the function of deliberation in the Lebel's framework, questions are raised about the consultation, discussion, and involvement of stakeholders in the decision-making/planning process of each phase. But in the response phase, Lebel's framework raises the question "what special directives or resolutions were invoked?" It is changed to "what special directives or resolutions are/were discussed?," as it is the assessment of deliberation function rather than the assessment of implementation function of the institutional capacity.
- Under the function of coordination in the Lebel's framework, questions are asked about the plans, policies, coordination setup,

and responsibilities assignment. But, Lebel's framework wrongly asks questions regarding implementation of these plans and policies. This shortcoming is removed, by changing the words "implemented," "undertaken," and "mobilized" with suitable words like "planned" and "allocated."

Apart from these improvements, the whole framework of Lebel *et al.*[8] is adapted in this study, with due acknowledgment. Pakistan's flood management institution is assessed against this modified framework (Table 1) using the case studies from the literature, and authors' personal knowledge of working in water sector (since 1993). After in-depth analysis of the current situation, each of the function for each of the phase is categorized into two levels, satisfactory (S) and not-satisfactory (NS), while the overall performance of each phase and each function and the whole institutional capacity is categorized based on three levels, namely satisfactory (S), average (A), and not-satisfactory (NS), as shown in Table 2.

4.1. *Deliberation*

Deliberation requires discussions and involvement of all the stakeholders, especially under-privileged and more vulnerable ones. Deliberation is required to do informed risks-taking and to decide that interests of which of the group should be preferred. In the case of Pakistan, where democratic institutions are not well developed, the process of planning and devising legal processes is much centralized and bureaucratic, and large-scale deliberations are not norms. After the earthquake of 2005, when the need is felt for a permanent disaster management umbrella agency, National Disaster Management *ordinance* is promulgated by the President of Pakistan,[14] although parliament should have taken the lead to discuss the need and present a bill in the national assembly and senate for consensus development and deliberation. Similarly, the decision about the location of breach section is said to be done based on engineering studies, but there are examples when the engineering designs are modified to accommodate the powerful groups, at the cost of already vulnerable groups.[12] Also, the benefits of this practice (breaching the embankment, strategically) are not well distributed, and usually the sufferings and benefits are segregated as per political/economic status and power of the groups. For example, in the case of extremely high floods in the Ravi river in 1988, right embankments were breached to save the provincial capital Lahore; similarly; right banks of

Table 1. Modified Lebel's (2006) framework for flood institution's assessment.

Institutional functions	Phases of disaster cycle			
	Mitigation (Well before)	Preparedness (Before)	Response (During)	Rehabilitation (After)
Deliberations: What should be done?	(a) How were decisions made about what and who should be at risk? (b) Whose knowledge was considered and whose interest were represented?	(a) Was the public consulted about the disaster preparation? (b) How were decisions to give special powers to particular authorities made?	(a) How were decisions made about what and who should be saved or protected first? (b) What special directives or resolutions were *implemented*?	(a) How were decisions made about what is to be on the rehabilitation agenda? (b) Whose knowledge was considered, whose interest were represented?
Coordination: Who is responsible?	(c) What national and basin level policies, strategies, or legislation were in place to reduce risks of disasters? (d) What structural measures were *planned* to reduce likelihood of severe flood events?	(c) How were responsibilities divided among authorities and public? (d) Was an appropriate early warning system *planned*? (e) Were public authorities well prepared?	(c) How were specific policies targeting emergency response operations planned? (d) Were there gaps between stated responsibilities and performance of key actors? (e) Who was in charge?	(c) Were the resource *allocated* for the recovery adequate? (d) Were they allocated effectively? (e) How was rehabilitation integrated into community, basin, or national development?

(*Continued*)

Table 1. (*Continued*)

Institutional functions	Phases of disaster cycle			
	Mitigation (Well before)	Preparedness (Before)	Response (During)	Rehabilitation (After)
Implementation: How was it done?	(e) To what extent were laws and regulations regarding land-use in flood-prone areas implemented? (g) What measures were taken to improve coping and adaptive capacities of vulnerable groups?	(h) Was the public well-informed? (g) How were specific national or basin-level policies targeting disaster preparedness implemented?	(f) How were emergency rescue and evacuation operations performed? (g) Were special efforts made to assist socially vulnerable groups? (h) Was there any measures taken to prevent looting?	(f) Did the groups who most need public assistance get it? (g) Who benefited from reconstruction projects? (h) Was insurance available and used, and if so how were claims processed? (i) Was the compensation process equitable and transparent?
Evaluation: Was it done well?	(h) How is the effectiveness of risk reduction measures assessed?	(h) How is the adequacy of preparedness monitored?	(i) How is the quality of emergency relief operations evaluated?	(j) How is the effectiveness of the rehabilitation program evaluated?

Table 2. Performance of Flood Management Institution in Pakistan.

Function/Phase	Mitigation	Prepared	Response	Rehabilitation	Overall
Deliberation	NS	NS	NS	NS	NS
Coordination	S	S	S	NS	S
Implementation	NS	S	S	NS	A
Evaluation	NS	S	S	NS	A
Overall (Phase)	NS	S	S	NS	A

embankment at Trimmu Barrage (the Chenab river) were breached in 1992, to save the Jhang city. But, no transfer of benefits (at least in economic terms) was seen in the rehabilitation stage from the beneficiary (urban dwellers) to the sufferer (rural dwellers). A similar message is conveyed by WCDR,[10] mentioning "who and what should be at risk" are decided based on economic, political, and strategic importance at the cost of less important and less significant people and areas. Also, the review of a district flood fighting plan[15] shows that the first objective is to reduce the economic damages of the governmental infrastructure, and the secondary objective is to preempt the danger to life and property of people. If deliberations were practiced, the objective would have been the other way round.

These examples show the prevailing situation of lack of deliberation process in the flood management institution. The consultation and deliberation during any stage of the flood management is not a norm in Pakistan, and as such the performance of the institution with respect to this function is categorized as not-satisfactory.

4.2. *Coordination*

Regarding coordination, Lebel *et al.* raised the question whether the national and basin level policies/strategies and legislature are formulated. Also, whether the structural measures are taken to avert the risk of severe flooding, and above all whether the role and responsibilities of each of the flood management entity is clear and who has to do what, at the time of crises, is clear? In this respect, the literature shows that although Pakistan does not have an agreed Water Policy (draft is ready since 2002) yet Pakistan has a strategy to manage floods through structural and nonstructural measures.

Since 1978, the improvement of flood management organizations and infrastructure in Pakistan is mainly carried out through two 10-year plans, National Flood Management Plan-I (NFMP-I) and National Flood Management Plan-II (NFMP-II). The second plan NFMP-II (with the original duration of 1989–1998) contained two major projects, called FPSP-I and FPSP-II (flood protection sector project). A recent report[16] shows that the Government is working on a third project under the extended second plan, with an outlay of Rs. 10 billion to be spent on the flood sector in the next few years. Coordination and planning for the phase of preparedness is done by Pakistan Meteorological department (plans to improve the flood forecasting and warning infrastructure and capacity), and that for the emergency response is done by district governments, through flood fighting plans under the supervision of relief commissioners.[15] Overlaps and gaps in the roles of the organizations are not prominent, at least in the phases of preparedness and response to the flood hazard. While the policy-making and coordination in the phase of rehabilitation are not yet brought into the spectrum of flood management institutes, the rehabilitation planning is done, after the crisis is over through routine development organizations such as WAPDA, PIDAs, and NHA. As such, we can say that satisfactory plans and coordination mechanisms are available for all the phases except the rehabilitation phase.

4.3. *Implementation*

Deliberations and coordination without implementation are useless. In this respect, the two flood management plans at national level (mainly for mitigation of flood risks), with their phased projects have resulted into successful implementation of many flood protection structures, although with reduced scope of work and considerable delays. Within the plans, the implementation of the nonstructural parts lags behind as compared to the structural parts. Such as when the plan's scope was reduced in case of NFMP-II, due to the financial crunch in late nineties, non-structural schemes like capacity building of the institutions were dropped at the first instance.[9] While on structural interventions, the implementation is relatively better, and till 2006 there are 6700 km length of protective embankments (for confinement of floods) in the country, supported by 1360 numbers of spurs[3] fortifying nearly whole lengths of the major rivers in the lowland areas. The usual delays also affect the rehabilitation projects such as the project to rehabilitate the damages of 1988 flooding was time overrun by 50%.[17] Implementation of certain bylaws such as "the Punjab land preservation act-1900" could not restrict the encroachment of flood plains by urban developers. Due to these facts, it is considered that the level of implementation of mitigation plans and rules is not-satisfactory.

Implementation of projects related to flood preparedness and response (such as modernization of flood forecasting and warning system) is completed by the Pakistan Meteorological department through acquisition and installation of seven weather radars at the strategic locations to improve the flood warnings in the country through real-time meteorological data acquisition across the international boundary. Recent pre-cyclone warnings (with the help of meteorological satellite data) issued with respect to cyclone Gonu and then Yemyin (both in June 2007) are well commended by the international organizations like WMO.[18] Gaps in the implementation phases are of sociopolitical nature, such as political pressure to modify the schemes in the benefits of powerful,[10,12] and will be rectified with further improvement of the internal evaluation process and external economic and social condition of the country by large. Considering these facts, the preparedness and response phases implementation is considered satisfactory and that of rehabilitation implementation is considered not-satisfactory.

4.4. *Evaluation*

In order to reduce the flood risk, the institution must have the capability of self-evaluation. A good practice in Pakistan, going on for the last few years is a pre- and then a post-flood meeting of all the stakeholders (especially Government agencies).[3] These meetings help in pointing out the gaps and weaknesses of the plans, their implementation, and the infrastructure. Other kinds of evaluation, available within the individual organizations (Army, WAPDA, Flood Forecasting Division, FFC, District administration, etc.) are as per overall culture of the governmental organizations. Recent development of the electronic media (private TV channels and Internet) in Pakistan and the shortcomings of any institutes are not likely to be covered up. The evaluation process after the 1992 flood event, has resulted in a vast-scale upgradation of flood warning system in the country and preparation of flood warning manual.[19] The manual has clarified the duties and responsibilities of each organization and the particular individual involved in the incident of flooding. Similarly, international organizations do keep a watch on the working of the institution. WMO's commendation letter[18] to Pakistan's Meteorological department about early warnings and reports[11] are just a few examples. As such, the evaluation culture is working satisfactorily.

5. Conclusions and Recommendations

This study, one of its kind, has evaluated flood management institutions of Pakistan by going through weaknesses and strengths of the flood management institution during each phase of the flood management. The results of the assessment are summarized in Table 2. It has been found that two of the phases (preparedness and response) performed satisfactory (S), while the phases of mitigation and rehabilitation are not-satisfactory (NS), if evaluated as per the improved assessment framework of Ref. 8. Considering the performance of the flood institution with respect to each function of the institution, the analysis shows that the functions of coordination is satisfactory, while the functions of implementation and evaluation performed average.

Overall performance of the flood management institution may be ranked as average (A). Although no other similar study could be found to compare the results, yet the study of APFM[7] regarding the level of implementation of integrated flood management (IFM) concepts in 17 participating countries shed some light on the flood institution in Pakistan. APFM has ranked the level of Pakistan at the lowest (Level-I), which signifies that Pakistan flood management institution is being run on "adhoc approaches to flood management." APFM study result (regarding IFM) and the current study result show that not-satisfactory results for the function of deliberation, and phases of rehabilitation and mitigation are in conformance with each other.

The current study's results will serve as a guide for planners and decision-makers to find the gaps and put more emphases to fill these gaps. Better deliberation and improved mitigation and preparedness is a must for improved institution capacities in Pakistan. For future studies, it is recommended that the same study should be carried out with input from a wide panel of experts to remove the professional biasness. Also, it is recommended to carry out similar studies for flood management institutions in other countries of the region.

References

1. EM-DAT, The OFDA/CRED International Disaster Database, 2006 disasters in number, www.em-dat.net, Université catholique de Louvain, Brussels, Belgium, 2007 (accessed on 13.6.2007).
2. EM-DAT, The OFDA/CRED International Disaster Database, Pakistan Country Profile: Natural Disasters, www.em-dat.net, Université catholique de Louvain, Brussels, Belgium, 2007 (accessed on 30.7.2007).

3. FFC, Annual flood report 2006, Federal Flood Commission, Ministry of Water and Power, Islamabad Pakistan, available at http://www.pakistan.gov.pk/ministries/water-power-ministry/media/Final.pdf, 2007 (accessed on 30.6.2007).
4. Dartmouth, http://www.dartmouth.edu/~floods/Archives/2007sum.htm, 2007 (accessed on 27.7.2007).
5. ADPC, *Integrated Flood Risk Management in Asia-2: A Primer* (The Asian Disaster Preparedness Center, Bangkok, Thailand, 2005).
6. G. B. Lamoree, L. E. Garcia, R. Perez and E. Castro, Methodology for the assessment of institutional frameworks for water resources management: Experiences from Latin America, *IWRA Water Int.* **30**(3) (2005) 283–293.
7. APFM, Overview situation paper on flood management practices, by Associate program on flood management, 2005, http://www.apfm.info/pdf/osp_full.pdf (accessed on 1.8.2007).
8. Y. L. Lebel, E. Nikitina and J. Manuta, Flood disaster risk management in Asia: An institutional and political perspective, *Sci. Culture* **72**(1–2) (2006).
9. Ministry of Water and Power (2002), *Pakistan Water Sector Strategy: Detailed Strategy Formulation*, Vol. 4 (The Chief Engineering Advisor/Chairman Federal Flood Commission, Islamabad, Pakistan, 2002).
10. WCDR, A review of disaster management policies and systems in Pakistan, Islamabad, Pakistan, 2005, www.unisdr.org/wcdr/preparatory-process/national-reports/Pakistan-report.pdf (accessed on 2.7.2005).
11. OCHA, Evaluation of disaster response agencies of Pakistan, National Disaster Management Advisor, Islamabad (available online on http://www.ndma.gov.pk/publications.html, (accessed on 25.7.2007).
12. D. Mustafa, Structural causes of vulnerability to flooding hazard in Pakistan, in *Econom. Geogra.* **74**(3) (1998) 289–309.
13. D. Mustafa, Reinforcing vulnerability? Disaster relief, recovery, and response to the 2001 flood in Rawalpindi, Pakistan, *Environ Hazards* **5** (2003) 71–82.
14. Dawn, News Report: Disaster management ordinance promulgated, 2006, http://www.dawn.com/2006/12/24/nat11.htm (accessed on 25.1.2007).
15. LFFP, Layyah food fighting plan for year 2003, by district administration Layyah district, 2003, available on www.punjb.gov.pk (accessed on 15.7.2007).
16. K. Mustafa, News Report: Rs 10 bn. flood protection plan on the anvil, 2007, http://www.thenews.com.pk/print3.asp?id=9013 (accessed on 19.7.2007).
17. ADB, *A Project Performance Audit Report: Flood Damage Restoration Project (loan no. 957-Pak(SF) in Pakistan* (Asian Development Bank, 1996).
18. NDMA, A presentation by Pakistan meteorological department to NDMA, 2007, http://www.ndma.gov.pk/Briefings/Cyclone/14-07-2007/8AM/PMD/HMir140830.ppt (accessed on 27.7.2007).
19. H. Rehman and A. Kamal, *Indus Basin River System — Flooding and Flood Mitigation*, 2007, available on http://www.riversymposium.com/index.php?element=38 (accessed on 25.7.2007).

Advances in Geosciences
Vol. 11: Hydrological Science (2007)
Eds. Namsik Park *et al.*

POST FLASH FLOOD HYDROLOGICAL INVESTIGATION

SARINTIP TANTANEE
Civil Engineering Department, Naresuan University
Muang Phitsanulok 65000, Thailand
sarintipt@nu.ac.th

SAMAN PRAKARNRAT
Meteorological Development Bureau
Thai Meteorological Department
Bangkok, Thailand

In the past decade, the occurrences of flooding and flash flood have been increasing over the lower northern part of Thailand. Flash floods typically result from intense rainfall rates that occur in individual thunderstorms, and often cause land sliding over the mountainous area. The study was undertaken for post flooding investigation over the flooded area of Uttaradit after the big flash flood in 2006. By coupling the data from field investigation and information from satellite images, the changes of land use and watershed characteristics was then analyzed.

1. Introduction

Flooding potential is one of the critical factors for a hydraulic structure design.[1] To estimate flood discharge, flood elevation, and flood volume, the watershed characteristics should be identified. The watershed characteristics depend on the physical properties of watershed area. If there are changes in topography over the area due to the occurring of flash flood with landslide, the watershed physical parameters are deformed. Change of landuse within the watershed greatly affects the collection capacity and consequent runoff behavior of watershed.[2] The remote sensing (RS) data are particularly useful in watershed study of which land cover and soil information are principaly estimated.[3] As Geographic Information systems (GIS) are capable of providing the spatial analysis and visualization, several studies had been done to integrate the GIS to various hydrological models.[4–6] These studies usually utilized GIS to investigate the spatial distribution of hydrological parameters over the studied area which had been proved that GIS was the effective tool for spatial analysis. As this research's concept is to investigate the changes of watershed between before and after flash flood and landslides, the study of watershed characteristics therefore has been undertaken by exploiting the power of RS data and GIS.

In the past decade, the flash floods have often occurred over the lower northern part of Thailand which increases the vulnerability to floods. In year 2006, there was a severe flash flood over this part of the country, covering the area of Uttaradit, Phrae, Sukothai, and Nan provinces. Over the mountainous area of Uttaradit, not only flash flood but also landslide occurred which caused a lot of damages for both lives and assets. The event of this big flash flood also altered the topography of the area. As the watershed functions dominate the hydrological parameters, the objective of this study was set to post investigate the watershed characteristics after the 2006 flash flood at Uttaradit. Using the satellite images, and obtained relevant data from Land development department, the topography and land covering of the watershed prior to and after flash flood was then analyzed. Daily rainfall from four stations close to the studied watershed was also analyzed to investigate the return period of rainfall.

2. Studied Area

Considering the regularly flooded and also faced the flash flood with landslide in 2006, the watershed of Huai Nam Rit in Tha Pla district, Uttaradit province of Thailand was selected to study. The daily rainfall data from four stations of Thailand meteorological department (TMD) were also analyzed in this study. Figure 1 shows the location of studied watershed and Fig. 2 shows the four studied rainfall stations.

3. Watershed Characteristics

The physical characteristics of watershed include drainage area, watershed length, watershed slope, and watershed shape.[1,6,7] The watershed parameters that reflect factor for a hydraulic structure design includes basin shape indices, channel geomorphology, and physical hydrological parameters. The studied basin shape indices are: Shape Factor (L_l)[7]; Circularity Ratio (F_C)[7]; Lemniscate ratio (K)[8]; Basin circularity (c)[9]; Form factor (F)[10]; and Basin elongation (E).[11] These typical basin shape indices

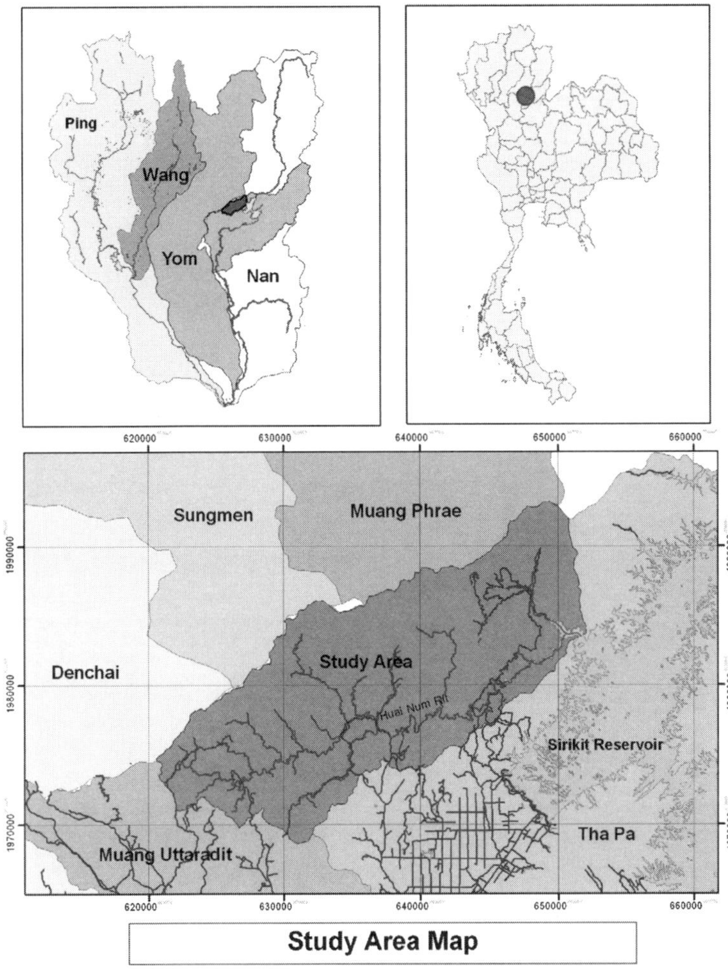

Fig. 1. The studied watershed and location of studied rainfall stations.

can be computed as follows:

$$\text{Shape factor } (L_l): \quad L_l = (L \times L_{ca})^{0.3} \tag{1}$$

$$\text{Circularity ratio } (F_c): \quad F_c = \frac{P}{(4\pi A)^{0.5}} \tag{2}$$

$$\text{Lemniscate ratio } (K): \quad K = \frac{L^2\pi}{4A} \tag{3}$$

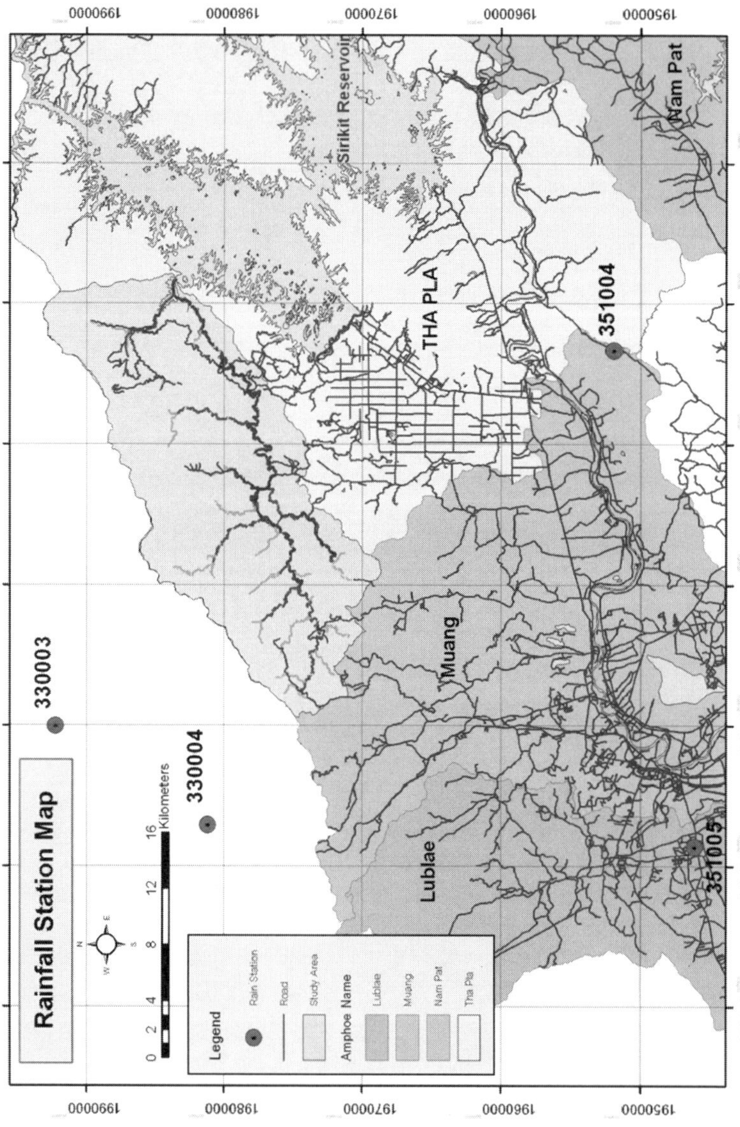

Fig. 2. Location of four studied rainfall stations.

$$\text{Basin circularity } (c): \quad c = \frac{4\pi A}{P^2} \tag{4}$$

$$\text{Form factor } (F): \quad F = \frac{A}{L^2} \tag{5}$$

$$\text{Basin elongation } (E): \quad E = \frac{2\sqrt{A}}{L\sqrt{\pi}}, \tag{6}$$

where L is the length of the watershed; L_{ca} is the length to the center of watershed area; P is the perimeter of the watershed; and A is the area of the watershed.

The channel geomorphologic parameters that were analyzed in this study comprise of Channel length,[8] Channel slope,[8] Drainage density[7] and Bifurcation ratio.[10] The parameters of Drainage density and Bifurcation ratio can be given by

$$\text{Drainage density,} \quad D = \frac{L_t}{A} \tag{7}$$

$$\text{Bifurcation ratio,} \quad R_b = \frac{N_i}{N_{i+1}}, \tag{8}$$

where L_t is the total length of stream channels in the watershed; A is the watershed area; N_i is the number of stream at any order, and N_{i+1} is the number of stream of the next highest order.

The studied physical hydrological parameters are Runoff Curve Number (CN)[12] and Time of Concentration (t_c). Three equations were utilized to obtain Time of Concentration as follows:

$$\text{US Bureau of reclamation equation}^{13} \quad t_C = 60\left(\frac{11.9L^3}{H}\right)^{0.385} \tag{9}$$

$$\text{Kirpich's equation}^{14} \quad t_C = 0.0078L^{0.77}S^{0.385} \tag{10}$$

$$\text{Barnsby's equation}^{14} \quad t_C = \frac{21.3L}{A^{0.1}S^{0.2}}, \tag{11}$$

where L is the length of channel; H is the elevation difference between upstream and downstream point, and S is the average watershed slope.

4. Studied Process

The process of study comprises of rainfall analysis, GIS based watershed map development and the watershed characteristics analysis. The daily

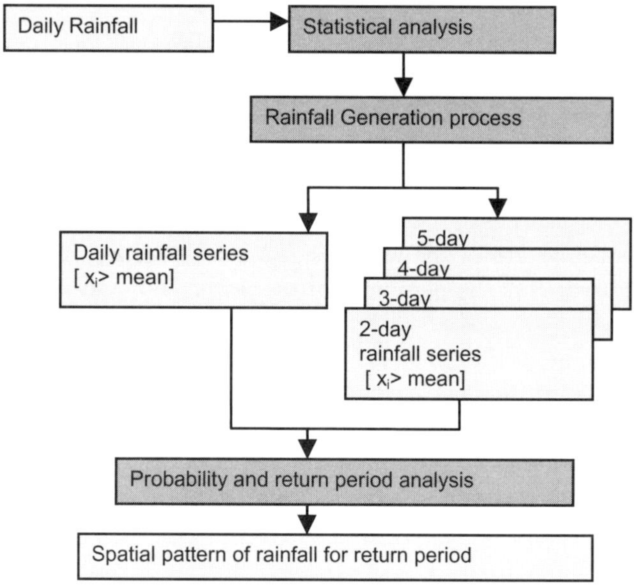

Fig. 3. Rainfall analysis process.

rainfall records of 1976–2006 from four studied stations were analyzed for statistical characteristics. The series of daily, 2, 3, 4, and 5 day accumulated rainfall which exceed mean values were then generated to analyze for rainfall at various return period of occurrences. The details of the rainfall analysis process are shown in Fig. 3.

The return period of an event of a given rainfall can be defined as the average recurrence interval between equaling or exceeding a specified magnitude.[6] Thus, the return period T_r can be estimated from the probability $p = P(X \geq x_T)$ of occurrence of the event $X \geq x_T$ in any observation, where X is the random variable and x_T is the specified level. The return period or recurrence interval can be estimated by Eq. (12).[7] The spatial patterns of rainfall in each return period were then constructed by plotting on the studied GIS watershed map with the interpolation function of Inverse Distance Weighting (IDW) of ARCGIS.

$$T_r = \frac{1}{P_r(X \geq x_T)}. \tag{12}$$

In order to compare the watershed identification between prior to and after flooding, the GIS based watershed map of these two periods were then developed with ARCGIS. Using ERDAS, the satellite images from year 2002

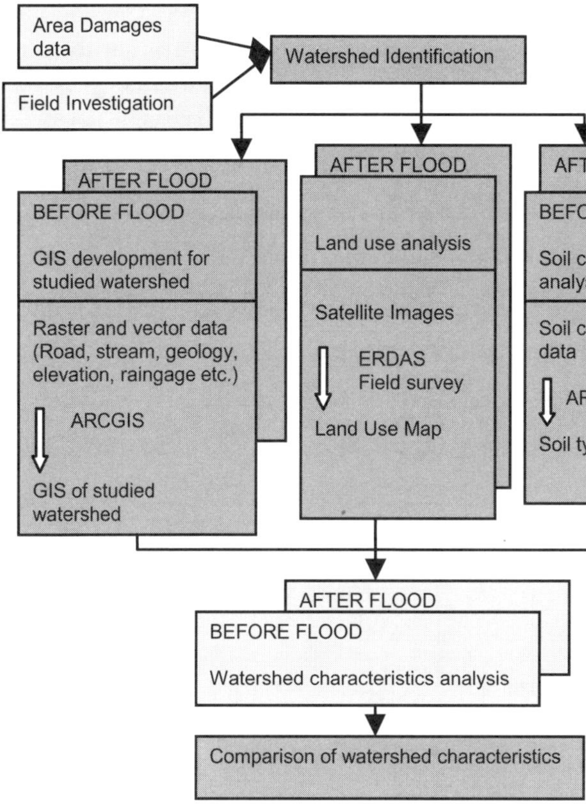

Fig. 4. GIS based watershed development.

and 2007 were analyzed for landuse of before and after flooding. The soil covering type data obtained from Land Development department were also utilized in the process of *CN* evaluation. The process of GIS based watershed map development is shown in Fig. 4. The watershed characteristics were then considered by basin shape indices, channel geomorphology, and physical hydrological parameters.

5. Results and Discussion

5.1. *Rainfall analysis*

The maximum daily, 2-day, 3-day, 4-day, and 5-day accumulated rainfall were analyzed for rainfall occurrences at several return periods as shown in Fig. 5. Table 1 shows maximum daily and 2–5 day accumulated rainfall for the year 2006 compared with the records of 1976–2006. Considering 2006 as the year of big flash flood and landslides, it is shown that the obtained return period of occurrence of 2–5 days accumulated rainfall at 351005 are in the range of only 15–17 years, which means that there was not the heavy rain event in 2006. Moreover, determining the location of the station, it is found that 351005 is the farthest station from the studied basin. Thus, 351005 is not the appropriate point to detect the rainfall for the studied basin. Considering the 2006 maximum rainfall for the three remaining stations, the return periods of occurrences of 2–5 days accumulated rainfall are very large which are in range of 318–2324 years. Therefore, further study should be done to analyze the accumulated rainfall as a significant factor of flash flood with landslide.

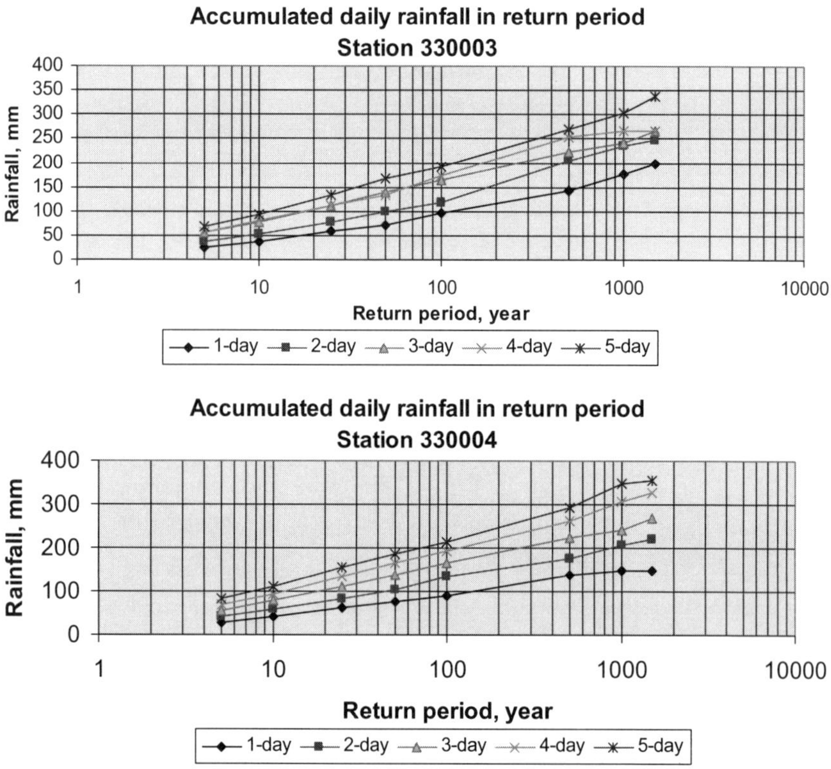

Fig. 5. Return period of occurrences for daily rainfall of four studied stations.

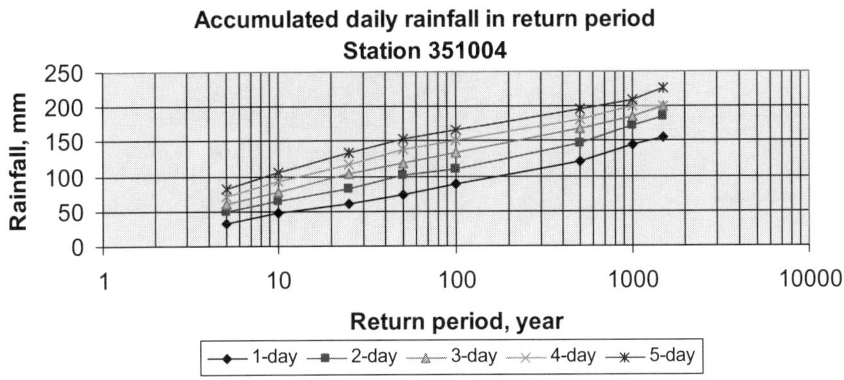

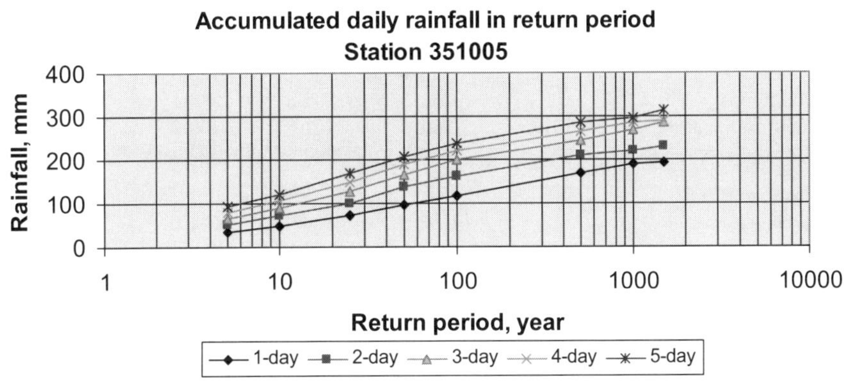

Fig. 5. (*Continued*)

5.2. *Watershed identification*

Geographic Information Systems based watershed was developed to analyze the difference of characteristic between before and after flooding. The Soil type map of the studied basin was developed from the data obtained from Land Development Department, shown in Fig. 6. Figure 7 demonstrates the stream order analysis of the study. Applying the satellite images of 2002 and 2007 with the ERDAS and ARCGIS, the land use map of before and after flooding were constructed as shown in Fig. 8. Comparing between before and after flooding, the area of the open forest decreases from 44.27% to 24.62% whereas the dense forest shows slight increase from 16.69% to 23.66%.

Coupling these developed maps with the topography of the watershed, the physical hydrological parameters were then analyzed. Tables 2–4 show

Table 1. Comparing 2006 rainfall with maximum and mean of records (1976–2006).

	Daily rainfall	Accumulated rainfall			
		2-day	3-day	4-day	5-day
Station 330003					
Mean	90.8	123.1	135.9	151.1	164.4
Maximum	212.0	265.7	265.7	305.9	393.8
2006 max. rainfall	118.4	201.9	208.5	231.6	235.9
Return period, yr	294	491	400	390	318
Station 330004					
Mean	91.9	122.9	145.1	160.7	172.3
Maximum	150.0	223.0	310.0	345.0	363.0
2006 max. rainfall	123.0	220.0	248.0	268.0	293.0
Return period, yr	401	1993	2347	1596	1379
Station 351004					
Mean	85.1	113.7	125.1	143.0	155.4
Maximum	161.2	201.3	201.3	230.4	230.4
2006 max. rainfall	141.8	201.3	201.3	230.4	230.4
Return period, yr	453	2324	2000	1068	1016
Station 351005					
Mean	115.5	152.2	180.0	194.7	210.5
Maximum	198.0	252.6	286.1	318.2	381.8
2006 max. rainfall	52.5	80.4	96.6	114.1	129.4
Return period, yr	12	17	15	17	17

the results of basin shape indices, channel geomorphology, and physical hydrological parameters, respectively. It is found that after flooding most of the basin shape indices and physical hydrological parameters are changed, which means that the flooding potential of the watershed is also altered. Considering the channel geomorphology, there are changes in channel length and channel slope whereas the drainage density and bifurcation ratio are constant.

The fundamental watershed functions in term of hydrology are: to collect the water from precipitation, to store water at various amounts and durations and to discharge water as runoff.[2] From the results in Tables 2 and 3, it is obviously seen that there are changes in channel shape indices and channel geomorphology which caused changes in the physical hydrological parameters, as well. These parameters are the significant factors for flooding design. *CN* demonstrates how much the cumulative precipitation becomes the cumulative direct runoff.[13] The result shows that *CN* value slightly decreases after flooding because of the increase in forest covering the studied area from 2002 to 2007. Less *CN* means that for the

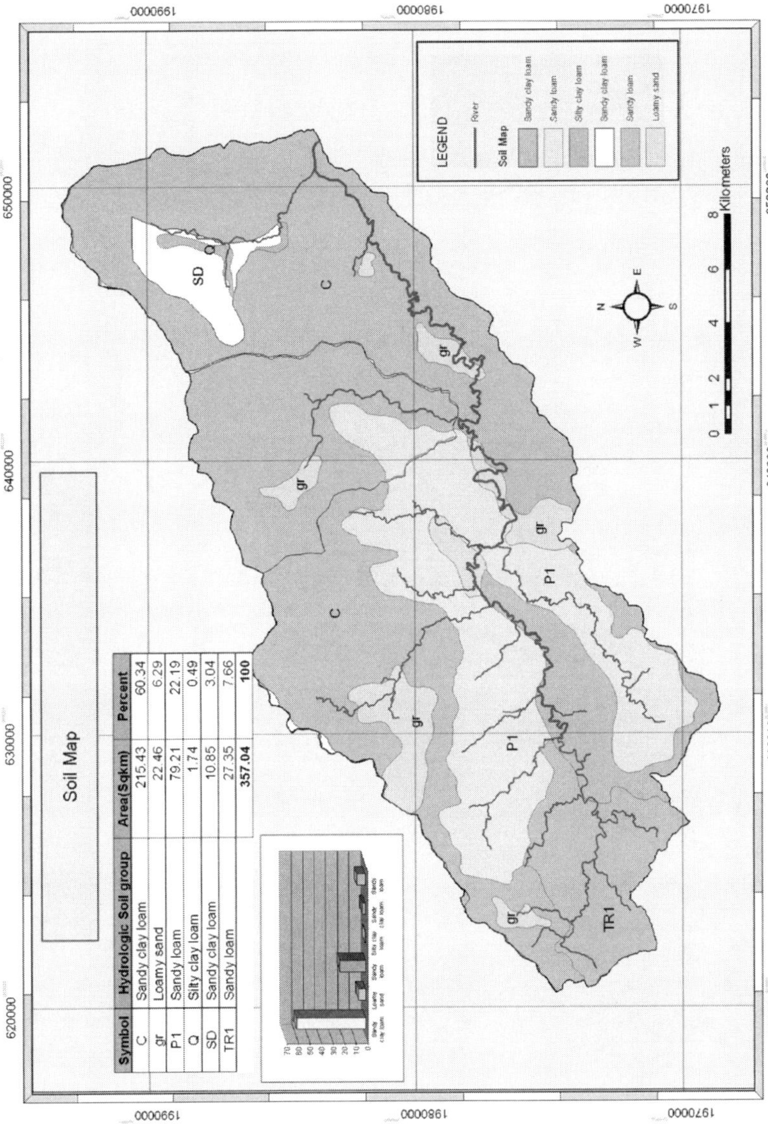

Fig. 6. Soil type map of the basin.

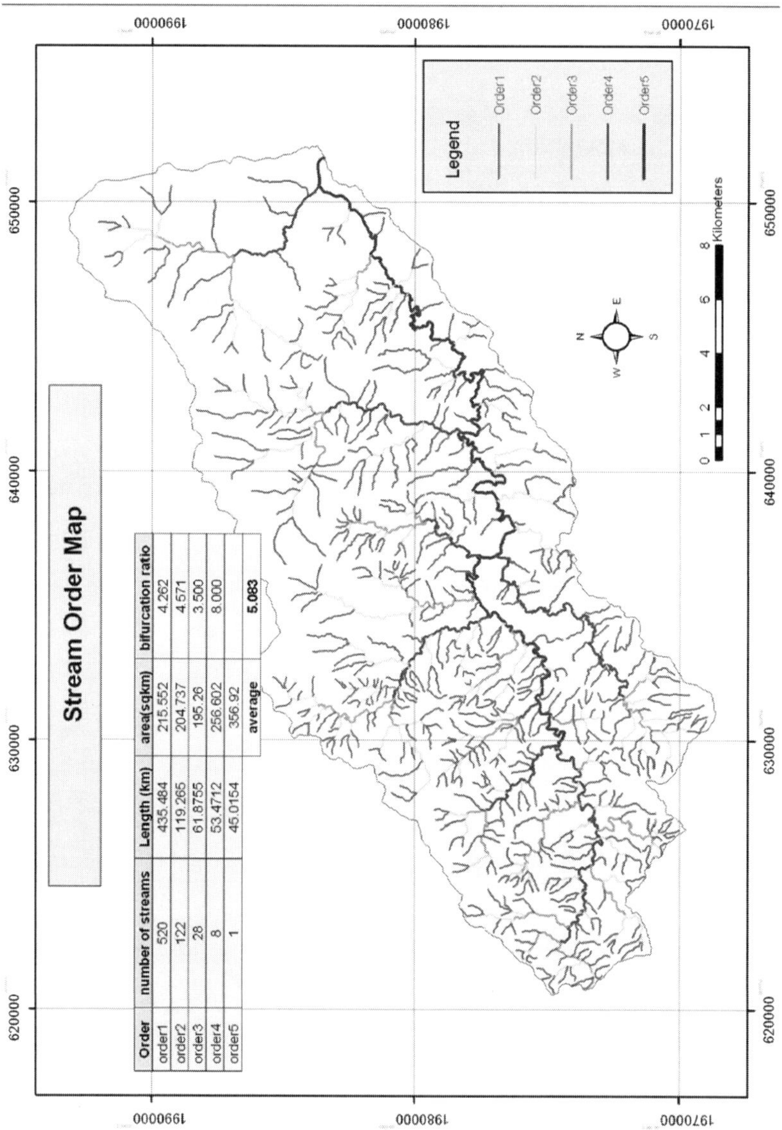

Order	number of streams	Length (km)	area(sqkm)	bifurcation ratio
order1	520	435.484	215.552	4.262
order2	122	119.265	204.737	4.571
order3	28	61.8755	195.26	3.500
order4	8	53.4712	256.602	8.000
order5	1	45.0154	356.92	
			average	**5.083**

Fig. 7. Stream order analysis.

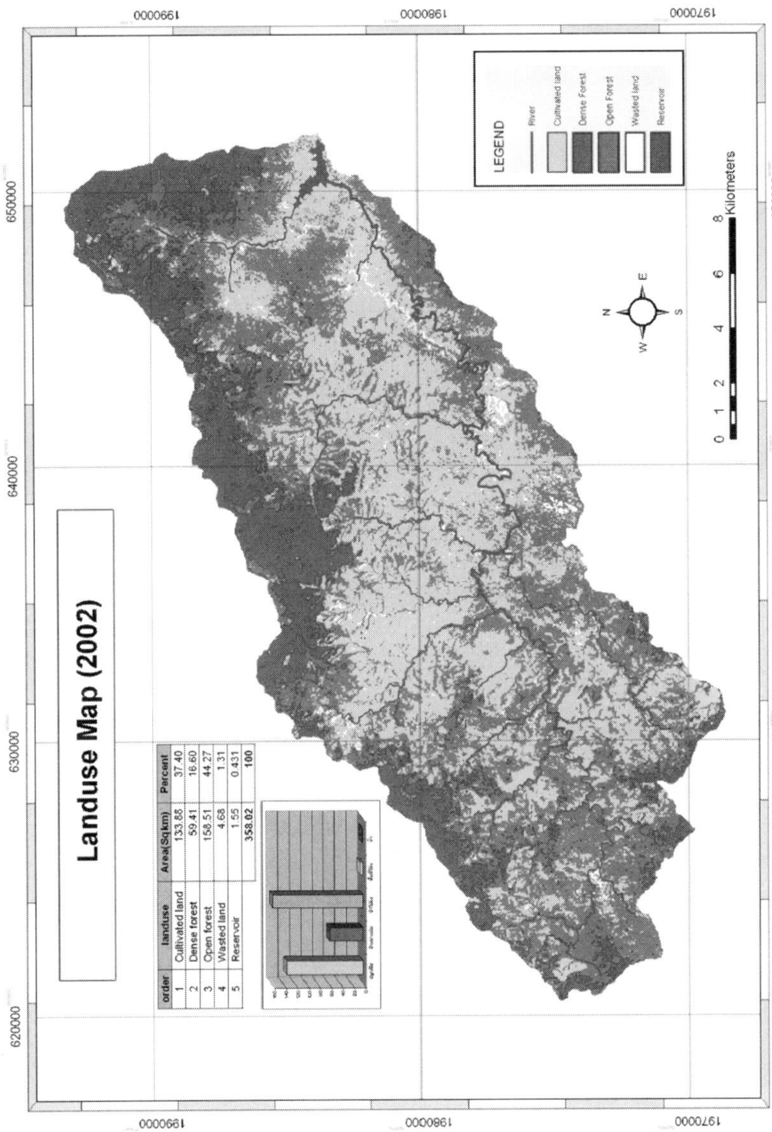

Fig. 8. Landuse analysis for 2002 and 2007.

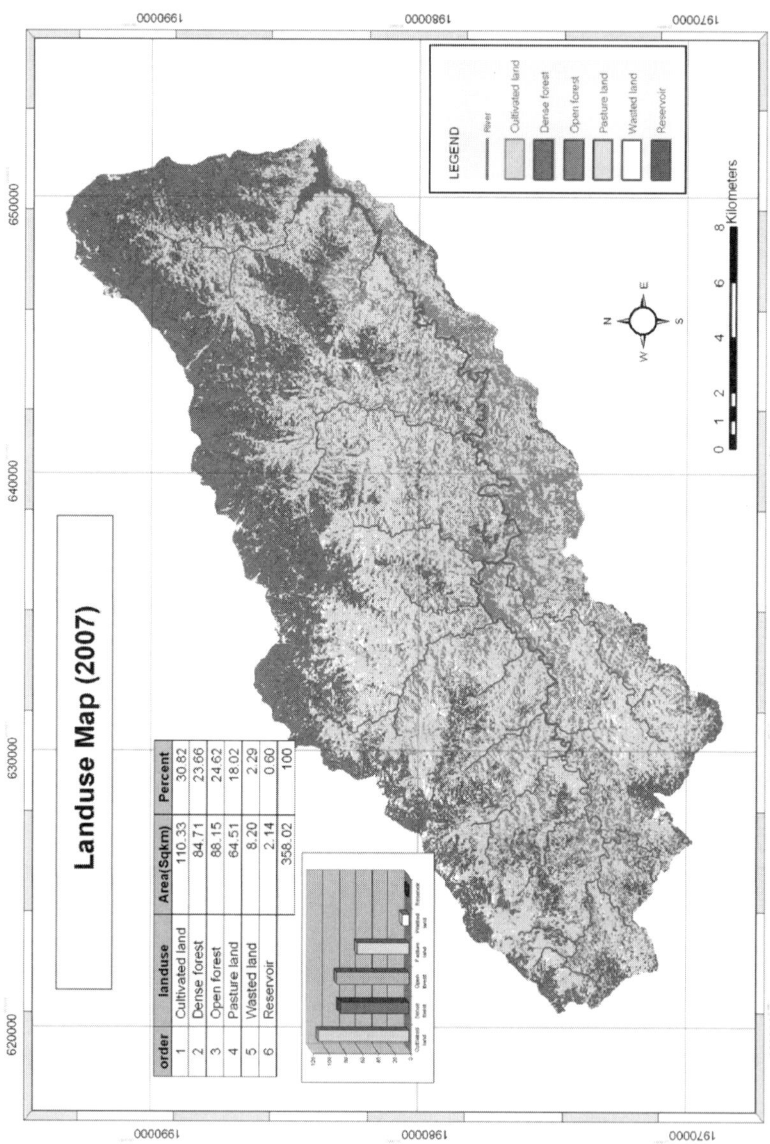

Fig. 8. (Continued)

Table 2. Basin shape indices.

Basin shape index	Before flooding	After flooding
Shape factor (L_l)	6.987	6.953
Circularity ratio (F_c)	1.518	1.518
Lemniscate ratio (K)	6.889	6.666
Basin circularity (c)	0.434	0.434
Form factor (F)	0.114	0.118
Basin elongation (E)	0.381	0.387

Table 3. Channel geomorphology.

Channel index	Before flooding	After flooding
Channel length, mile	33.892	33.326
Channel slope,%	0.678	0.689
Drainage density	3.235	3.235
Bifurcation ratio	5.083	5.083

Table 4. Physical hydrological parameters.

Curve Number (CN)			Time of Concentration (t_c)		
Antecedent moisture classes (AMC) Group	Before flooding	After flooding	Equation	Before flooding	After flooding
I	63	60	— US Bureau of Reclamation	10.04	9.85
II	80	78	— Kirpich	9.87	9.68
III	90	89	— Barnsby	8.13	7.97

same antecedent moisture condition, any particular cumulative rainfall will become less direct runoff. Time of concentration (t_c) is also the important parameter for flooding structure design that demonstrates time of flow from the farthest point on the watershed to outlet.[15] The results show the decrease in t_c which means that runoff needs less time to travel to the outlet of the watershed after a flooding event.

6. Conclusions

As the flooding potential of watershed depends on the watershed characteristics, it is necessary to analyze these watershed properties for the hydraulic structure design. In general, the watershed characteristics can be identified by basin shape indices, channel geomorphology, and physical hydrological parameters. Flooding and flash flood with landslide is the important factor on changes in the topography of the area, which cause the deformation of watershed physical characteristics. From this study, it is shown that the changes of these characteristics affect the hydrological parameters which mean the change in obtained runoff and time of flow, as well. Thus, the watershed properties analysis over the flooding area should be considered when the flooding design is needed.

References

1. R. C. Sorrell, *Computing Flood Discharges for Small Ungaged Watershed* (Michigan Department of Environmental Quality Geological and Land Management Division, 2003).
2. B. E. Black, Runoff from watershed models, *Water Resour. Res.* **6**(2) (1970) 465.
3. G. W. Kite and A. Pietroniro, Remote sensing application in hydrological modeling, *Hydrol. Sci.* **41**(4) (1996) 563–591.
4. D. J. Abel, P. J. Kilby and J. R. Davis, The systems integrate problem, *Int. J. Geographic Inf. Syst.* **8**(1) (1994) 1–12.
5. Z. X. Xu, K. Ito, G. A. Schultz and J. Y. Li, Integrated hydrological modeling and GIS in water resources management, *J. Comput. Civil Eng.* **15**(3) (2001) 217–223.
6. S. R. Satti and J. M. Jacobs, A GIS-based model to estimate the regionally distributed drought water demand, *Agriculture Water Management* **66** (2004) 1–13.
7. V. T. Chow, D. R. Maidment and L. W. Mays, *Applied Hydrology* (McGraw Hill, 1988).
8. C. T. Haan, B. J. Barfield and J. C. Hayes, *Design Hydrology and Sedimentology for Small Catchments*, 2nd Edn. (Academic Press, 1994).
9. R. J. Chorlev, D. E. G. Malm and H. A. Poaorzelski, A new standard for estimating basin shape, *Am. J. Sci.* **255** (1957) 138–141.
10. V. C. Miller, A Quantitative Geomorphic Study of Drainage Basin Characteristics in the Clinch Mountain Area Virginia and Tennessee, Tech. Report No. 3 (Office of Naval Research, Geography Branch, Project no. 389-042, 1953), 1–30.
11. R. E. Horton, Erosion development of streams and their drainage basin; hydrophysical approach to quantitative morphology, *Bull. Geol. Soc. Am.* **56** (1945) 275–370.
12. S. A. Schumm, The evolution of drainage systems and slopes in badlands at Perth Amboy, New Jersey, *Bull. Geol. Soc. Am.* **67** (1956) 597–646.

13. Soil Conservation Services, *National Engineering Handbook, Section 4 Hydrology* (US Dept. of Agriculture, 1972).
14. US Bureau of Reclamation, *Design of Small Dam*, 2nd Edn. (Washington D.C., 1973).
15. M. Wanielista, R. Kersten and R. Eaglin, *Hydrology Water Quantity and Quality Control*, 2nd Edn. (John Wiley & Sons Inc., 1997).

Advances in Geosciences
Vol. 11: Hydrological Science (2007)
Eds. Namsik Park *et al.*

FLOOD IMPACT ASSESSMENT IN THE SURROUNDING AREA OF SUVARNABHUMI AIRPORT, THAILAND

TAWATCHAI TINGSANCHALI* and D. ENG
School of Engineering and Technology, Asian Institute of Technology
P.O. Box 4, Klong Luang, Pathumthani, Thailand 12120
**tawatch@ait.ac.th*

The existence of the Second Bangkok International Airport (SBIA) or the Suvarnabhumi International Airport induces more adverse effect to the flooding situation in its surrounding area. Due to limited drainage capacity, during a heavy storm, flooding in the surrounding area occurs over the area. The objective of the study is to find the most suitable flood control and drainage system that can drain floodwater from the surrounding area of 624 sq. km with minimum flood damages and impact to social and living conditions of the people in the study area. This study involves the application of MIKE FLOOD hydrodynamic model for determining the relative effects of flood control and drainage system in the surrounding area of the airport. The results of the study show that flood damages mostly occur in the central and downstream parts of the study area where drainage is insufficient. Flood depth and duration are main parameters used for the estimation of flood losses.

Flood mitigation and management in the surrounding area of SBIA is planned by pumping water of 100 m^3/s from Klong Samrong canal inside the study area through the proposed drainage channel to the Gulf of Thailand. The existing dikes along boundaries of the study area can protect water from the outer area to enter into the surrounding area of the airport. Flood simulation shows that a canal with capacity of 100 m^3/s and a pumping station at the downstream end of the canal are required to cope with the drainage capacity for the flood of 100 years return period. A flood drainage channel of capacity of 100 m^3/s is designed and will be constructed to drain flood from Klong Samrong to the sea. On the other hand, the embankment along the proposed drainage canal project improves traffic flow in the vicinity of the airport. On economic benefit, the project investment cost is Baht 8,410 million. The project benefit cost ratio is 2.12 with the economic internal rate of return of 15.61%. The construction period is 4 years.

Environmental and social impacts are investigated and counter measures are proposed to reduce the impacts. The study considers compensating scheme for people who are directly affected by the flood drainage project and those who will lose their lands or their professions. Considerations are also extended to people who are indirectly affected by the project. Institutional framework is recommended to be established to manage flood control and drainage and water resources in the surrounding area of the airport.

1. Introduction

The Second Bangkok International Airport (SBIA) or the Suvarnabhumi International Airport (Figs. 1 and 2) reduces the flood storage and drainage capacity of the surrounding area hence inducing more adverse effect to the flooding situation. Because of the limited drainage capacity, during a heavy storm, rainfall will accumulate causing flood inundation inside the area for a few months. Flooding will cause more significant losses and inundation over a long period of time if flood drainage capacity is not sufficiently provided, especially in the surrounding area of the airport.

Causes of flooding in the eastern part of the Lower Chao Phraya River Delta in the vicinity of the airport are mostly due to heavy local rainfall in the area, storm runoff from surrounding areas and limited flood drainage capacity. The limitation of flood drainage capacity is due to (1) general reduction of drainage canal capacities due to sediment deposition, vegetation and encroachment by human activities; (2) insufficient drainage openings of the east–west highways, particularly Bang Na-Trad highway; (3) insufficient drainage canals in north–south direction from Klong

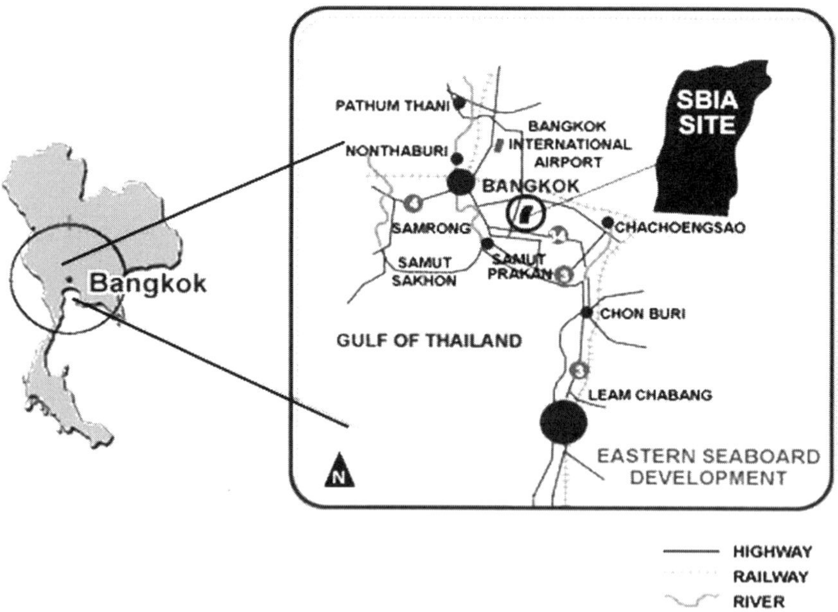

Fig. 1. Location map of Second Bangkok International Airport (Suvarnabhumi Airport).

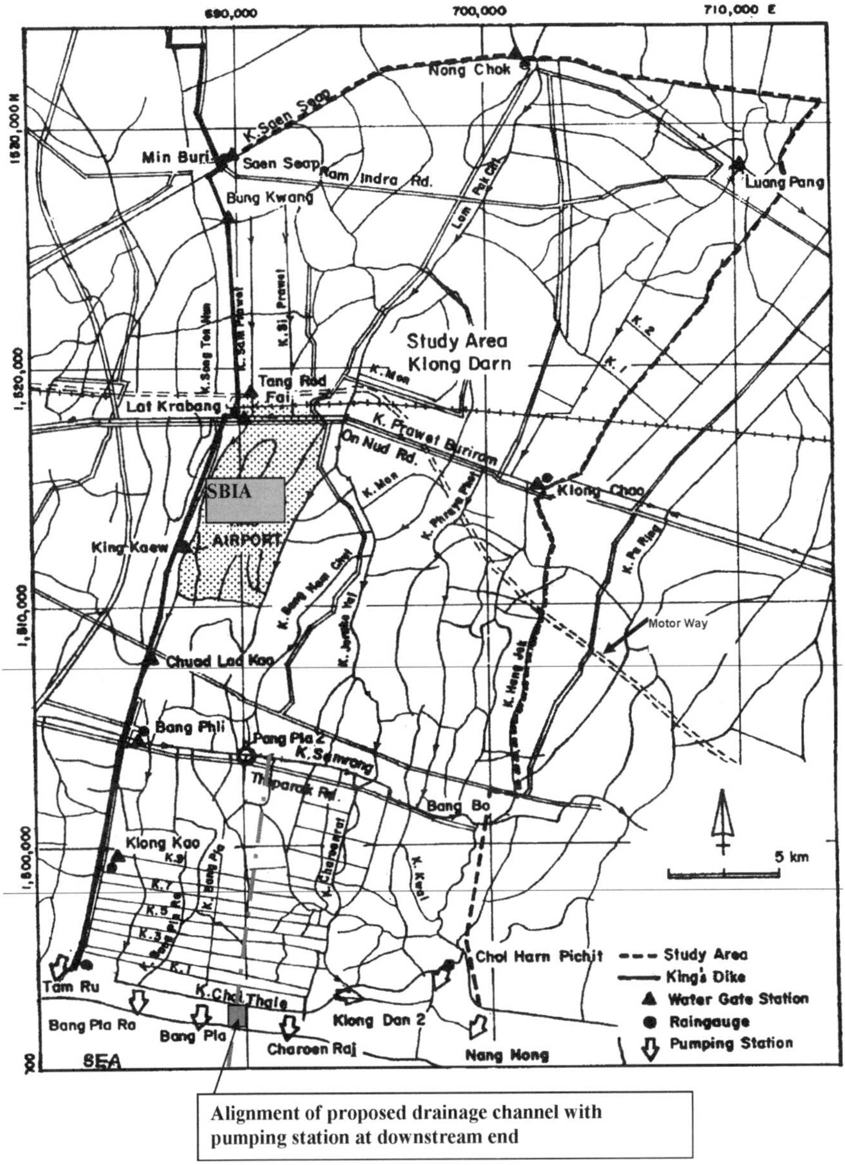

Fig. 2. Detailed location map of SBIA and its surrounding area.

Samrong canal to the Gulf of Thailand; and (4) topology of the land is low-lying depression area which does not provide drainage by gravity. Runoff from surrounding areas flow into the depression area which causes prolonged flooding and considerable damages, especially in the downstream part of the study area. The overall study area covers an area of 624 sq. km in the east of Bangkok.

The objective of the study is to find the most suitable flood control and drainage alternative that can drain floodwater, in the surrounding area of SBIA effectively with minimum impact to environment, social, and living conditions of the people living in the area. It is desired to have the project decision-making involve the government, the people, the community, and other stakeholders through public participation.

2. Description of Study Area

The study area covers an area of 624 sq. km in the east of Bangkok. As shown in Fig. 2, this area is surrounded by Klong Saen Saeb canal and Klong Nakorn Neung Khet canal in the north, Klong Phra-Ong Chao Chaiya Nuchit canal in the east, and by the Rom Klao-King Kaew Road (or the King Dike) in the west. The south of the study area is Klong Chai Talae which runs parallel to the coastline of the Gulf of Thailand. The average ground level is about +0.40 m to +1.00 m (msl). Topography of the area is a flat terrain with nearly horizontal slope in which the water can drain out only slowly.

The airport occupies a part of the eastern flood plain of the Lower Chao Phraya River (Fig. 2). Land use in the study area can be mainly classified to agricultural area, residential area, and non-residential area. Most of the study area is covered by agricultural area, about 56.57% of overall area. The Royal Irrigation Department (RID) plans to drain flood in the study area from Klong Samrong canal to the Gulf of Thailand by improving the existing drainage canals and construction of a main drainage channel with a pumping station at the downstream end to pump water to the Gulf of Thailand.

3. Previous Studies on Flood Control and Drainage

Previous studies concerning flood control and drainage in the study area included three alternatives namely: drainage channel, underground tunnel, and elevated water flume. The previously proposed drainage channel by RID[1] was found to face problem in implementation because the channel route proposed for construction is now fully occupied by people and communities, industrial factories and public facilities. This totally obstructed land acquisition and thus the implementation of the previously proposed flood drainage channel. Other flood control and drainage alternatives proposed in previous studies, by the Royal Irrigation Department, include an underground tunnel or an elevated water flume. These were found to have practical limitations with high operation and maintenance cost. In the present study, three alternatives of previously proposed flood control and drainage system namely, drainage channel, underground tunnel, and elevated water flume were considered and discussed including the newly proposed drainage channel in the public forum of the concerned authorities, stakeholders, communities and people concerned in 2003. Public hearings were organized to allow these concerned sectors to express their opinion and suggestions on the project.

4. Hydraulic Model Study and Design of Flood Control and Drainage System

MIKE FLOOD[2] is a hydrodynamic model that integrates the one-dimensional hydrodynamic model MIKE 11 and the two-dimensional hydrodynamic model MIKE 21 into a single, dynamically coupled modeling system. Using a coupled approach enables the best features of both MIKE 11 and MIKE 21 to be utilized, whilst at the same time avoiding many of the limitations of resolution and accuracy encountered when using MIKE 11 or MIKE 21 separately. Connection between MIKE 11 and MIKE 21 is controlled by using a broad-crested weir.

The MIKE 11 model is used to simulate flow in canal networks. In this study, the flood in 1995 which was the largest flood in the last decade is considered. External boundary conditions are required at all upstream and downstream ends of model branches which are not connected to junction.

In this study no inflow is considered at each of these model branches. The initial flow conditions for the hydrodynamic model computation is the bank full water level and corresponding discharge in each canal according to Manning equation. The model was calibrated and verified satisfactorily by using the data of daily water level for 30 days at Klong Samrong canal (Bang Phli regulator station) in 1995 and 2001, respectively.[3]

In the calibration, the canal Manning roughness coefficient was adjusted by trial and error and was found to be 0.035, while in the flood plain the

Manning coefficient is 0.05. It was found that to cope with flood of 100 year design return period; a canal capacity of $100\,m^3/s$ with a downstream pumping station are required. The flood drainage channel of capacity of $100\,m^3/s$ is designed and constructed to drain flood from Klong Samrong through this canal to Gulf of Thailand. The channel has a trapezoidal cross-section with a bottom width of 48 m, a side slope of 1:3.5, a depth of 3.36 m and a bottom slope of 1:25,000. A pumping station of $100\,m^3/s$ is installed at the downstream end of the drainage channel to pump floodwater from this channel to the sea. It is found from flood simulation study that for the 100 year design flood, the new drainage channel can effectively drain the floodwater in the critical areas around the airport especially in the south of the airport and will help to mitigate water pollution and water weeds in that area. With cleaning of the existing drainage canals, the newly proposed drainage channel can reduce water level in Klong Samrong canal as much as 1.63 m on average after 30 days. In the inundation areas mostly in the south of the airport, the flood depth of 0.3 m can be almost eliminated (Figs. 3 and 4). The proposed drainage channel can improve drainage capacity of Klong Samrong canal and acts as a shortcut for drainage of floodwater out of the area to the sea. The flood control and drainage project can drain floodwater of 8.64 mcm from the project area and reduce the inundation area from 140 sq. km to almost nil and flood duration from 10 to 2 days.

5. Flood Control Project Benefit and Flood Damages

On economic benefit, the project investment cost is Baht 8,410 million. The project benefit cost ratio is 2.12 with the economic internal rate of return of 15.61%. The construction period is 4 years from 2005 till 2008.[4]

In this study, only tangible damages are considered for analysis. The survey is mainly targeted for three main categories of land use, i.e. residential area, non-residential area, and agricultural area. Based on the categories used in the standard loss estimation methodologies,[5,6] questionnaire surveys were carried out in all districts to cover all main types of land use which are mostly affected by flood.

The questionnaires are made simple for the people to understand. The questions are limited to the past 10 years in order not to be too long for the human memory factor. The selected size of samples for questionnaire surveys is 350 samples for each category of land use. From the surveys it was found that flood damages mostly occur in the central and downstream parts

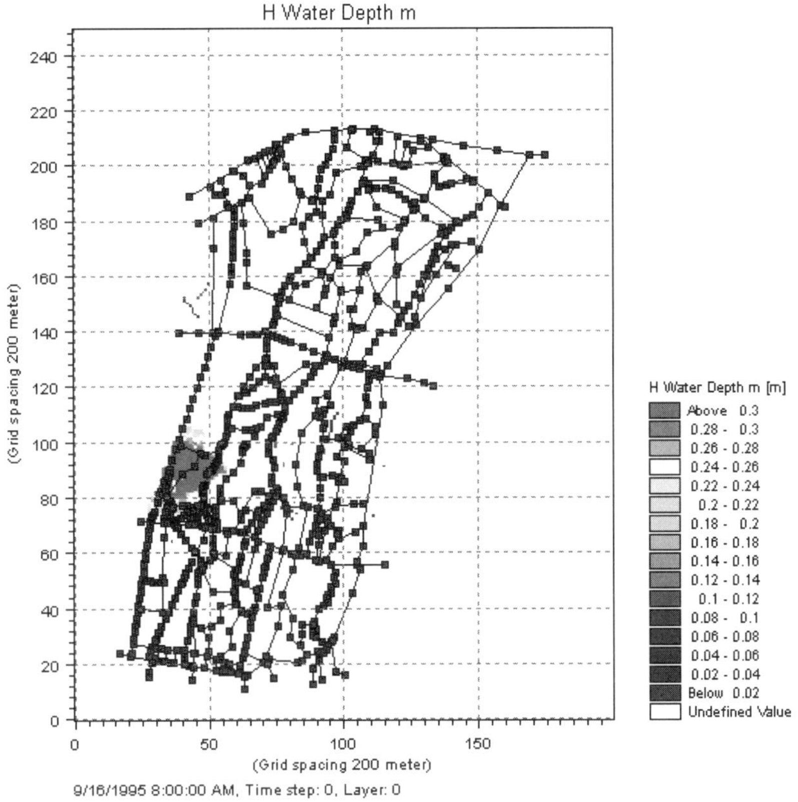

Fig. 3. Observed inundation area due to flood depth of 0.30 m in study area.

of the study area between Klong Pravet, Klong Samrong, and Klong Chai Talae, where drainage is insufficient (Fig. 2). This area is mostly agricultural with some residential area and non-residential area. The obtained data from questionnaire surveys also show that the flood warning systems are unavailable for the people. If the flood warning systems are provided, the people can prepare themselves and their properties to avoid flood and reduce flood damage on average by more than 40%.

From survey results, information of flood depth/duration and flood damage was plotted. Curve fitting method was applied in order to make a preliminary of flood damage functions. The average damage values of each flood depth were plotted for different flood durations.

The obtained data are statistically analyzed to develop the relationships between the parameters that affect flood damage and economic damage

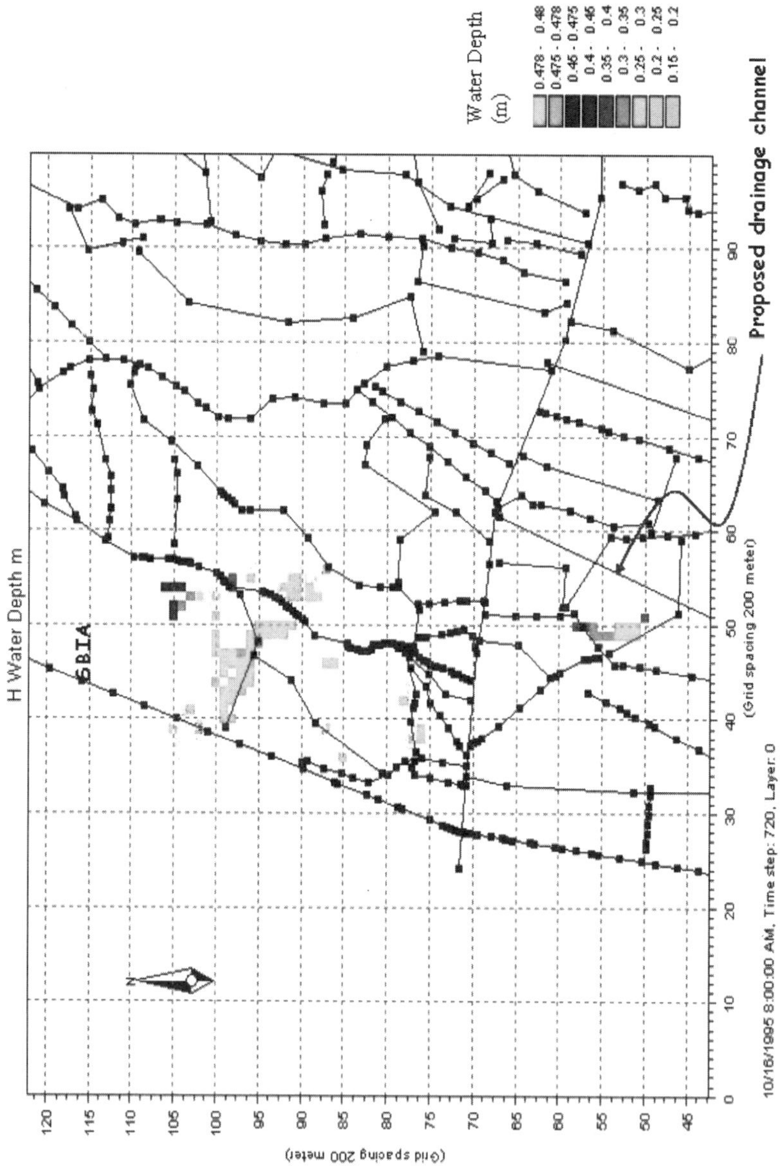

Fig. 4. Predicted flood depth and inundation area after dredging of new drainage channel.

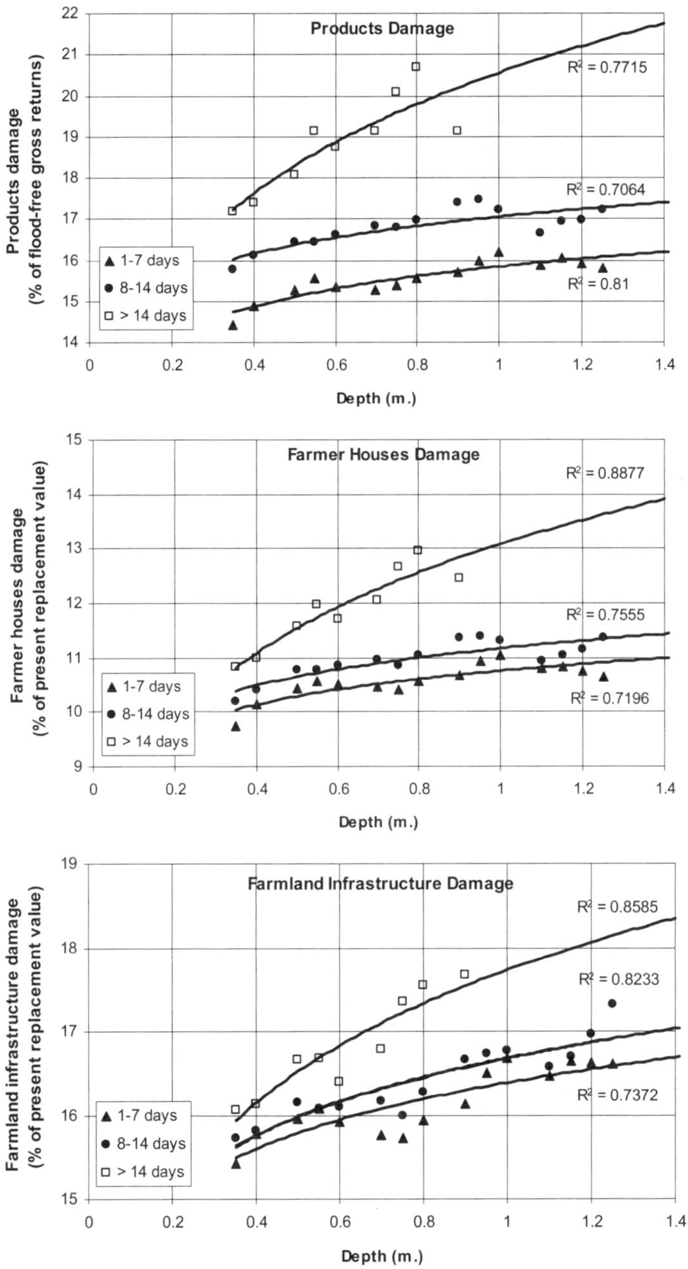

Fig. 5. Average depth–duration–damage curves for agricultural area.

caused by floods. An example of flood depth–duration–damage relationship for agricultural area is shown in Fig. 5. The factors considered in the flood damage relationships are product damage, farmer house damage, and farmland infrastructure damage.

6. Preliminary Environmental Impact Assessment

The environmental impact assessment covers physical resources, biological resources, utilization of resources, and quality of life. From the assessment, it is found that the surface water quality in the study area is under deteriorating condition due to high turbidity, high concentration of pollutants, high coliforms, and low dissolved oxygen of less than 2 mg/l. The water quality in the canal within the project area is suitable for agricultural use and general use for industry but it is not suitable for domestic use and drinking purposes. The groundwater in this area is generally not suitable for domestic use and drinking purpose but it is suitable only for irrigation and industrial uses. The impact due to project implementation on groundwater quantity and quality are expected to be insignificant. On the geological condition, soil stability problem may exist. The thickness of soft soil layer is about 12–16 m. It is necessary to improve the soft soil layer to prevent instability and subsidence of canal slope and embankment.

There are six to eight tourist locations in the project area which have good potential. All public health measures must be taken to eliminate water storages where mosquitoes breed. Mosquito bites may lead to dengue fever, which is fatal.

7. Social Impact Assessment

The flood control and drainage in the surrounding area of the Suvarnabhumi Airport (Second Bangkok International Airport) has certain impacts on socio-environmental and living conditions of the people and local residents, both directly and indirectly. Mostly 75% of the families in the area have lived in the area surrounding the airport from their childhood for more than 30 years. About 95% of all houses are permanently built and 87.2% of the residents are Buddhist while 12.5% are Muslims by religion. Mostly 68.8% of local transportation is by private cars and motorcycles. Only 16.3% of local transportation is carried by boats, trucks and buses. 40.3% of water supply for domestic consumption is from groundwater, 35.1% from canals in the area, and 24.3% is from piped water supply system in the area.

Public participation is required to avoid conflicts, confrontations, problems, and protest in the future.[7] All stakeholders are given the opportunity to express their opinion or giving suggestions in selecting the most appropriate alternative of the flood control and drainage system. These suggestions are meant to minimize the negative socio-environmental impacts of the flood control and drainage system on the people living in the project area. The process of encouraging the public participation was carried out in six steps as shown in Fig. 6 as follows:

Step 1: Coordination with local officers concerned, which includes the provincial governors and district officers. (32 local leaders have been identified).

Step 2: Organize seminar for the 32 representatives/local leaders.

Step 3: Local leaders collected basic survey data at community level. For the secondary data they obtained from villages, sub-districts offices, and from Internet resources. The primary data are obtained by using questionnaires and group discussions with local people, in particular with the religious leaders and heads of villages. Apart from getting data and information from communities, discussions and exchange of ideas with local leaders yielded other suggestions on the original design as follows:

1. They are in agreement with dredging of a new main drainage canal but it was suggested to have road embankments on both sides of the canal.
2. They disagree to construct an underground tunnel or elevated water flume to convey floodwater from the study area to the sea due to high operation and maintenance cost and their limited utilization for other purposes.
3. They suggest cleaning up water weeds in the existing drainage canals including enlarging the cross-sections of the existing canals, in particular Klong Bangpla and Klong Trong.
4. They agree that extensive work in cleaning and rehabilitation of other drainage canals in the study area surrounding the Second Bangkok International Airport should be done.

Step 4: Organization of meetings of local representatives/local leaders to collaborate in verifying and analyzing the data and information of the communities. This enables the local representatives/local leaders to confirm or suggest improvement in the design and implementation of the flood control and drainage system.

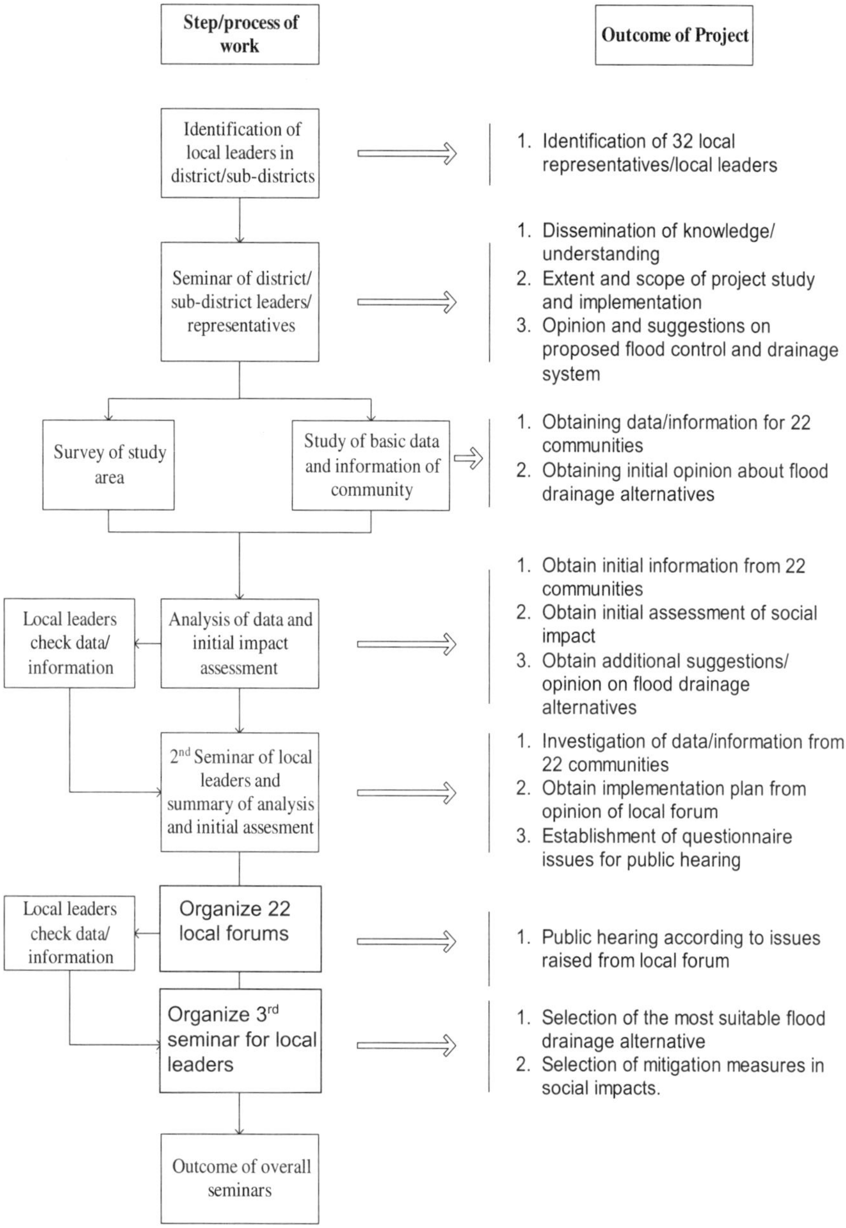

Fig. 6. Flowchart of procedure and expected outcomes of social impact assessment.

Step 5: Organization of local forum at community or village level to get opinion or suggestion according to the above-mentioned four issues.

Step 6: Organization of seminar for local representatives or local leaders to summarize and conclude the overall output of local forum and recommendations on proposal of most suitable flood control and drainage alternative.

From public participation, it can be concluded that the local residents agreed with the necessity of having a new main drainage canal of $100\,m^3/s$ capacity. This will give an ultimate benefit to the area surrounding the airport and the airport itself as a whole. Finally, the whole community agreed with the dredging of a new canal with roads along both sides of the canal. It is also agreed that the canal length can be 11–12 km long and there should be at least one to two bridges crossing the canal to allow local people to cross from one side to another.

For those who are directly affected by dredging of the new canal, as their lands are acquired, it is suggested that a reasonable compensation should be given to those people. Moreover, there should be a recovery program to help the people to develop their new profession and to promote job opportunities for the affected people as the first priority. For those people who are indirectly affected, they suggested to have a good control and protection measure of wastewater and water pollution due to urbanization and industrial development in the study area. Also it was suggested to utilize the new canal as water storage for agricultural water use and domestic consumption, especially in dry seasons.

8. Discussion and Conclusions

The research project is initiated by the Royal Irrigation Department of Thailand. Apart from technical investigations, the project study also considers initial assessment of environmental impact and social impact based on the people and community in the project area. The outcome of the study is the most suitable alternative of the flood control and drainage system for the study area.

Flood mitigation and management in the surrounding area of the airport is done by draining water from Klong Samrong canal into the proposed flood drainage channel and pumped out at the end of the channel to the Gulf of Thailand. The dikes along the southern bank of Klong

Prawate canal can block the inflow from the north into the study area. The existing dikes along the west and east boundaries of the study area can protect the water from the outer area to enter into the surrounding of the airport.

The accumulation of floodwater in the study area is caused by direct rainfall and the discharge of $12\,m^3/s$ pumped out from the airport. The proposed drainage channel canal of $100\,m^3/s$ capacity can reduce flood depth and inundation area. The study results are based on the design 100 year rainfall. The flood level in Klong Samrong canal can be reduced on average by 1.63 m. The inundation area is mostly in the south of the airport in which the flood depth of 0.3 m can be almost eliminated. Improvement of the upstream drainage canals north of Klong Samrong canal will effectively feed more discharges into the proposed drainage channel.

Most of the obtained depth–damage curves have logarithmic relationship or power equation relationship with approximately an average correlation coefficient of 0.7 to 0.8. From the analysis it is also found that the factors which affect the flood damage are flood depth and flood duration. The flood events considered in the questionnaire surveys have floodwater depth of more than 0.10 m and flood duration of more than 1 day on average.

We can conclude that the proposed drainage channel from Klong Samrong canal to the Gulf of Thailand with improvement of the upstream drainage channels and building of the southern dike along Klong Prawate can significantly improve the drainage capacity of the surrounding area of the airport. The embankment along the proposed drainage canal will improve traffic flow of Samut Prakarn province especially in the vicinity of the airport and in the south.

The flood damages, environmental, and social impacts are investigated and countermeasures are proposed to reduce the impacts. Compensation to affected people and recovery program has been established to minimize the impact due to the presence of the new airport.

Acknowledgments

Grateful acknowledgements are expressed to the Royal Irrigation Department of Thailand in allowing the author to use the data and information from the project study report in writing a part of this paper. Sincere thanks are due to the two co-investigators namely: Associate Professor Dr. Viraphol Taesombat of the Kasetsart University and

Associate Professor Chob Khameglud of the Thammasat University for their full effort in carrying out this study.

References

1. RID, Study on integrated flood mitigation in Chao Phraya River system, Report for Royal Irrigation Department, Thailand, 1998.
2. DHI Water and Environment, MIKE FLOOD: 1D-2D Modeling — User manual, DHI Software Technical University of Denmark, Denmark, 2003.
3. T. Tingsanchali and K. Keokhumcheng, Application of Mike flood model for flood drainage study of Eastern Bangkok Plain, *Proceedings of the International Conference on Modeling Tools for Environment and Resources Management*, AIT, Thailand, June, 2005, pp. 143–150.
4. T. Tingsanchali, V. Taesombat and C. Khameglud, Review and study of flood drainage project of surrounding areas of Suvarna Bhumi Airport, Feasibility Report for Royal Irrigation Department, September, Bangkok, Thailand, 2003.
5. D. I. Smith, T. L. Lustig and J. W. Handmer, Tangible urban flood damage, An Outline Manual, *Proceedings of the 2nd National Conference on Local Government Engineering*, Institute of Engineers, Canberra, Australia, 1983 pp. 376–381.
6. D. Dutta, S. Herath and K. Musiake, A mathematical model for flood loss estimation, *Journal of Hydrology*, **277**(1–2) 2003, pp. 24–49.
7. T. Tingsanchali, C. Khameglud and S. Prajamwong, Process in solving flood problems due to Second Bangkok International Airport and its impacts on way of life of people in a community, *International Conference on Urban Flood: Experiences for Sustainable Development*, Center for Urban and Community Development, Vietnam National University, Ho Chi Minh, 2006, pp. 39–50.

Advances in Geosciences
Vol. 11: Hydrological Science (2007)
Eds. Namsik Park *et al.*

INUNDATION ANALYSIS ON ROAD-ORIENTED MESH WITH DIGITAL CITY AND SEWERAGE MAPS

HIROMITSU BABA
Graduate School of Civil Engineering
Tohoku University, Aramaki Aoba06, Aoba-ward Sendai-shi
Miyagi-ken, 980-8579 Japan

AKIRA MANO
Group, Laboratory of Disaster Potential, Aramaki Aoba06
Aoba-ward Sendai-shi, Miyagi-ken, 980-8579 Japan
mano@civil.tohoku.ac.jp

In this study, for an area which does not put up city block data, calculation technique to calculate city block is developed by combining digital map 2500 with city region digital map in Sendai City. We can obtain from digital sewerage map the information and the technique which can connect the road network to the ground. Furthermore, the method to divide into pervious area and impervious area which has an effect on the landside inundation with laser profiler data is developed. Based on the above result, analysis model is applied to the rainfall of August 1986, and the result is compared with the questionnaire result.

1. Introduction

Recently, the inundation damages in the city have been increasing. Furthermore, accumulation of valuable properties such as electronic industry together with urbanization makes the area fragile against flood. Water retention and retarding function of urban land decrease with the increase in the impervious area. While heavy rainfall over 100 mm/h occurs frequently, flooding in urban areas causes large damage to various properties and urban function. Besides, street flooding can limit or completely hinder the functioning of traffic systems and has indirect consequences such as damage to buildings and economic activities. As these disasters have been kept off, the accurate forecasting model of the urban flood is very effective. In a densely built up area, there is a limit to only structural action such as development of river channel and detention basin in view of the present situation. Therefore, it is necessary to build up analysis of inundation method and others. So, we need an establishment of high-accurate analysis of inundation method.

Many researchers[1–3] have been working on the construction of analytical model from such a current state and improving the accuracy of the analysis model. For example, there are models such as analysis of one-dimensional flow assuming flow channel and two-dimensional flow with divided square grid in the study area of the present study. In addition, inundation studies using unstructured mesh models are becoming widespread. However, there are issues that mesh setup of unstructured mesh takes much trouble, and the model which can infiltrate underground has not been established yet. Thus, it is important to develop a model which enables reproduction of actual phenomenon by partitioning the road and city block, taking into account the pervious area.

So we have come up with a new method to overcome these difficulties by the recently set up digital databases and integration.

2. Study Area

The area is Haranomachi, located in the east drainage district or Sendai city, Miyagi prefecture Japan (Fig. 1). The area is about 11 km^2. The circulation

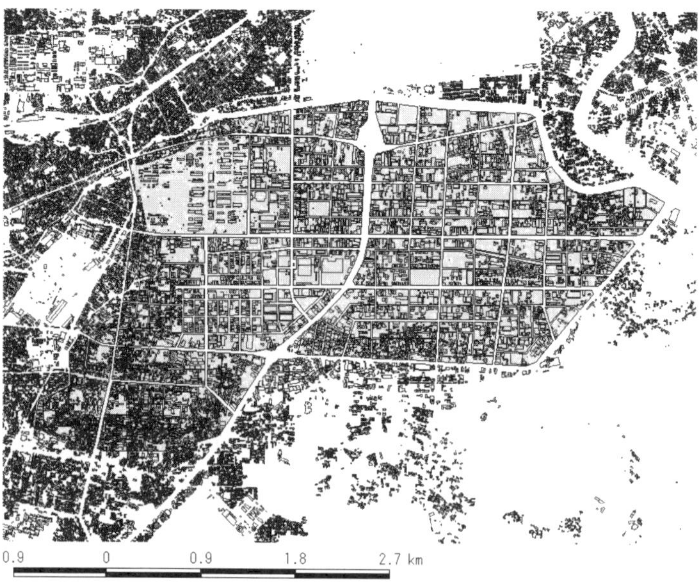

Fig. 1. Haramachi east part drain district.

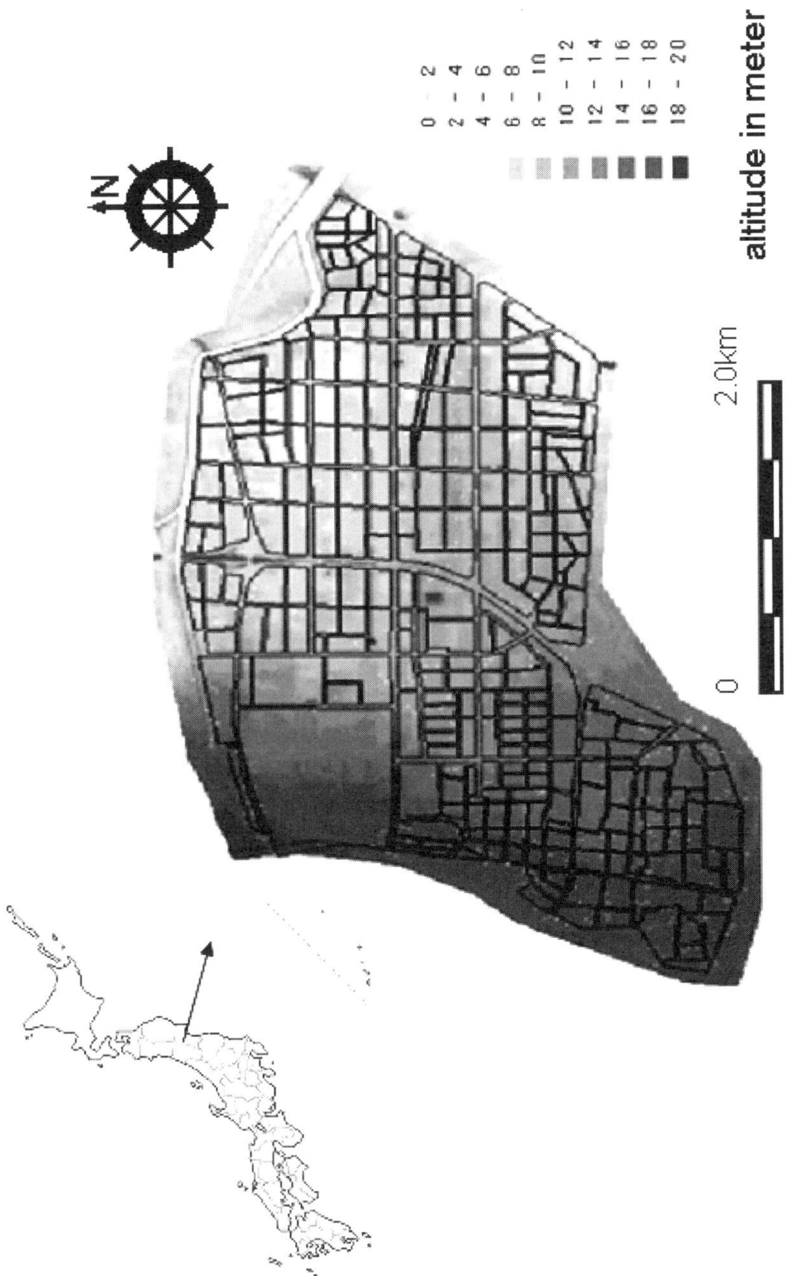

Fig. 2. Elevation of study area based on the laser profiler data with 2 m mesh.

and industrial area and the market such as of perishable food are assembled in this object area. But, there is a record that the area was flooded several times in the past years. So, disaster prevention action is very important in this area. You can see that the elevation decreases from the southwest to northeast. The ground elevation data was obtained by eliminating buildings and trees from the original profiler data using numerous filters (Fig. 2).

3. Analytical Model

The city regional digital map of Sendai City was originally made in 2003. The map is formed of GIS database, and has the data of a city block and the main building combined from the aerial photograph field survey information. In taking out city block data from this city regional digital map, and overlapping the GIS software with digital map 2500, city block polygon was related to the corresponding road element. Digital map 2500 is the map data made of basic data of GIS based on 1/2500, which was made by each municipal government. It was confirmed that the road part of both came in succession accurately, and the result of overlapping a city region digital map with the digital map 2500 is shown in Fig. 3.

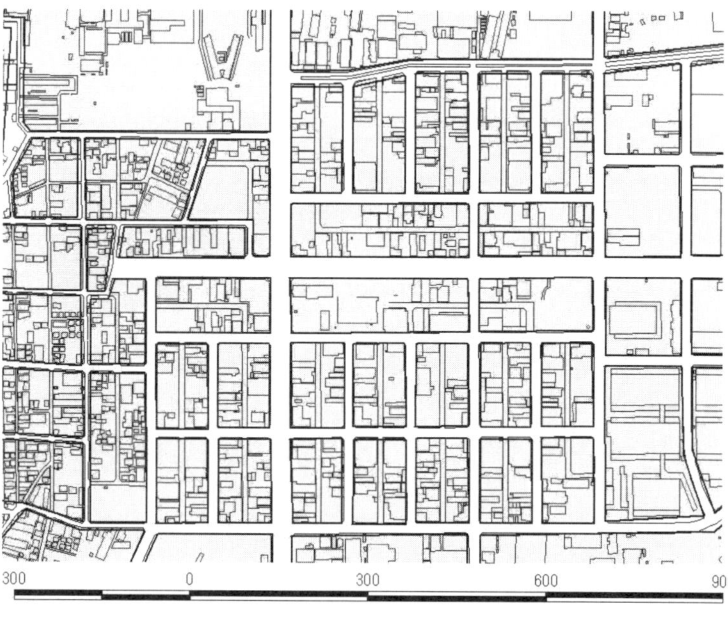

Fig. 3. Building from city regional digital map and roads from digital map 2500.

3.1. *Grid setup*

There is a variety of sizes and shapes for an actual city block. Therefore, it is difficult to expect improvement of precision even if a small city block is taken into calculation accurately, and therefore the number of calculations has to be increased. Therefore, a small city block is united and simplified. Big city blocks united into a heavy line is shown in Fig. 3. It is united by using the polygon making use of the GIS software, and is simplified. In this model, mesh in the object area consists of city block, road, and crossing mesh. Coordinates can be given to each city block by using GIS software. Moreover, the area is calculated to find the flowing quantity calculation. The number is allocated to all the polygons, line and point acquired from numeric maps 2500. We can recognize the intersection that exists at both ends of each road with the GIS tool on the basis of the number. Then, a line of a road nearest to the polygon is set to be adjacent to the polygon of the city block. Thus, all the related roads adjacent to the polygon are established to form the complete network.

3.2. *Surface runoff model*

In the surface runoff, the water level and the flow discharge of each time step of each grid are calculated from an equation of continuity and equation of motion. In the calculation of the city block, road and street, the coefficient of roughness was decided based on an earlier study (Table 1). Moreover, the flow discharge penetration was calculated from the building occupation rate:

$$\beta = \sqrt{1 - \lambda}, \tag{1}$$

where β shows the flow discharge penetration and λ shows the building occupation rate.

3.3. *Drainage outflow model*

The setting of sewer pipe network was made based on the Sendai City digital sewage map (6). The diameter of the considered pipe was only 2000 mm or

Table 1. Coefficient of roughness.

	City block	*Road*
Coefficient of roughness (n)	0.067	0.043

Fig. 4. Drainage network of 2000 mm or more.

more in this research (Fig. 4). Falling off flowing quantity to manhole from the ground level is calculated from Honma's overflowing formula, and if the depth of the manhole is less than the elevation of the ground level, it is assumed as complete overflow, and if the depth of the manhole is not less than the elevation of the ground level, it is assumed as incomplete overflow and diving overflow, and we calculate the reverse flow discharge to the ground level.

In the case of complete overflow ($h_1/h_2 < 0.45$),

$$q = 0.35h_1\sqrt{2gh_1}. \qquad (2)$$

In the case of imperfect overflow ($0.45 < h_1/h_2 < 0.8$) and the diving overflow ($0.8 < h_1/h_2$),

$$q = 0.91h_2\sqrt{2g(h_1 - h_2)}, \qquad (3)$$

where q is the unit width flow discharge, h_1 is the overflow depth, and h_2 is the downstream depth.

3.4. *Consideration of infiltration capacity characteristic*

This model divides every city block into pervious area and impervious area, and applies the effective rainfall model which is considered infiltration capacity characteristic. Finally, the volume of runoff from one city block is assumed by bringing together the volume of runoff of the pervious area and

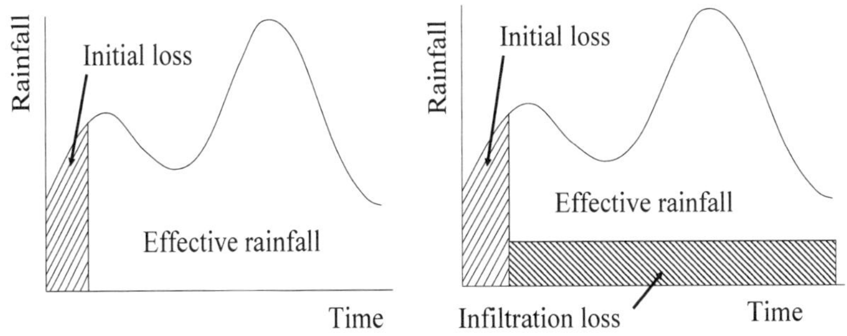

Fig. 5. Schematic diagram of effective rainfall.

the impervious area. In the calculation of an effective rainfall, the model (Fig. 5) proposed by Ando *et al.*[4] is considered. This model considered the initial loss and ultimate infiltration capacity infiltration according to the land use.[5]

3.4.1. *Impervious area*

Abstraction is only set as initial loss L_{imp} as initial water absorption on the surface of asphalt, and the asphalt side and boghole storage in impervious area. After the rain water fills the initial loss L_{imp} as shown in Fig. 5, the whole becomes an effective rainfall. Time series of rainfall $R(t)$ and the time series of an effective rainfall in impervious area can be denoted R_{eimp}.

$$\mathrm{R}_{eimp}(t) = \begin{cases} 0 & \left(\sum R(t) \leq L_{\text{imp}}\right) \quad (4) \\ R(t) \times \dfrac{60}{DT} & \left(\sum R(t) \geq L_{\text{imp}}\right). \quad (5) \end{cases}$$

$R(t)$ is the time series of rainfall (mm) and R_{eimp} is the time series of effective rainfall (mm/h). L is initial loss (mm) and DT is rainfall observation interval (min). Initial loss L_{imp} in impervious area is set as 2 mm in an earlier study.

3.4.2. *Pervious area*

About the effective rainfall in the pervious area, abstraction is considered the initial loss and ultimate infiltration capacity for every land use. Abstraction in pervious area is initial loss $L_p^{(k)}$ and ultimate infiltration capacity $I_c^{(k)}$, for every land use k.

To be more precise, the initial loss $L_p^{(k)}$ corresponds to the sum of the infiltration of initial rainfall before infiltration capacity reaches ultimate infiltration capacity $I_c^{(k)}$ and boghole loss, as shown by Fig. 5. So, the rainfall which goes over ultimate infiltration capacity $I_c^{(k)}$ after the initial loss ends, becomes an effective rainfall. Therefore, time series of effective rainfall $R_{ep}^{(k)}$ depending on land use k is given by the following expression:

$$\sum R(t) \leq L_p^{(k)}, \quad \mathrm{R}_{ep}^{(k)}(t) = 0$$
$$\sum R(t) \geq L_p^{(k)},$$

$$\mathrm{R}_{ep}^{(k)}(t) = \begin{cases} 0 & \left(R(t) \leq I_c^{(k)}\right) \quad (6) \\ R(t) \times \dfrac{60}{DT} - I_c^{(k)} & \left(R(t) \geq I_c^{(k)}\right). \quad (7) \end{cases}$$

k is number of land use division and I_c is the ultimate infiltration capacity (mm/h).

3.4.3. *Calculation of impervious area*

In this research, it was assumed that impervious area is detected and calculated in the building by finding the remainder of the elevation distribution data before and after the numerous filter processing of the laser profiler data. Here, the place with 2 m or more difference of elevation was judged in the building. Figure 6 shows the ratio of the area of the building distribution of each city block in unit %. The residential area extends to the southwest part in the drain district and we can find the distribution of a comparatively high rate of the building area.

4. Questionnaire Survey

The damage caused by this heavy rainfall was very large. But the survey data about inundation condition had not been left. So, in advanced calculation, we carried out the questionnaire survey to examine the inundation depth in October 2006. The questionnaire survey set the investigation area with a focus on the flood damage area, and the object buildings were selected at random from the survey area by mail collection and visit survey. The number of questionnaire distributions was 84 votes, the collection number was 39 votes, and the recovery percentage was 46%. Inundation distribution is shown in Fig. 7. Nonshaded blocks means no data.

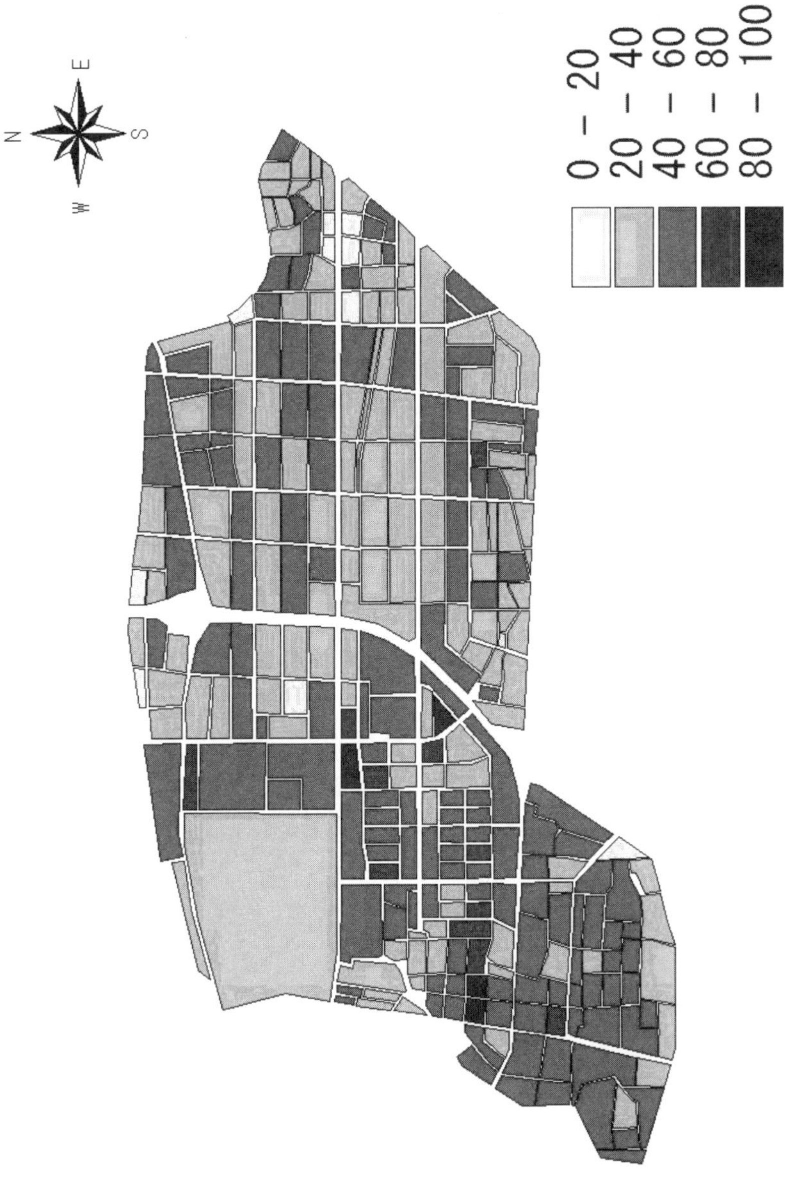

Fig. 6. Impervious area ratio in percent.

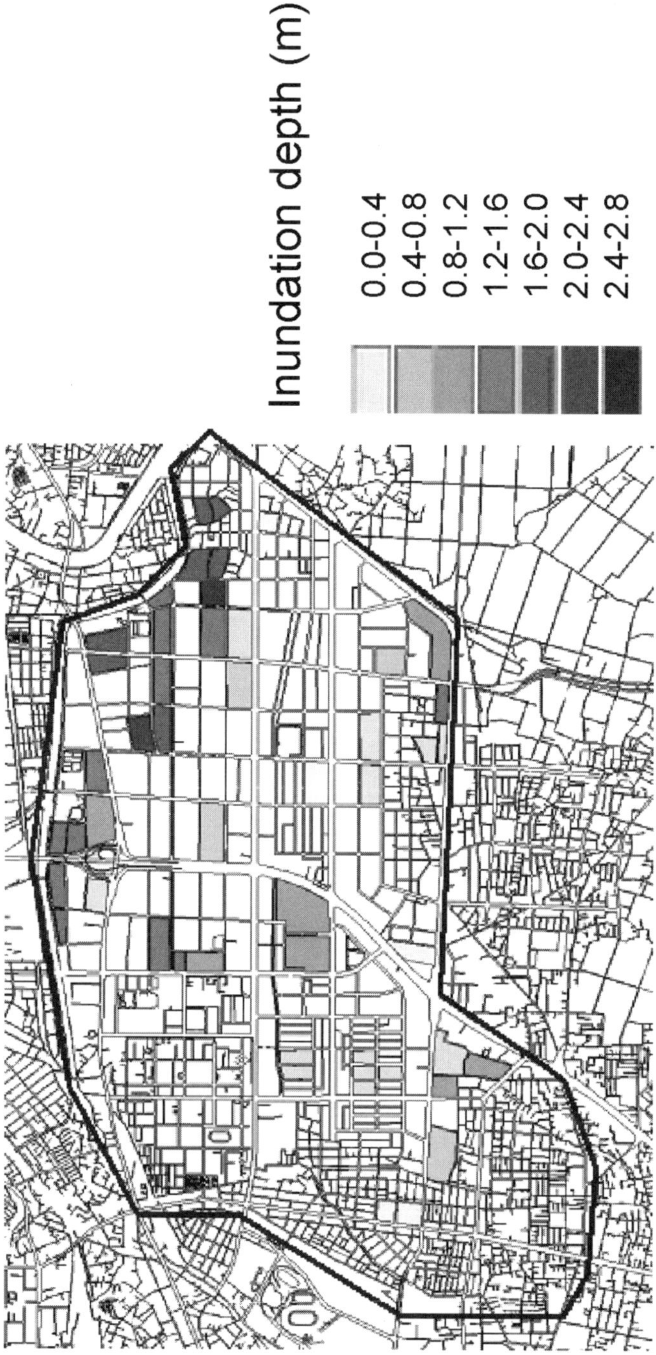

Fig. 7. Questionnaire survey result.

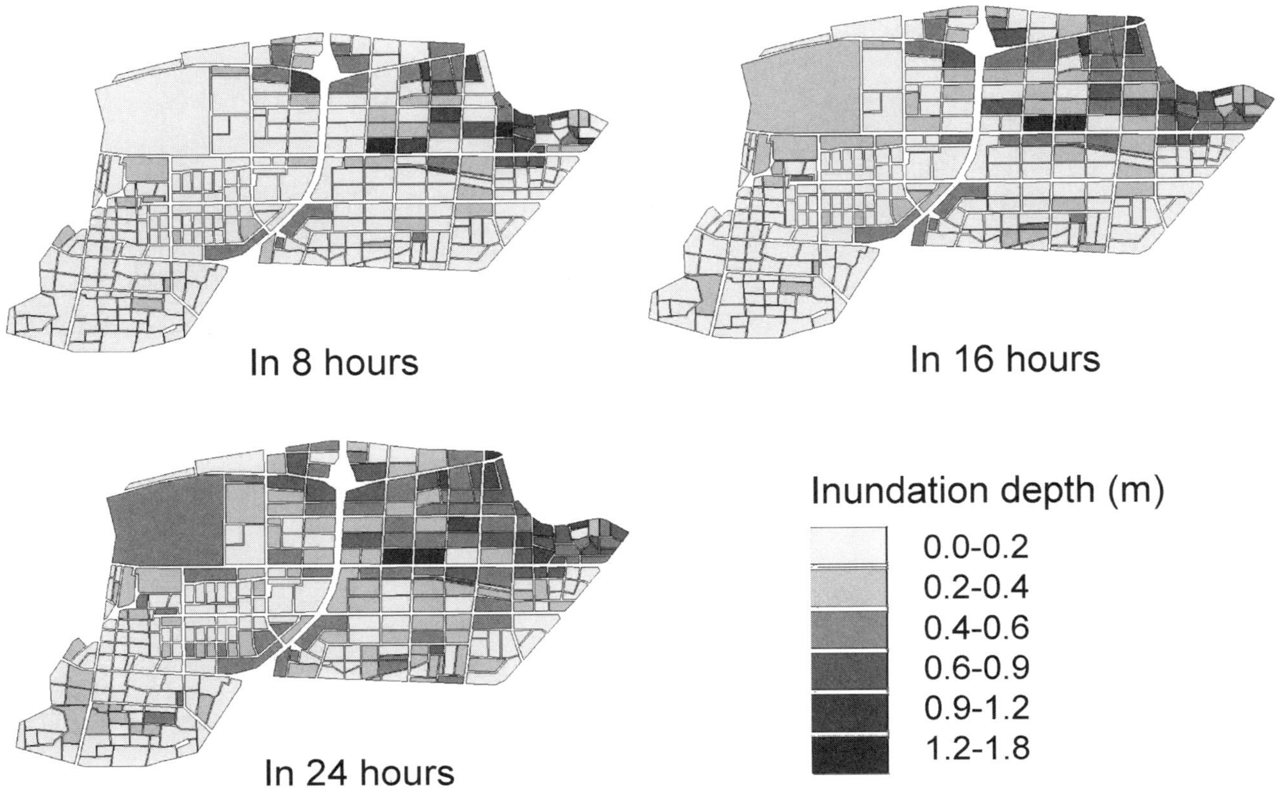

Fig. 8. Calculation result which did not consider the infiltration.

Fig. 9. Calculation result which considered the infiltration.

5. Analysis Result and Discussion

The calculation results are shown in Figs. 8 and 9. These figures show the inundation distribution from the start of calculation up to 24 hs. Figure 8 corresponds to the calculation results that considered the infiltration and Fig. 9 considers those without infiltration. From Fig. 8, it is seen that when infiltration capacity was not considered, inundation depth is large and inundation is widespread by distribution. On the other hand, when the infiltration capacity is considered, the deep inundation appears on the northeastern lowland area and inundation depth was about 1.7 m. Although the inundated part is locally seen also in other areas, the inundation depth and the spread are not large. From here onward, initial loss is an important parameter to understand the abstraction phenomenon, and in principle, it has a much greater impact on the outflow of the initial rainfall. But, the questionnaire result may lack credibility, because it was the event over 20 years ago. Therefore, we cannot assess its value quantitatively, but it is thought that the calculation result which considers the infiltration can reproduce the result of the survey well on intensive northeastern inundation and water depth.

6. Conclusion

We have developed a way to construct the unstructured mesh effectively on GIS from the digital city map, and digital sewage map, and the method for accounting impervious area from laser profiler data. As the result of these calculations, the model calculations well reproduced inundation water depth close to the questionnaire results. In an earlier study, when water circulation is considered, the model which calculates the amount of surface retention, infiltration, and direct runoff is established, though it is very few in number about urban runoff analysis. In the future, using examples from it, we are going to develop the runoff model close to actual phenomenon considering the tank model which has the infiltration to the underground and more in-depth data of sewage system.

Acknowledgment

This work was partially supported by the 2005, 2006 Grants by the Foundation of River and Watershed Environmental Management.

References

1. I. Otsubo, K. Sakurai, M. Takeda and N. Matsuo, *Ann. J. Hydraul. Eng. JSCE* **45** (2001) 887 (in Japanese).
2. K. Toda, K. Inoue, S. Murase, Y. Ichikawa and H. Yokoo, *Ann. J. Hydraul. Eng. JSCE* **44** (2000) 479 (in Japanese).
3. S. Satou, *Tohoku J. Natural Disaster Sci.* **41** (2005) 105.
4. Y. Andou, Y. Takahashi, K. Izumi and K. Knao. *Ann. J. Hydraul. Eng.* **29** (1985) 49.
5. K. Fujimura and Y. Ando, *J. Jpn. Soc. Hydrol. Water Res.* **11**(4) (1998) 360.

Advances in Geosciences
Vol. 11: Hydrological Science (2007)
Eds. Namsik Park *et al.*

INTEGRATED WATER RESOURCES MANAGEMENT (IWRM) — NEEDS AND TRAINING FOR THE OCEANIA REGION

JAMES P. TERRY
Department of Geography, National University of Singapore
1 Arts Link, Singapore 117570
geojpt@nus.edu.sg

KIFLE KAHSAI
School of Biological, Chemical and Environmental Sciences
The University of the South Pacific, Private Mail Bag, Suva, Fiji

V. GROVER and C. MAYFIELD
United Nations University — International Network on Water
Environment and Health (UNU-INWEH)
175 Longwood Road South, Suite 204
Hamilton, Ontario L8P OA1, Canada

Limited water resources on many islands in the Pacific are highly vulnerable to overexploitation and pollution, and are affected by climatic change and variability. Island physical environments, together with isolation and remoteness, and a range of socio-cultural traditions, all combine to present many challenges to good governance for sustainable water management. In response, this paper outlines the need for an Integrated Water Resources Management (IWRM) training program in the South Pacific Islands and the establishment of a Water Virtual Learning Centre as a regional focus for advanced training in island water resource management issues. An overview of the 10-unit IWRM curriculum is presented, and challenges discussed in terms of delivering the training through flexible and multi-modal teaching to water/environment professionals across the vast Oceania region.

1. Introduction and Background

The global community recognized the importance of sustainable water management at the 2002 World Summit on Sustainable Development, complementing the existing Millennium Development Goal to halve by 2015 the proportion of people without dependable access to safe drinking water. The "Water in Small Island Countries" sessions during the 3rd World Water

Forum in Kyoto, Japan in 2003, provided the South Pacific island nations an opportunity to put forward their priority water issues. Three priority concerns identified were:

- Uniquely fragile water resources due to lack of natural storage, small size of island water resources, competing land use, inadequate water resources management during climatic extremes, and vulnerability to anthropogenic hazards such as urban pollution.
- Water service providers are constrained by lack of human and financial resource bases, restricting the availability of experienced staff and investment, leading to poor cost-recovery.
- Highly complex water governance due to the disconnectedness between traditional communities and national administration practices.

The Oceania region is characterized by widely scattered countries composed of numerous islands that vary considerably in their size, types, and hydrologic characteristics. There are large, high volcanic islands such as those in Fiji and Solomon Islands, raised limestone islands like Niue, Nauru, and Tongatapu, and tiny, low-lying coral atolls of Kiribati, Tuvalu, and the Marshall Islands.

Small or low-lying islands have very limited water resources because of the small size of stream catchments or underground aquifers. Figure 1 illustrates the special problems of water supply on an atoll. On Tarawa, the main atoll of Kiribati, the small island of Betio in the south west of the atoll is the main administrative center and has an extremely high population density — approximately 15,000 people/km^2. Betio's groundwater aquifer is unable to meet water demands, so water must be piped from larger aquifers on Bonriki and Buota islands, which lie more than 30 km away from the south east of the atoll.

Limestone islands, like coral islands on atolls, also have no natural fresh surface water systems at all, and are completely dependent upon rainwater harvesting and groundwater. Figure 2 shows the shape and thickness of the groundwater lens on Niue, which is a raised limestone island, 259 km^2 in area. Although the population on Niue is small, approximately 1,800, and the freshwater resources are normally sufficient to meet demands, the porous nature of the bedrock means that the natural aquifer must be carefully protected against pollution by human or agricultural effluent.

Contrary to popular belief, even high volcanic islands with surface drainage (rivers and streams) are vulnerable to prolonged drought. Figure 3 illustrates the point, showing the behavior through time of river discharge (flow) in the major Ba River in northern Viti Levu island in Fiji. This

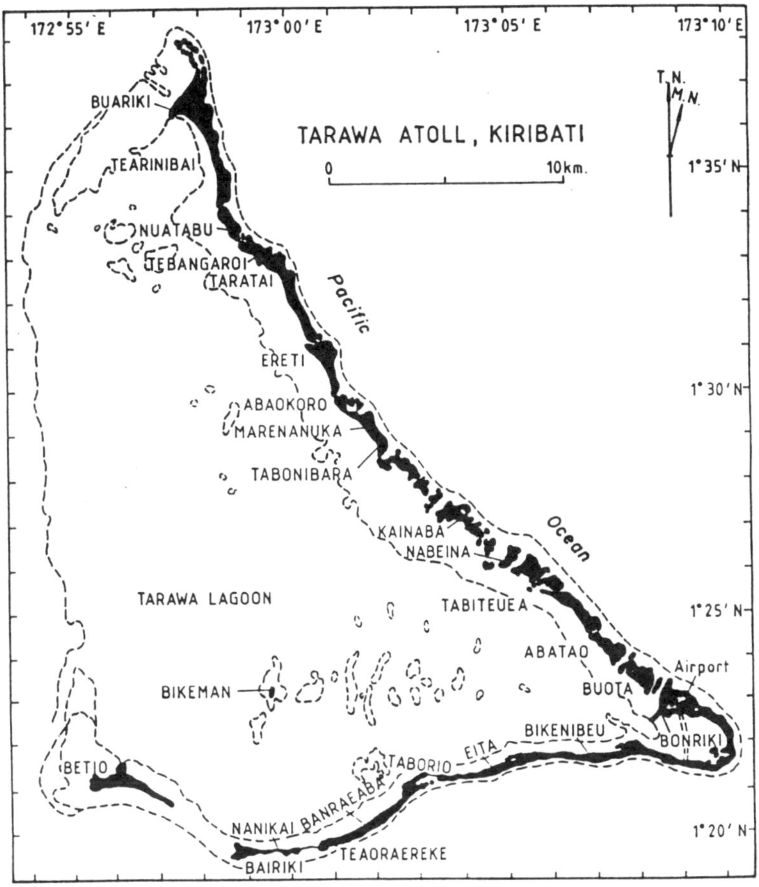

Fig. 1. Tarawa Atoll in Kiribati. Betio in the south west of the atoll is the most populous island. Water must be supplied from Bonriki and Buota islands (labeled), over 30 km distant. From White *et al.*[1]

river is typical of most watercourses on volcanic Pacific islands in the sense that it is used for domestic water supply by many villages located along the banks of its main channel and tributaries. The river is also utilized by the Fiji Public Works Department as the source of urban water supply in Ba Town located on the river delta. Periods of rain failure and associated minimum flow in the Ba River therefore affect a large population living within the river basin.

The hydrograph in Fig. 3, plotted over a 20-year period from 1980 to 2000, is drawn above a graph of fluctuations in the Southern Oscillation Index (SOI) for the same period. Prolonged low values of the SOI indicate

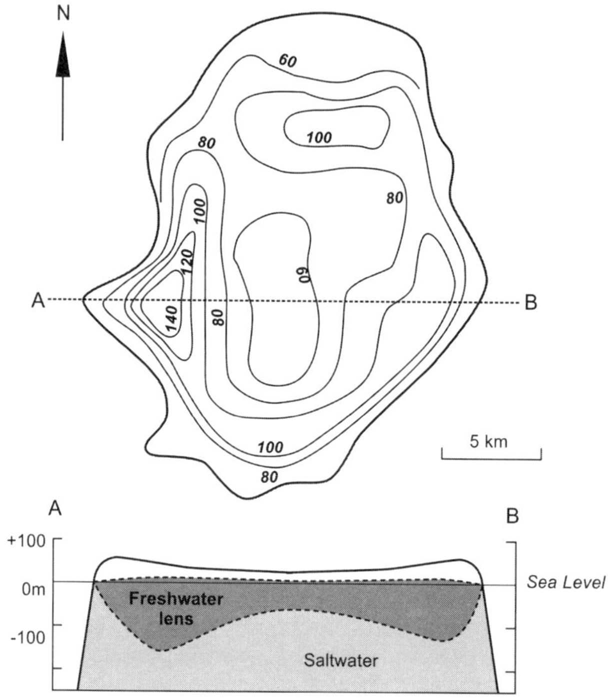

Fig. 2. Plan view and cross-section of Niue island's freshwater lens. Niue is a single island nation with a population of 1,800. The "contours" in meters show the thickness of the lens within the limestone bedrock. Adapted from Jacobson and Hill[2] and Nunn.[3] Note that the vertical scale exaggerates the apparent thickness of the lens.[4]

El Niño episodes. The patterns in the two graphs demonstrate how times of very low average water flow in the Ba River correspond with strong El Niño phases (e.g. 1983, 1987, 1998) because these normally cause significant reductions in precipitation across extended areas of the south west Pacific for periods of a year or more.

Projections of future climate change in the Pacific also mean that surface and groundwater supplies could be affected by variations in rainfall, higher surface temperatures (leading to an increase in evapotranspiration), and sea level rise.[6] At the same time, there are competing activities in watersheds. Water is needed for a wide range of sectors: urban and rural water supplies, commercial forestry, tourist developments, and subsistence agriculture. These all combine to present many challenges to good water governance at different scales.

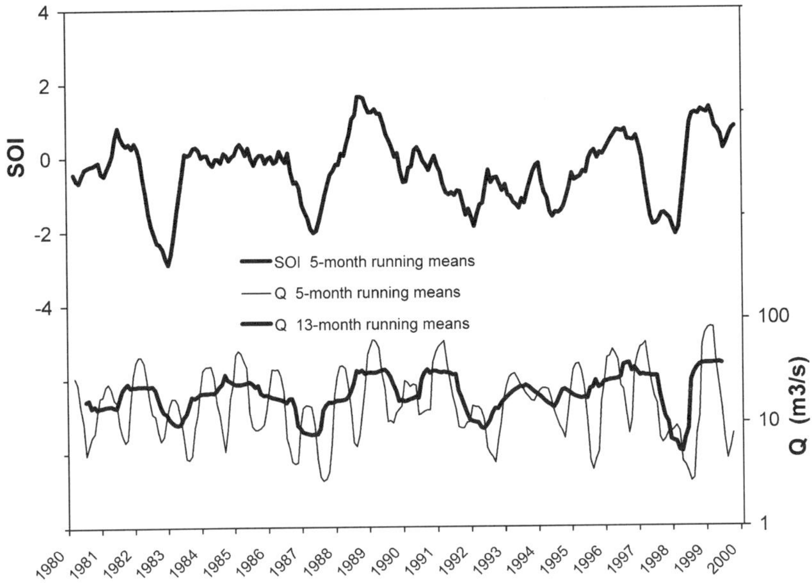

Fig. 3. The behavior of river discharge (Q) in the Ba River on Viti Levu island in Fiji, illustrating how surface water resource availability is strongly influenced by El Niño events, measured by episodes of low SOI values (Southern Oscillation Index). From Terry.[5]

2. Building Institutional Capacity — The Water Virtual Learning Centre

For the above-mentioned reasons, sustainable water resources management is not being fully achieved in many Pacific Islands. In light of this, the South Pacific countries are now committed to the long-term goal of enhancing national capacities that will promote the development and implementation of sustainable water management at local, regional, sub-regional and basin levels, and contribute to strategic planning in the Water for Life Decade (2006–2015). One of the agreed main activities is to improve water governance using Integrated Water Resources Management (IWRM).

The first step in this process has been the establishment of a regional training center for the South Pacific by the United Nations University International Network on Water, Environment and Health (UNU-INWEH). This was achieved in partnership with the University of the South Pacific (USP) which is the main provider of tertiary education to 12 island nations (Cook Islands, Fiji, Kiribati, Marshall Islands, Nauru, Niue, Samoa, Solomon Islands, Tokelau, Tonga, Tuvalu, and Vanuatu). For this project, the USP set up the "Water Virtual Learning Centre" (WVLC) in 2005 on the university campus in Suva, Fiji. The objective of the WVLC is to provide professional training in IWRM to assist managers, scientists, and policy-makers, to improve sustainable water resource management practices in the Pacific Islands. Two other WVLCs have also been established with UNU-INWEH support at the Asian Institute of Technology in Bangkok, Thailand, and at the University of Ghana in West Africa where similar IWRM training is being trialed.

The second step is delivery of a core curriculum on IWRM, that offers opportunities for continuing education and upgrading skills. The WVLC is flexible in terms of delivery mechanisms and procedures, so as to accommodate the widely varying schedules, work requirements, technology access, and backgrounds of the participants.

3. "Customizing" IWRM Training for Tropical Islands

Although IWRM is a relatively new "brand" in the Pacific islands, the concepts and approaches it embodies are not new. Figure 4 shows that there is a need to take a holistic integrated approach to water

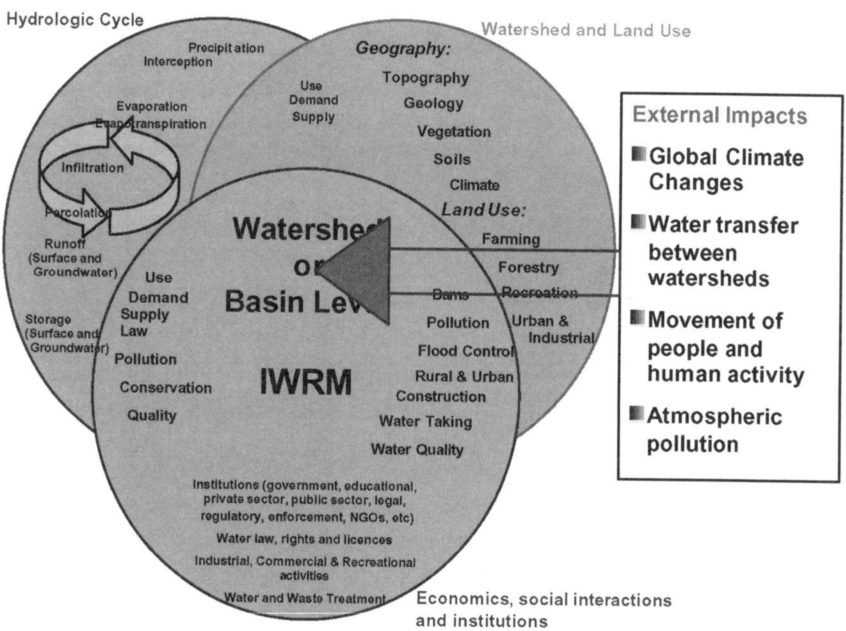

Fig. 4. Elements of Integrated Water Resources Management.

resources management, to ensure that socio-cultural, technical, economic, and environmental factors in the development and management of water resources are considered, and that traditional practices in the Pacific islands are acknowledged and utilized. For example, the problem of competing land pressures — choosing whether to use limited available land for a plantation, a water reserve, a school or recreation area, must be appreciated at the household, village and community levels. In particular, every community must understand that activities on the land have an impact on freshwater storage.

A curriculum that provides the core knowledge-base for IWRM was developed by UNU-INWEH in Canada, to be taught through the three WVLCs in Fiji, Thailand, and Ghana. The framework of IWRM curriculum, prepared by the Advisory Committee, consists of 10 courses, totaling 250 learning hours, directed at new and practising professionals in the water sector. The curriculum commenced for the first time in late 2007, and will take approximately 18 months to complete. United Nations University certificates will be given for each successfully completed course, and a Diploma in IWRM from the UNU will be awarded for the satisfactory completion of the entire program. The diploma also serves as a foundation to other degrees, such as Masters level in Environmental Studies or Science. The 10 courses comprising the IWRM program have been specifically designed to be easily modified and updated, and use non-proprietary standards for formatting and presenting materials. The contents of each course relate back to the concepts of IWRM, and are as follows:

1. *An Introduction to Integrated Water Resources Management*: A broad examination of critical concepts and knowledge needs related to IWRM, including essential human and institutional capacities.
2. *Water Transfer*: The natural components and processes of the hydrologic cycle, temporal variance, global and regional processes, and impacts of anthropogenic alterations on the water cycle.
3. *The Terrestrial Ecosystem and The Impacts of Land Use Changes*: Changes and impacts that occur as a result of land use change and development.
4. *The Aquatic Ecosystem*: Concepts, processes, analytical approaches and procedures that relate to our understanding of the structure and function of the aquatic ecosystem.
5. *Aquatic Ecosystem Health and Impact Assessment*: Aspects of anthropogenic impact and change to the physical, chemical, biological, and ecological components of the aquatic environment.

6. *Water Use*: The many human uses of water, both consumptive and non-consumptive, and their implications for the hydrologic cycle, ambient water quality, and both ecosystem and human health.
7. *Wastewater*: The discharge of wastewater to natural and artificial catchments, potential impacts, methods of treatment and mitigation, the urban water cycle and methods to evaluate and choose appropriate technologies.
8. *Governance and Community Based Approaches*: Legislative, regulatory, legal, jurisdictional, community, and individual responsibilities of IWRM. Gender issues, the requirements and benefits of capacity building, public education and community involvement.
9. *Organizational Infrastructure and Management*: Building and maintaining water management infrastructure, various public and private-sector organizational models. Auditing, reporting, and long-range planning.
10. *Applying Integrated Water Resources Management*: Customized case studies, practical illustrations of the concepts and procedures of IWRM, and investigative techniques for students to assess their own IWRM needs.

A special feature is that according to the requirements of the South Pacific nations, the courses are "customized" by introducing significant local and regional materials and data. Essentially, customization means using appropriate island case studies and examples, based on research carried out in the region, to make the core program relevant to the special needs of the Pacific. The academic staff teaching courses in the curriculum are responsible for customizing the courses in their area of expertise, for example hydrology, development economics, water management and governance, after which courses are delivered to participants through the WVLC at USP in Fiji. For example, in Course #2 focusing on water transfer, available case studies and existing research materials on atoll lens recharge processes, factors influencing hillslope hydrology and stream flow behavior on small tropical islands were provided as illustration. Student understanding of these processes also formed part of the course assessment.

Another important element of the program is Course #10, which is a field-based module in Fiji. The aim of this module is to reinforce the basic principles already learnt through two to three weeks of fieldwork on IWRM practice. The objective is to bring all participants together at the WVLC, in order to visit a range of selected sites that are relevant as case studies, for example, in relation to water resources issues, drainage basin management, water harvesting, treatment, and storage. In this way, IWRM is applied in a "hands-on" way to find solutions to real water-based problems and to evaluate a range of resource development options.

4. IWRM Training by Distance and Flexible Learning

The vast geographical distances across Oceania make it difficult for all IWRM students to come to study at one of the three major campuses of the University of the South Pacific in Fiji, Vanuatu, and Samoa. Fortunately, USP has been a pioneer in "Distance and Flexible Learning" (DFL) since the 1970s. The IWRM course will therefore be offered by DFL, by providing multi-modal course delivery to participants across the South Pacific islands, who would otherwise not be able to study face-to-face. DFL is ideal for students who wish to study independently, live in remote areas of the South Pacific, or have heavy family or professional responsibilities which prevent them from studying on-campus.[7]

To support multi-modal course delivery, DFL students are provided with printed materials and CDs. In addition, students have access to online learning at their regional USP centers through "USPNet." This is a state-of-the-art satellite communications network, owned and operated by USP (Fig. 5). USPNet provides the opportunity to participate in audio tutorials, communicate by email with a lecturer or another student, access the Internet, watch live video broadcasts of classes, and take part in video conferences (and tutoring) with other course participants. Video conferencing allows the presentation of PowerPoint slides, the use of overhead camera for transparencies and other visual material, and video playback. Via USPNet, IWRM course participants are, therefore, able to interact effectively with other students, lecturers or support staff, no matter where the students are located.

The early indications are that 17 students who registered for the first pilot program are performing well. Sixty-five percent of the students from six island countries were able to complete the first four courses successfully. However, 24% never participated fully from the start of the program due to personal reasons and work commitments. The remainder have shown mixed results. The main challenge for most of the Pacific regional students (i.e. those outside Fiji) is the lack of fast Internet connectivity on their home islands. The best solution to this problem, at least until Internet service providers are able to improve broadband access, is to provide remote

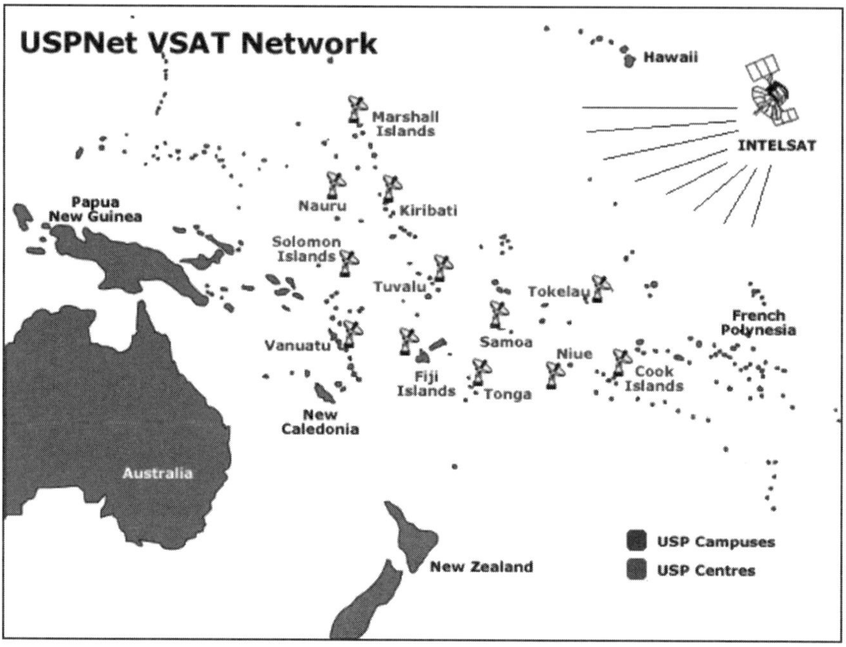

Fig. 5. The "USPNet" Regional Telecommunications Network — supporting Distance and Flexible Learning through multi-modal education delivery.

participants with as much information (reading materials, illustrative data, etc.) as possible on CD or in hard copy, so that their reliance on the Internet is reduced. The principal challenges for the IWRM program coordinators are the limited availability of relevant resource materials on island water issues, owing mainly to the small amount of scientific research being carried out in this area in most Pacific island states, and also overcoming the limited experience of some students with regard to water issues on islands that are physically different from where they live, for instance teaching stream hydrology to students from atolls where there are no surface watercourses. Course #10 was specifically designed to address these issues.

5. Conclusions

Across the developing island nations of the tropical South Pacific, there is a wide diversity of water resources characteristics. The region has scattered countries composed of numerous islands that vary considerably in their size, physical geography, and hydrology, including high volcanic islands, emerged limestone islands, and coral atolls. Limited water resources are under threat from overexploitation by growing human populations and are vulnerable to climate change. Island environments, remoteness, and socio-cultural traditions all combine to present many challenges to good governance for sustainable water management.

In this context, IWRM is a relatively new concept in the region, but has been recognized by the University of the South Pacific, which established the Water Virtual Learning Centre in 2005 for advanced training in IWRM, through partnership with the United Nations University–International Network on Water, Environment and Health. A curriculum in IWRM is offered for the first time in 2007 through distance and flexible learning modes, using an advanced satellite communications network for audio tutorials, interactive video broadcast, and access to online learning through the Internet. The training ensures that socio-cultural, economic, and environmental features of different islands are properly considered in the management of water resources. There is a special focus on course customization for regional needs, and to acknowledge traditional practices carried out in the islands, so that these are included where appropriate in IWRM training. The IWRM curriculum can enable the South Pacific nations to enhance the implementation of sustainable water strategies at local, basin, island, and regional scales. This will assist overall in the sustainable development of Pacific Island communities.

References

1. I. White, A. Falkland and D. Scott, Droughts in small Coral Islands: Case Study, South Tarawa, Kiribati, UNESCO IHP-V, Technical Documents in Hydrology, No. 26, UNESCO Paris (1999), 55 pp.
2. G. Jacobson and P. J. Hill, Hydrogeology of a raised coral atoll — Niue Island, South Pacific Ocean, *BMR J. Australian Geol. Geophys.* **5** (1980) 271–278.
3. P. D. Nunn, *Oceanic Islands* (Blackwell, Oxford, UK, 1994), 413 pp.
4. L. Mosley and C. Carpenter, Niue coastal water quality and groundwater resources assessment, SOPAC Technical Report 372, Suva (2005), 35 pp.
5. J. Terry, Hazard warning! Hydrological responses in the Fiji Islands to climate variability and severe meteorological events, in *Regional Hydrological Impacts of Climatic Change—Hydroclimatic Variability*, eds. S. Franks, T. Wagener, E. Bøgh, H. V. Gupta, L. Bastidas, C. Nobre and C. de Oliveira Galvão, International Association of Hydrological Sciences (2005), Publication No. 296, pp. 33–41.
6. S. Bettencourt, J. Campbell, N. de Wet, A. Falkland, J. Feresi, R. Jones, P. Kench, G. Kenny, W. King, P. Lehodey, L. Limalevu, R. Raucher, T. Taeuea,

J. Terry, N. Teutabo and R. Warrick, The impacts of climate change in Pacific Island economies: policy and development implications, *Asia Pacific J. Environ. Develop.* **9** (2002) 142–165.
7. Distance and Flexible Learning, The University of the South Pacific website, accessed from <www.usp.ac.fj/> (2008).

Advances in Geosciences
Vol. 11: Hydrological Science (2007)
Eds. Namsik Park *et al.*

THE INFLUENCE ON THE WATER USAGE PLANNING IN THE NEW THEORETICAL AGRICULTURAL SYSTEM UNDER THE STOCHASTIC RAINFALL

SUWATANA CHITTALADAKORN* and EAKAWIT JORNPRADIT†
Department of Water Resources Engineering, Faculty of Engineering
Kasetsart University, 50, Phaholyothin Rd., Lardyao, Chatuchak
Bangkok 10903, Thailand
**j.suwatana@gmail.com*
†eakawit36@gmail.com

A major input of precipitation occurring to become water resources for agricultural systems is the daily rainfall. The historical data record in the hydrology and climatology integrated reflected the underlying climate, the stochastic impact of climate variability on water resources, and land-water usage or storage planning in cultivation systems. The initiative policy of the New Theoretical Agriculture System has been given beneath the aim of sufficiency philosophy by His Majesty King Bhumibol Adulyadej affected under stochastic rainfall as well. With the basic criteria of management for land division in the New Theoretical Agriculture System, the land is divided into pond area:paddy field:crops area:living area by percentage of 30:30:30:10. For assessing water and land management all regional parts of Thailand, displaying the variability in quantity, time and space, the spatial dependence between the weather inputs at different regions has to be accommodated. In this paper, it is proposed to study the influence over water resource usage under the stochastic of daily rainfall affecting land portions divided into the New Theoretical Agriculture System by the differential evolution algorithm; a numerical programming for searching solution applied in this study. A mathematical model is designed for water balance calculation between water input and water consumption in the system. The calculation consists of the excess daily rainfall, effective daily rainfall, and daily consumption for both household and cultivation. For analysis, specific cropping pattern was determined corresponding to the cultivation behavior in each regional area. The 10% of the 1.6 and 2.4 hectares farm sizes was determined for living area. For the rest (90%), it would be divided into three portions: an area for a pond of 4.0 m in depth, an area for paddy field, and an area for cultivating other crops both in wet and in dry seasons. The decision variables for optimization were defined by the relationship of those area portions. The studied results found that the appropriate portions of divided areas had relationship with the average annual rainfall amount (Rf_{av}) of each regional area as follows:

The portion in % for pond area is given as

$$A(pond) = -31.38\ln(Rf_{av}) + 257.33$$

The portion in % for paddy field in wet season is given as

$$A(paddy_{wet}) = 31.338\ln(Rf_{av}) - 187.02$$

The portion in % for paddy field in dry season is given as

$$A(paddy_{dry}) = 51.319\ln(Rf_{av}) - 344.80$$

The portion in % for abandon area in dry season is given as

$$A(abandon) = 225.3 - 28.01\ln(Rf_{av})$$

1. Introduction

The kingdom of Thailand has approximately 20.8 million hectares of agricultural areas in which 4.8 million hectares are of irrigated areas. Non-irrigated areas are counted to be 76.92% of agricultural land holding in the country. Most of the agricultural areas are facing droughts affecting both domestic usage and cultivation. Drought is often faced by farmers who constitute 70% of the population in the country. His Majesty King Bhumibol Adulyadej has given the initiative policy to farmers for ultimate profitable usage of land and water resources of small farmland known as the New Theory Integrated Farming, for small farmland size of 1.6–2.4 hectares which is the average size range of the farmer's land holding. The objectives to approach sufficiency philosophy are

- self-reliance,
- self-sufficiency, and
- sustainable agriculture.

In the first precept, the land portion is divided into four parts consisting of pond area; paddy field; upland and vegetable crops area; and living area by percentage of 30:30:30:10 as shown in Fig. 1:

- the first part approximately 30% for pond;
- the second part approximately 30% for paddy field;
- the third part approximately 30% for upland crops, orchard, vegetable, etc.; and

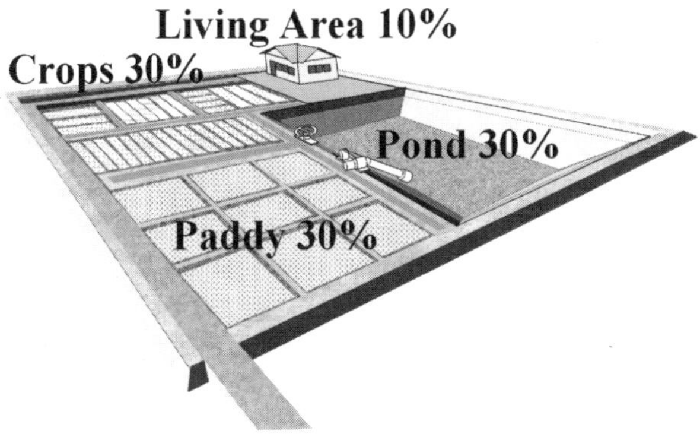

Fig. 1. Land portions are divided into four parts of the New Theory Integrated Farming System.

— the fourth part approximately 10% for living area, road, infrastructure, etc.

In this portion, the farmers can adjust appropriately depending on the rainfall quantities in any local area.[1]

2. Objectives and Boundary

The objectives of this study are the planning of land usage appropriated with water resources. Proportional setting of pond portion, agriculture portion, and living portion in the New Theory Integrated Farming System is the research matter suitable for five local regions in Thailand consisting of the northern, northeastern, eastern, central, and southern parts. The case study representatively is defined in each part of the regions by comparing daily rainfall data with the closest to average annual rainfall, which are ChiangMai, KhonKhen, ChonBuri, Bangkok, and Nakorn-Si-Thammaraj province, respectively. The area of study covers especially for land size of 1.6–2.4 hectares. The study is based on the balancing between the water gained daily and used only, but it did not include other sources of water budget. The water usage is analyzed by using cropping patterns defined in each region and calculates the water consumption used by weighting crop coefficient (Kc) from average actual ratio of 28 kinds of crops[3] such as (1) Kor-Khor rice, (2) DorkMaLi-105 rice, (3) BasMaTi rice, (4) wheat, (5) maize, (6) sweet corn, (7) soeghum, (8) soybean, (9) mungbean, (10) sesame, (11) sunflower, (12) watermelon, (13) cabbage, (14) terminalia vegetable, (15) tomato, (16) huge onion, (17) onion, (18) momordica, (19) zinnia, (20) cotton, (21) sugarcane, (22) caster, (23) asparagus, (24) taro, (25) lemon (1–5 years), (26) mango, (27) pamelo, and (28) vetiver. They are simplified into four kinds of crop coefficient directed by section group of Irrigation water use research, Hydrology and Irrigation water management division section, Royal Irrigation Department Thailand as (1) paddy in wet season, (2) paddy in dry season, (3) upland crops in wet season, and (4) upland crops in dry season (analyzed by Cropwat 7.0 Program). Reference Potential Evapotranspiration (ETo) is analyzed by the Modified Penman Monteith method by the Royal Irrigation Department.[3]

3. Methodology

The study of the water usage planning is on the basis of water budget balancing.[4] Daily climatology–hydrology data were used for evaluating water budget in order to balance with daily consumption dealing with livestock, fishery, and various domestic water usage. In analyzing proportion of land usage appropriately, a mathematical model simulation using differential evolution optimization algorithms is presented in the following sections.

3.1. *Mathematical model*

Creating the mathematical model is composed of decision variables in the objective function and constraints to seek suitable answer of decision variables for achieving the objective function.

3.2. *Simulation techniques*

The mathematical model is calculated repeatedly in various situations by changing the group of decision variables in many iterations by selecting an algorithm, which is able to rapidly approach the solution.

3.3. *Optimization algorithmic technique*

Many algorithms are available; in this study, the differential evolution algorithm or DE is selected in seeking the best solution responding to the objective function. This algorithm is capable of searching the nearest solution rapidly and directly.[5] It is classified to be the stochastic search space algorithm as same as genetic algorithms (GAs), but it has less complexity than GAs. Furthermore, it uses real number instead of binary number transformation and needs no bit — String encoding, fit to non-linear problems, and non-differentiate function. Principle procedure for approaching the solution by differential evolution created by Price and Storn[2] can be generally described in the following section.

3.3.1. *Initial population*

The first step is randomly created a set of numbers of initial population. Each of those has its cost value or fitness value.

3.3.2. *Mutation*

The next step is mating couple of population, calculating the different values between its populations, multiplying the different values by weighting factor F or mutation factor F to get a magnitude of a newly different population vector that consists of subprocedure as follows:

(1) Define the target vector (X_i, G), where $i = 1, 2, 3, \ldots, NP$.
(2) Randomly select three vectors $(Xr_1, G, Xr_2, G, Xr_3, G)$ from the initial population, which are different from target vector.
(3) Calculate to find out the mutant vector $(V_i, G+1)$ from the following relation:

$$V_i, G+1 = Xr_1, \quad G + F(Xr_2, G - Xr_3, G) \tag{1}$$

where X_i, G is the target vector; $V_i, G+1$, the mutant vector; Xr_1, G, Xr_2, G, Xr_3, G, the random vector; and F, the weighting factor shown in (Fig. 2).

3.3.3. *Crossover or recombination*

It is the mating step in order to generate the better population or the better solution and to select the best group of the population, as shown in Fig. 3, which is called trial vector $(U_i, G+1)$, where

$$U_i, G+1 = (U1_i, G+1, U2_i, G+1, \ldots, UD_i, G+1) \tag{2}$$

and

$$\begin{gathered} Vj_i, G+1/if/(randb(j) \leq CR)/or/j = rnbr(i), \\ Xj_i, G/if/(randb(j) > CR)/or/j \neq rnbr(i), \end{gathered} \tag{3}$$

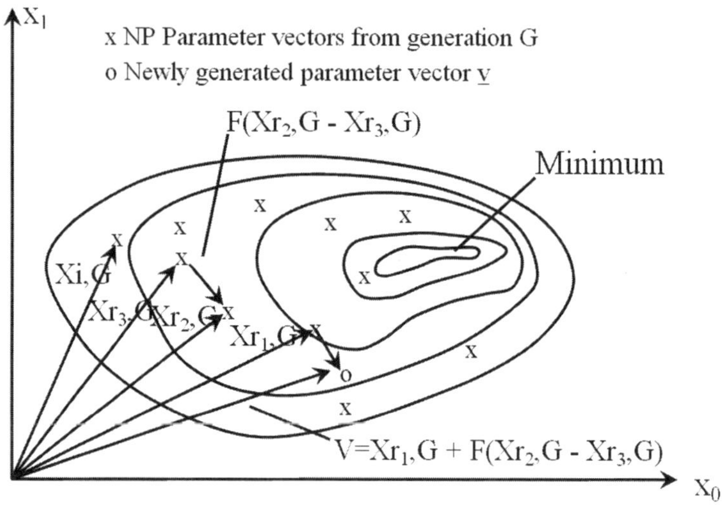

Fig. 2. Procedure to search the mutant vector from two variable populations.
Source: Price and Storn.[2]

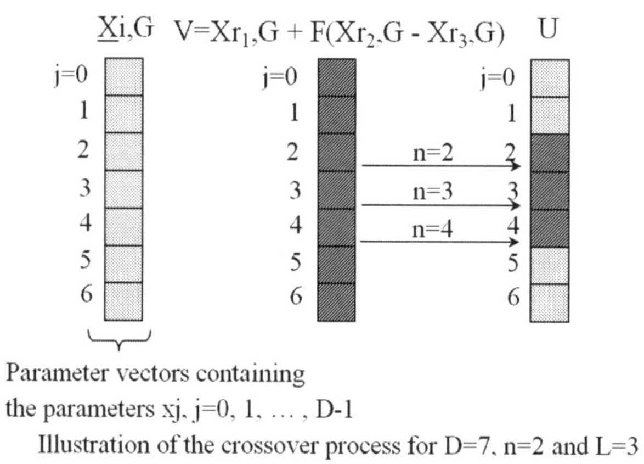

Fig. 3. Procedure in crossover step of the target vector and mutant vector for which the value $D = 7$.
Source: Price and Storn.[2]

where Uj_i, $G + 1$ is the trial vector; Vj_i, $G + 1$, the mutant vector; Xj_i, G, the target vector; $rndb(j)$, the random on real number from value equal to 0 to 1 for iteration j; CR, the crossover constant on real number from value equal to 0 to 1; $rnbr(i)$, the index randomly getting from fixed point number $1, 2, \ldots, D$; and j, the $1, 2, \ldots, D$.

3.3.4. *Selection*

It is the selection procedure for the next population generation $(G+1)$ from the newly better population by comparing the function value or the cost value of trial vector with target vector. In the case where the function value of trial vector is better than target vector, the target vector will be replaced by trial vector in the next generation.

3.3.5. *Evaluation and regeneration*

It is the repeating procedure from Secs. 3.3.2 to 3.3.4 by changing the target vector up to $i = \text{NP}$.

3.3.6. *Reach convergence tolerance*

By taking the target vector from Sec. 3.3.4 to repeat all of the population group, Fig. 4 shows the step in procedure for differential evolution optimization as applied to minimum optimization problem seeking.

4. Objective Function Setting

Decision variables are the interested parameters in this study, and the chosen five variables $(D1 \ldots D5)$ correspond to land portions in the New Theoretical Agricultural System: water budget flow into pond, water usage in agricultural activities, and domestic usage.

4.1. *Decision variables: D1–D5 definition*

The decision variables are defined in various proportion areas (Fig. 5) as follows.

$$D1 = A1/A(available),\ D2 = A2/A(agriculture),\ D3 = A3/A2,$$
$$D4 = A5/A4,\ D5 = V(init.)/V(max.),$$

where $A1 = A(pond)$, $A2 = A(paddy\ in\ wet\ season)$, $A3 = A(dry\ season\ paddy)$, $A4 = A(wet\ season\ crops)$, $A5 = A(dry\ season\ crops)$, $V1 = initial\ volume\ in\ pond$, and $V(max.) = maximum\ pond\ volume$.

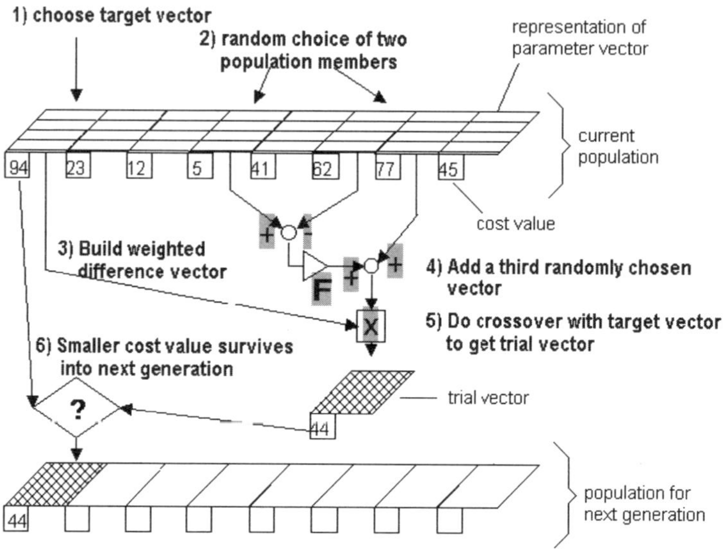

Fig. 4. Procedure of solution seeking by differential evolution algorithm from the steps in Secs. 3.3.2 to 3.3.4.
Source: Price and Storn.[2]

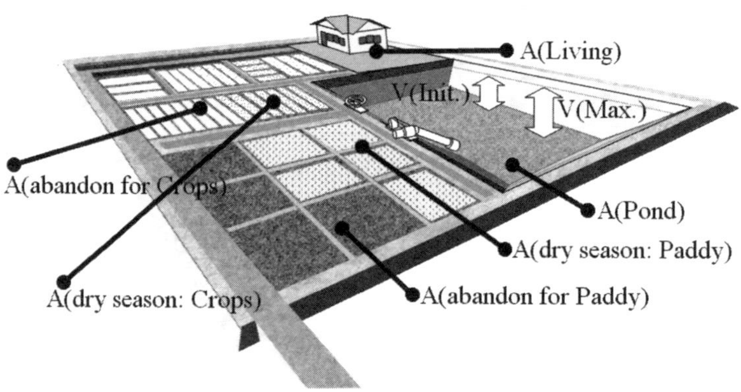

Fig. 5. Land portions dividing (Area: A) into various land use sorts for decision variables setting.

Hence,

$$\begin{aligned} A(total) &= A(living) + A(pond) + A(abandon\ for\ paddy) \\ &\quad + A(dry\ season\ paddy) + A(abandon\ for\ crops) \\ &\quad + A(dry\ season\ crops), \end{aligned} \tag{4}$$

where $A(available) = A(total) - A(living)$, $A(agriculture) = A(available) - A(pond)$, $A(wet\ season\ paddy) = A(abandon\ for\ paddy) + A(dry\ season\ paddy)$, and $A(wet\ season\ crops) = A(abandon\ for\ crops) + A(dry\ season\ crops)$.

4.2. *Objective functions and constraints*

The aim of objective functions to the maximum water usage can be expressed as

$$Max.Z = \sum_{i=1}^{365} V(use)_i \tag{5}$$

within the constraints

$$\left| \sum_{i=1}^{365} V(in)_i - \sum_{i=1}^{365} V(use)_i \right| = 0, \tag{6}$$

$$V(in)_i = (EX_i - ER_i)(A(agri)) + Rf_i(A1) - ETo_i A1), \tag{7}$$

$$\begin{aligned} V(use)_i = ((&Weighted\ Kc.\ of\ wet\ season\ paddy_i)A2 \\ &+ (Weighted\ Kc.\ of\ wet\ season\ crops_i)A4 \\ &+ (Weighted\ Kc.\ of\ dry\ season\ paddy_i)A3 \\ &+ (Weighted\ Kc.\ of\ dry\ season\ crop_i)(ETo_i)A5/1{,}000 \\ &+ Domestic,\ Fishery\ and\ Livestock\ daily\ use \\ &\quad and\ Daily\ Deep\ Percolation \\ &- (ER_i)A(agr_i))/Irrigation\ efficiency, \end{aligned} \tag{8}$$

$$Vstorage_{i+1} = Vstorage_i + V(in)_i - V(use)_i, \tag{9}$$

$$V(min.) = 5\%\ of\ V(max.), \tag{10}$$

$$Vstorage_i \geq V(min.), \tag{11}$$

$$Vstorage_i \leq V(max.), \tag{12}$$

where Rf_i is the daily rainfall (in year of the most closed average annual), ER_i is the daily effective rainfall, EX_i is the daily excess rainfall, and ETo_i is the daily reference evapotranspiration.

In calculating the water balancing between water budget and water usage, we will consider in holistic components of effective rainfall, percolation for compensating water pumping or daily demand in water consumption in order to estimate the storage need and define the minimum storage at 5% of maximum storage volume: $V(max)$. In addition, the daily deep percolation rate is defined on the basis of land preparation for percolation rate permitted, not exceeding the criterion of the Royal Irrigation Department for central, eastern and southern, and northeastern by the rate of 1.0, 1.5 , 2.0 mm/day, respectively.

5. Results

In setting the objectivc function and creating the mathematical model, simulation was performed to find appropriate decision variables for both lands of sizes 1.6 and 2.4 hectares. It is found that each region should set the plan for different lands used in dry season and wet season depending on the daily rainfall characteristics and quantities as shown in Tables 1–4. Nevertheless, for comparison, the obvious difference of the land proportion, shown in Figs. 6 and 7 for both land sizes are 1.6 and 2.4 hectares respectively.

Table 1. Land use plan portions in percent in the case of 1.6 hectares for wet season.

Land portion (%)	General	Northern	Northeastern	Central	Eastern	Southern
Pond area	30.00	35.04	30.99	38.83	34.24	11.78
Paddy in wet season	30.00	42.59	34.73	31.34	27.78	54.18
Upland crops in wet season	30.00	9.87	21.78	17.33	25.48	21.54
Living area, etc.	10.00	12.50	12.50	12.50	12.50	12.50

Table 2. Land use plan portions in percent in the case of 1.6 hectares for dry season.

Land portion (%)	General	Northern	Northeastern	Central	Eastern	Southern
Pond area	30.00	35.04	30.99	38.83	34.24	11.78
Paddy in dry season	30.00	24.74	8.55	21.78	15.64	53.78
Upland crops in dry season	30.00	2.75	8.18	10.28	12.25	15.97
Abandon area in dry season	—	24.97	39.78	16.61	25.37	5.97
Living area, etc.	10.00	12.50	12.50	12.50	12.50	12.50

Table 3. Land use plan portions in percent in the case of 2.4 hectares for wet season.

Land portion (%)	General	Northern	Northeastern	Central	Eastern	Southern
Pond area	30.00	36.31	32.09	40.36	35.54	12.09
Paddy in wet season	30.00	44.94	36.62	33.04	29.28	56.94
Upland crops in wet season	30.00	10.42	22.96	18.27	26.85	22.64
Living area, etc.	10.00	8.33	8.33	8.33	8.33	8.33

Table 4. Land use plan portions in percent in the case of 2.4 hectares for dry season.

Land portion (%)	General	Northern	Northeastern	Central	Eastern	Southern
Pond area	30.00	36.31	32.09	40.36	35.54	12.09
Paddy in dry season	30.00	26.11	9.02	22.96	16.48	56.52
Upland crops in dry season	30.00	2.90	8.63	10.84	12.90	16.79
Abandon area in dry season	—	26.35	41.93	17.51	26.75	6.27
Living area, etc.	10.00	8.33	8.33	8.33	8.33	8.33

6. Implementation

From the particular results, it has been found that the land divided into portions related to the amount of local rainfall behaviors. To apply these results for implementation, the recommended graphs have been created based on the relationship between local average rainfall and percentage of various land used such as pond area, paddy area in wet season, and paddy area in dry season. The rest of the portions are the percentage of the upland crops and the percentage of the abandoned area in dry season. On implementation, a graph is plotted, which expresses the relationship between annual average rainfall and percentage of all portions, as shown in Fig. 8.

7. Conclusion

The optimum land planning for water usage in the New Theoretical Agricultural System is the extension of the present study.[2] In particular, this study is analyzed for the relation of the area divided into portions in the New Theoretical Agriculture System against local rainfall (Rf_{av}). As

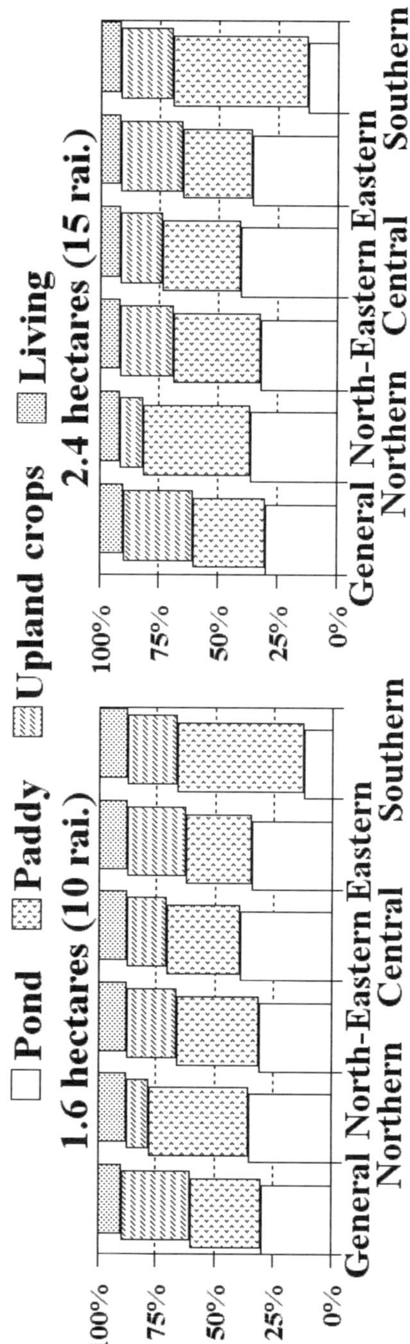

Fig. 6. Graph of land use plan by portions in percent compared with land sizes of 1.6 hectares (10 rai.) and 2.4 hectares (15 rai.) in wet season.

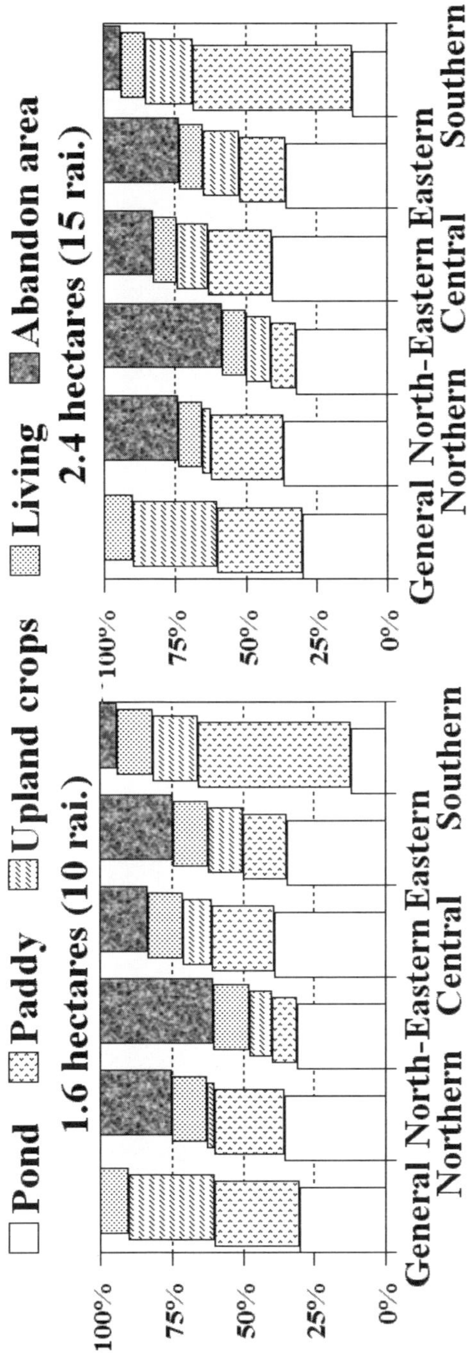

Fig. 7. Graph of land use plan by portions in percent compared with land sizes of 1.6 hectares (10 rai.) and 2.4 hectares (15 rai.) in dry season.

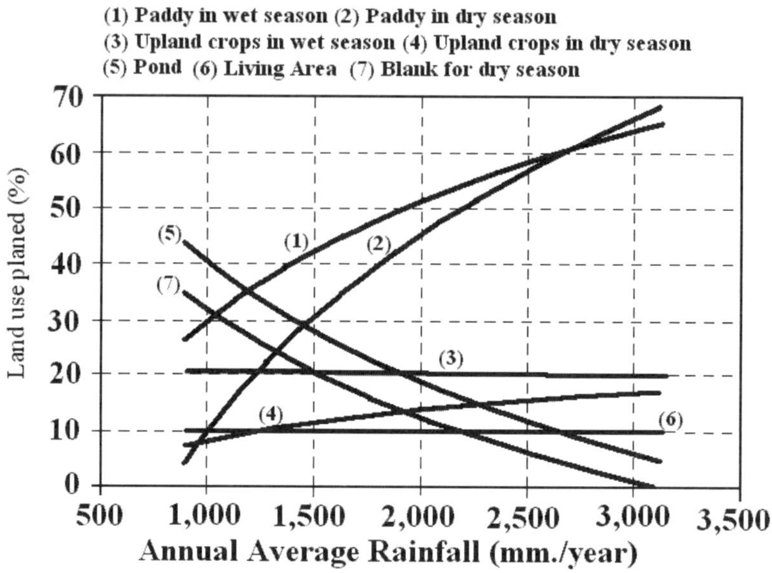

Fig. 8. Graph expressed relationship between annual average rainfall and all portions set for the recommended area in both wet and dry season, in percent.

the result shows, the relation is found to be significant and it could be concluded that the major area in percent should be appropriately planned for water resources usage effectively as follows.

Portion in percent for pond area is

$$A(pond) = -31.38\ln(Rf_{\text{av}}) + 257.33$$

Portion in percent for paddy field area in wet season is

$$A(paddy\ in\ wet) = 31.338\ln(Rf_{\text{av}}) - 187.02$$

Portion in percent for paddy field area in dry season is

$$A(paddy\ in\ dry) = 51.319\ln(Rf_{\text{av}}) - 344.80$$

Portion in percent for abandon area in dry season is

$$A(abandon) = 225.3 - 28.01\ln(Rf_{\text{av}})$$

The relation graph shown in Fig. 8 can be used for usage planning in different areas depending on localized rainfall characteristics. The differential evolution algorithm applied in this study to seek complicated solutions is a suitable method for rapidly searching tool to handle such problem efficiently.

Acknowledgments

Through this paper, the researchers would like to consecrate His Majesty King Bhumibol Adulyadej and dedicate to all the farmers in Thailand who always face droughts. It may be used as an opportunity to do sufficient agriculture in order to maintain a sustainable life.

References

1. M. Wanielista, R. Kersten and R. Eaglin, *Hydrology Water Quantity and Quality Control* (John Wiley & Sons, Inc. USA, 1997).
2. K. Price and R. Storn, Differential evolution a simple and efficient adaptive scheme for global optimization over continuous spaces, Technical Report TR-95-012, ICSI., 1998.
3. O. Chanvecha, 28 kinds of crop coefficients: Kc in Thailand, http://www.rid.go.th/attatch_branch/qrice.html, 1999.
4. Office of the Royal Development Projects Board, The New Theory, Integrated Farming System by H.M. Bhumibol Adulyadej initiatives, Thailand, 2003.
5. S. Chittaladakorn, E. Jornpradit, The optimum planning of water resource usage in new theory integrated farming system by differential evolution algorithm, *The 12th National Convention on Civil Engineering*, Thailand, 2007, p. 399.

Advances in Geosciences
Vol. 11: Hydrological Science (2007)
Eds. Namsik Park *et al.*

PUBLIC DOMAIN SYSTEM MODELING IN NORTHERN THAILAND

HANSA VATHANANUKIJ
Water Resource Engineering Department
Faculty of Engineering, Kasetsart University
P.O. Box 1032 Kasetsart, BKK10903, Thailand
hansa.v@ku.ac.th
http://gis.eng.ku.ac.th

Natural feature at northern region of Thailand has been regularly inundated and significantly encountered straight to central plain downstream. Rainy season in the year 1995, 2002, and 2006 caused land slide and serious flood damaged both northern and central plain of Thailand. It is important meaning for reducing these losses. This research adopted the public domain system model to simulate flood in northern Thailand.

Public domain system models have been introduced to academic researchers in Thailand (RM-GIS Center/Engineering/KU) since 2001. SWAT model was one of them. It unified continuous time and basin scale together with geoinformatic system interface. The model could simulate daily, weekly, and monthly data for both pre-process and post-process. Base flow methodology was employed on stream flow hydrograph separation and analysis program which leashed by sliding interval methodology. Data access was approached by basic standard query language.

Distant from the previous three prototype basins (Omkoi ~1,500 km^2, Upper Nan ~ 10,000 km^2, and Pasak ~ 15,000 km^2), a new research platform was on steep slope basin (Maewaang catchment in Chiangmai province). Drainage area was 533 square-kilometers, almost un-regulated with two observation stations.

Results indicated that the most sensitive parameter to model simulation is curve number (CN) while more relative parameter to land surface interpreter is NDVI (from spatial assessment through multi-resolution of satellite imageries). NDVI has much significant applicability to environmental monitoring, particularly for terrestrial dynamics. Their relationships have been affined and affirmed to represent fundamental parameters in both global scale and local scale such as temperature, plantation, evapo-transpiration, etc.

Assessment results on aspect and relative differences illustrated significant value for northern region of Thailand. Geoinformatic system structures on digital elevation model, basin delineation and model outputs could deliberately demonstrate digital terrain model and possible flood risk of the platform.

1. Introduction

Rainfall and discharge are strongly related and depended on topography, vegetation cover, soil and human activities which frequently derive distribution in different segments of spatial domain.[1] Trendy assessment through geoinformatic system has been enhanced capability and accuracy in managing enormous database algorithm.[4] Hundreds of hydrology/hydraulics/hydro-dynamic models in Thailand have been used to estimate and forecast normal and extreme flood since 1977. Most of them were much costly licensed and could not relief any flood in time. We all know that models need end-users to be acquainted with their specific parameters and simulation. Nowadays, freeware model could manage, develop, and inter-change all efficiency methodology. SWAT model has been introduced to this cyber world since 1990 through DOS system with FORTRAN77 language. It has been popularly developed into all systems by many researchers from all over the world. Recently, this model has full functions in GIS which could approach via Internet. This research studied and developed several interfaces together with calibration, verification, and evaluation to affirm the model efficiency. This paper summarizes the whole public domain model researches.

2. Assimilation

The public domain model, SWAT or SWAT/GIS model, is based on continuous time, basin scale, geoinformatic system interface, precipitation analysis, surface and sub-surface flow analysis, evapo-transpiration estimation and 500 sub-basins possible calibration. The model required database on topography, weather, soil, vegetation, land management practice, and watershed area. Methodology on flood and base flow analysis was engaged on hydrologic respond unit with stream flow hydrograph separation and analysis program.[17] This research was principally on the following formulae.

SCS curve number method is for surface discharge estimation

$$Q = \frac{(R - 0.2s)^2}{R + 0.8s}, \quad R > 0.2\,s, \tag{1}$$

$$Q = 0.0, \quad R \leq 0.2\,s, \tag{2}$$

$$s = 254\left(\frac{100}{CN} - 1\right). \tag{3}$$

Modified-rational formula is for peak discharge estimation

$$q_{\mathrm{p}} = \frac{\rho r A}{360}. \tag{4}$$

Continuity equation, momentum equation, and Manning's equation[2] are for channel routing;

$$G = \frac{\partial U}{\partial t} + \frac{\partial F}{\partial x}, \tag{5}$$

$$G = \begin{pmatrix} r - f \\ s_0 - s_f \end{pmatrix}, \quad F = \begin{pmatrix} uh \\ h \end{pmatrix}, \quad U = \begin{pmatrix} h \\ 0 \end{pmatrix}, \tag{6}$$

$$S_f = \frac{n^2 u^2}{h^{\frac{4}{3}}}, \tag{7}$$

$$n_i = n_0 \left(\frac{h_i}{\overline{h}} \right). \tag{8}$$

Soil water balance equation is for sub-surface routing

$$\mathrm{SW}_t = \mathrm{SW} + \sum (R_i - Q_i - ET_i - P_i - QR_i). \tag{9}$$

Lateral sub-surface discharge equation used mass continuity equation with soil profile-control volume and finite difference equation for kinematics storage.

$$\frac{\mathrm{SV}_2 - \mathrm{SV}_1}{t_2 - t_1} = iL - \frac{q_{\mathrm{lat1}} + q_{\mathrm{lat2}}}{2}. \tag{10}$$

Shallow aquifer equation is for capacity estimation

$$Vsa_t = Vsa_{i-1} + Rc - Rvap - q_{rf} - perc_{gw} - WU_{SA} \tag{11}$$

Standard methods for the evapo-transpiration estimation are selected from Penman–Monteith method

$$E_{\mathrm{p}} = \frac{\frac{\delta(h_0 - G) + 86.7AD(e_a - e_d)}{\mathrm{AR}}}{(\mathrm{HV})(\delta + \gamma)}, \tag{12}$$

Priestley–Taylor method

$$E_{\mathrm{p}} = 1.28 \left(\frac{h_0}{\mathrm{HV}} \right) \left(\frac{\delta}{\delta + \gamma} \right), \tag{13}$$

and Hargreaves–Samani method (1985)

$$E_{\mathrm{p}} = 0.0032\left(\frac{\mathrm{RAMX}}{\mathrm{HV}}\right)(T - 17.8)\,(T_{\max} - T_{\min})^{0.8}. \tag{14}$$

At present, remote sensing imageries did play an important role in model assessment. Relationship among fundamental *in situ* and remote sensing parameters such as land surface temperature, solar radiation, latent heat, soil, water, plantation, surface albedo, normalized difference vegetation index and etc., are significantly accepted.[16]

This research partially studies in energy balance, micro-meteorology characteristics, and evapo-transpiration through MODIS/Terra data set. Spatial analysis could be utilized for land use/land cover change interpretation and related to evapo-transpiration via normalized difference vegetation index as shown in the following equations:

$$\mathrm{NDV}I = \frac{\left(1 - \frac{\mathrm{NIR}}{\mathrm{RED}}\right)}{\left(1 + \frac{\mathrm{NIR}}{\mathrm{RED}}\right)}, \tag{15}$$

$$\mathrm{EV}I = \frac{G\,(\mathrm{NIR} - \mathrm{RED})}{(L + \mathrm{NIR} + C_1 R - C_2\mathrm{BLUE})}. \tag{16}$$

For low level vegetation; e.g. paddy field

$$\begin{aligned} \mathrm{ET} &= -1.93 + 1.42\mathrm{NDVI} + 0.2\mathrm{LST} + 7.44\mathrm{ALB}, \\ r^2 &= 0.85. \end{aligned} \tag{17}$$

For high level vegetation; e.g. teak plantation

$$\begin{aligned} \mathrm{ET} &= 3.7 - 0.60\mathrm{NDVI} + 0.07\mathrm{LST} + 4.57\mathrm{ALB}, \\ r^2 &= 0.64. \end{aligned} \tag{18}$$

This research developed data access program by basic standard query language to be convenient for the correction of spatial parameters, hydrograph separation analysis, and results comparison.

3. Assessment

3.1. *Research areas*

Figure 1 illustrates the four different research areas at Maewaang, Omkoi, upper Nan, and Pasak catchments whose topography enfolded by mountainous or hilly features and sloped down to outlets. Catchment areas were 533, 1,500, 10,200 and 14,500 square kilometers, respectively.[7,8,11–15]

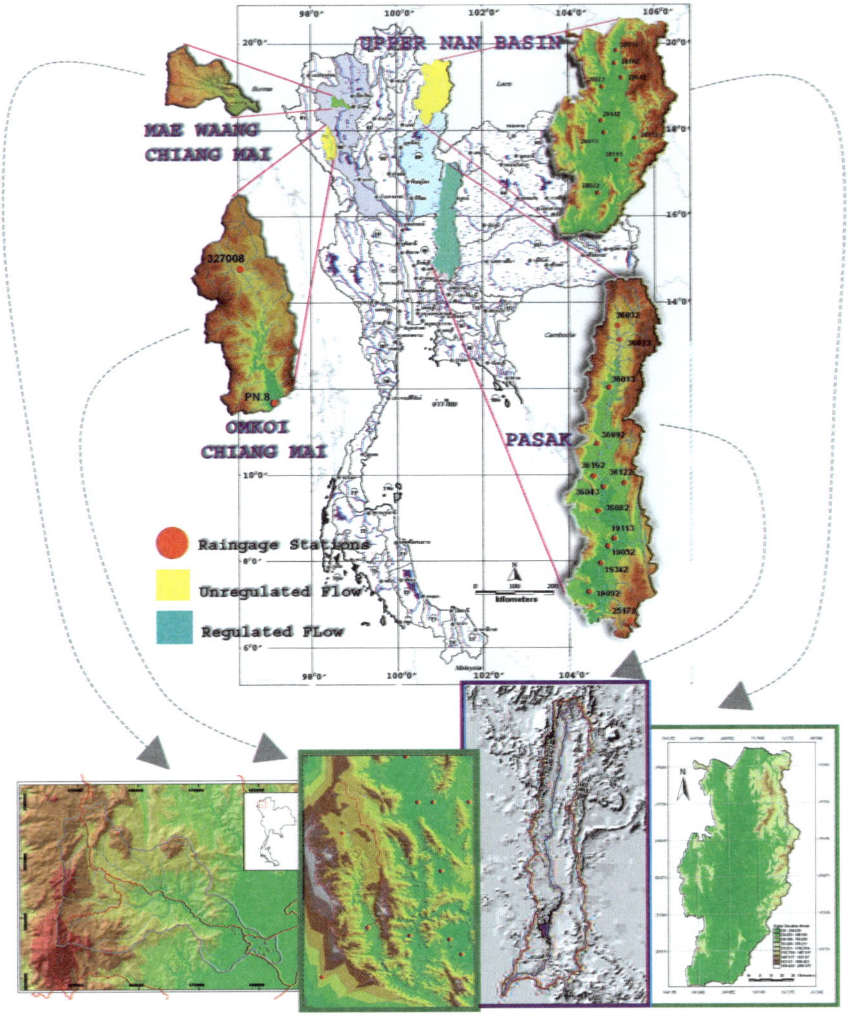

Fig. 1. Prototype areas.

3.2. *Model process*

Satellite imagery performance was elaborated on NDVI, land cover, and temperature through spatial analysis of geo-reference MODIS dataset. Land use/land cover and soil types were corrected and generated by Hydrologic Respond Unit (HRU). Horizontal and vertical resolutions produced digital elevation model (DEM) with finer-slope aggregation.

Hydrologic/hydraulic performance was evaluated through daily data (original from hourly data) on rainfall and discharge of 71 stations/ 16 catchments from various major zones of northern Thailand. Data were included with rainfall, discharge, temperature, evapo-transpiration, humidity, topographic details, soil types, land use, and land cover.

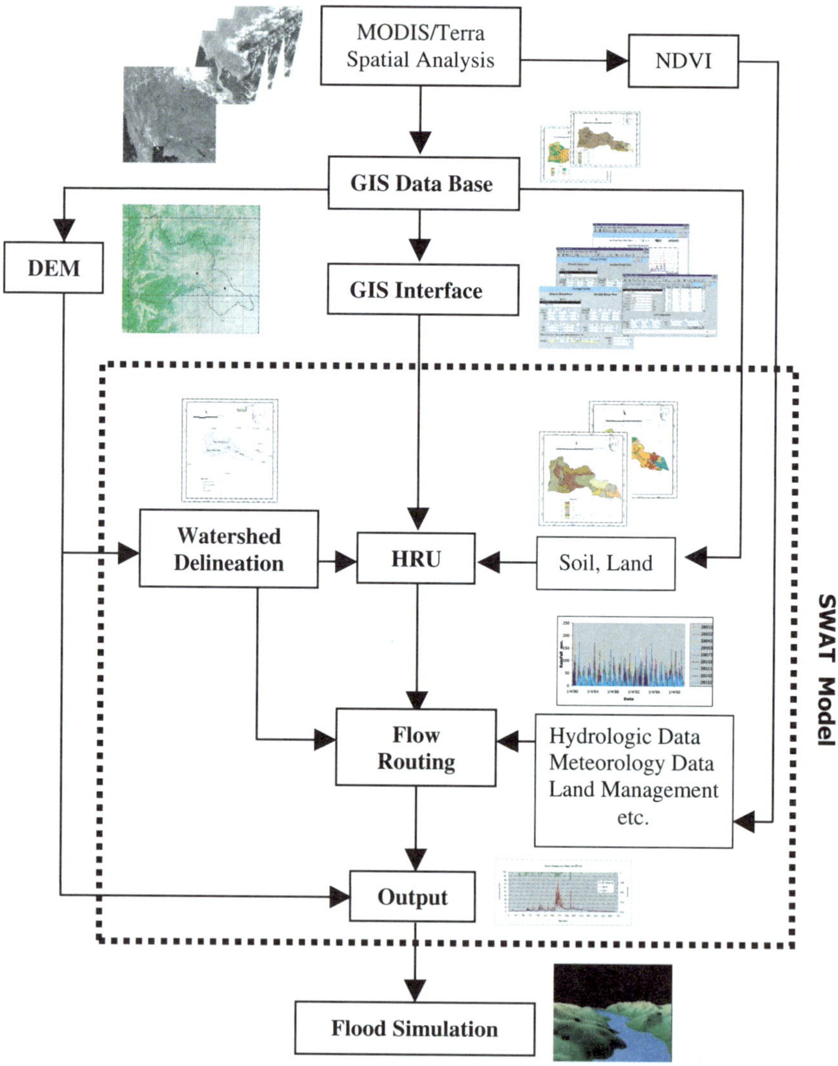

Fig. 2. Model algorithm (example on Maewaang basin).

SWAT model algorithm started from most upstream cells to downstream cells, determined cell-to-cell steepest descent (two-dimensional array as a cascading of DEM analysis) to outlet. Under GIS data base (1:50,000 scale), research area was divided into 500 sub-basins for calibration. Verification was on observation stations.[5,9,10]

Figure 2 illustrates model algorithm of Maewaang basin.

3.3. *Results*

According to the model calibration and verification, it was found that the most sensitive parameter to discharge in northern Thailand is curve number (CN). Sub-surface water and evapo-transpiration performed medium to low. Significant parameters on the output results are shown in Table 1.

The correlative efficiency comparison did best on mild aspect and good on steep aspect (regulated basin performed above 90% and unregulated basin performed over 70%). Figure 3 illustrates the example of best correlation between observed and simulated discharges. Figure 4 illustrates the example on good correlation of satellite parameters and hydrologic fundamental parameters.

Table 2 contains correlation at verified stations in slope, number of rain gauge intensity, and correlation coefficients of surface flow (between simulated results and observed data).

4. Conclusion

This research attempted to structure, verify, and modify public domain system model SWAT. Many parameters in the model had improved and approved via satellite imageries assessment. Sensitive input data were

Table 1. Five discharge sensitivities of model calibration for northern Thailand.

Parameter	Total discharge	Surface discharge	Base flow
Sensitivities			
CN	High	High	Low
AWC	High	Medium	Low
GW_REVAP	Medium	Low	Medium
REVAPMN	Medium	Low	Medium
GWQMN	Medium	Low	Medium

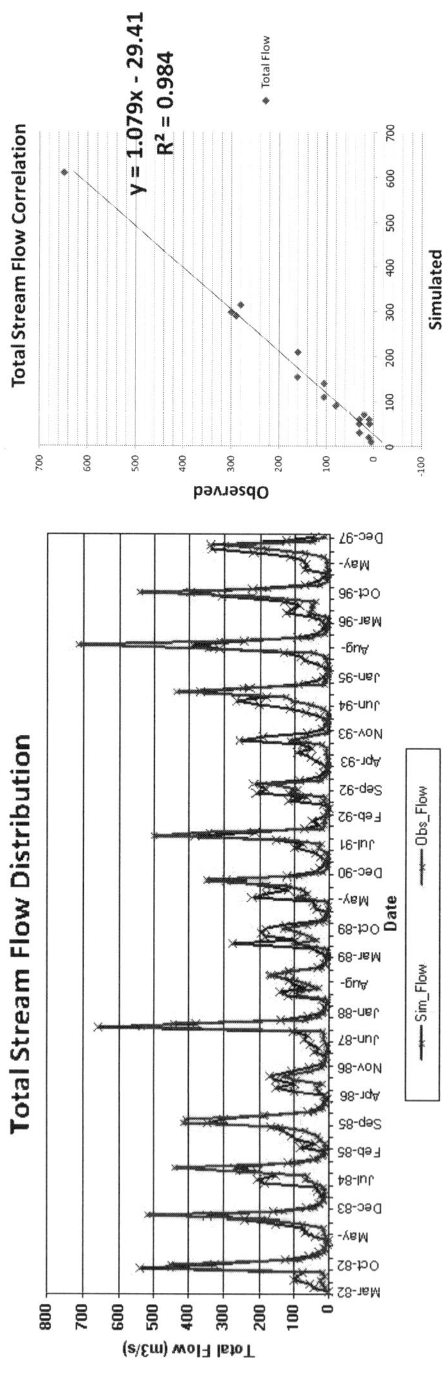

Fig. 3. Example of observed–simulated discharge comparison.

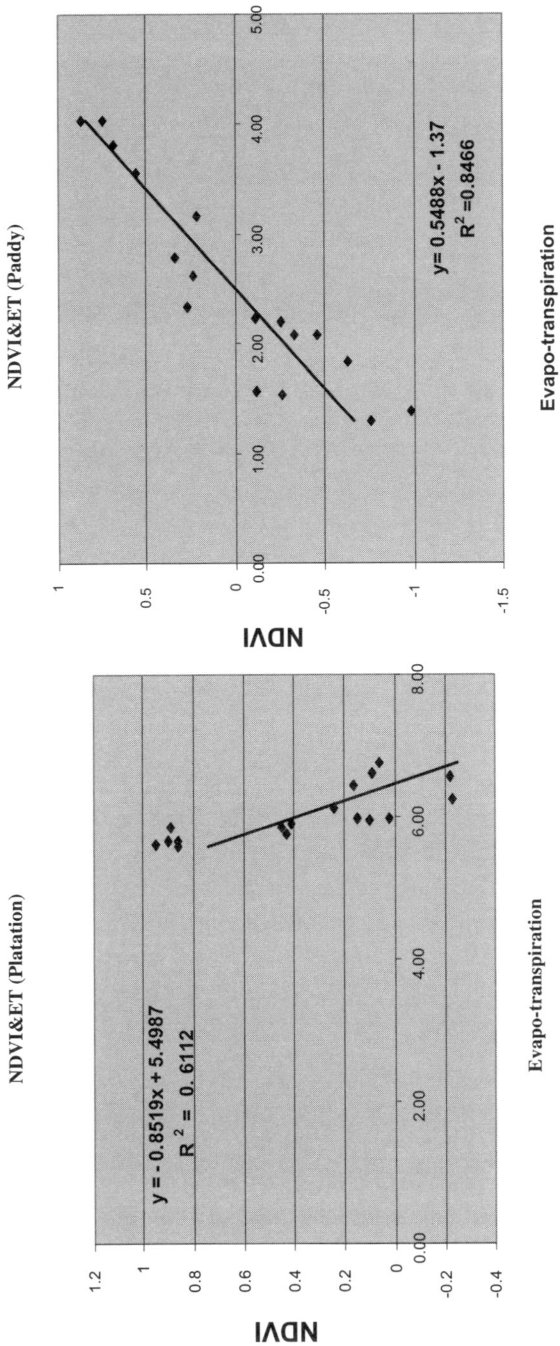

Fig. 4. Example of NDVI–evapotranspiration relationship.

Table 2. Characteristics of five indicators at verified stations.

Calibration area	Stream gauge	Basin area (km^2)	Basin slope	R-gauge intensity	C-coefficient (R^2)
Area 1	P.82	389.00	0.250	0	0.5600
Area 2	P.84	493.00	0.180	1	0.6400
Area 3	P.64	487.18	0.266	1	0.5795
Area 4	PN.8	1567.67	0.299	1	0.5361
Area 5	N.17	1091.15	0.306	0	0.3512
Area 6	N.42	2047.11	0.296	2	0.6219
Area 7	N.49	153.16	0.317	0	0.4784
Area 8	N.50	194.61	0.324	0	0.4775
Area 9	N.51	758.53	0.240	1	0.5284
Area 10	N.63	776.05	0.202	1	0.4099
Area 11	N.1	4495.06	0.241	8	0.6008
Area 12	N.13A	8566.91	0.237	8	0.7045
Area 13	N.35	10,156.01	0.230	11	0.7008
Area 14	S.10	300.79	0.100	0	0.6037
Area 15	SM.1	1132.70	0.248	2	0.7585
Area 16	S.4B	3321.00	0.181	3	0.8285
Area 17	S.12	476.11	0.230	0	0.5450
Area 18	SM.2	7329.49	0.158	6	0.8089
Area 19	S.13	395.25	0.100	0	0.7182
Area 20	S.14	1252.77	0.087	0	0.8572
Area 21	S9.(UnReg)	14,323.80	0.068	13	0.7397
Area 22	S9.(Reg)	14,323.80	0.068	13	0.9708

rainfall amount, basin slope, and land use. The most sensitive parameter to model simulation was curve number. The number of rainfall stations, evapo-transpiration data, and area dimensions did not signify to the model.

Results on both unregulated steep aspect and regulated mild aspect catchments have been much efficiently utilized to local operation. Combination of digital terrain model and model outputs was possible to demonstrate flood simulation which became essential for water resource decision supporting system.

References

1. M. B. Abbott, J. C. Bathrust, J. A. Cunge, P. E. O'Connel and J. Ramsussen, An introduction to hydrological system 1. History and philosophy of a physically based, distributed modeling system, *J. of Hydrol.* **87** (1986), 45–59.
2. V. T. Chow, *Applied Hydrology* (Mc-Graw Hill, New York, 1988), p. 273
3. H. Vathananukij, *River Engineering* (Physics Center Co. Ltd. 1998), p. 120.
4. H. Vathananukij, *Geoinformatic System/Geographic Information System* (*GIS*) (Physics Center Co. Ltd. 2003), p. 151.

5. H. Vathananukij, *Water Resource Engineering Innovation: Public Domain System Model* (Aksornsobhon Co. Ltd. 2007), p. 135.
6. H. Vathananukij, Geoinformatic public domain system model SWAT in Thailand, *Kasetsart Journal: Nat. Sci.* **40**(1) (2006) 264–272.
7. H. Vathananukij and K. Thanasiriyakul, Thailand tributaries relativity assessment through geoinformatic system and SWAT modeling, *Proceedings International Workshop on GAME-T and Hydro-Meteorological Studies in Thailand and Southeast Asia* (2002), pp. 62–68.
8. H. Vathananukij, Public domain model (SWAT-GIS) development on Nan River Basin, in *Proceedings International Symposium on the Climate System of Asian Monsoon and Its Interaction with Society* (2003), pp. 15–23.
9. H. Vathananukij and T. Paewpisakul, Design flood application through geoinformatic system and AutoCAD land development desktop (LDT) model, *Proceedings International Fourth Regional Symposium on Infrastructure Development in Civil Engineering* (2003), pp. 1087–1093.
10. H. Vathananukij and C. Thongpond, River model evolution intimacy on geoinformatic system, *Proceedings of International Seminar on Innovation Model for Sustainable Water Resource Management* (2003), pp. 139–149.
11. H. Vathananukij, *Portentous flood exposition on geoinformatic system, Proceedings International Symposium on Water Resource and Its Variability in Asia in the 21st Century* (2004), pp. 55–28.
12. H. Vathananukij and RM-GIS Team, Regulated and unregulated river reach perception through geoinformatic public domain model particularized upon The Chaophraya Tributaries, *Proceedings of the 6th International Study Conference on GEWEX in Asia and GAME* (2004), T7PHV165014-1/7-7/7.
13. H. Vathananukij, Sustention models for water resource management, *Proceedings of 2nd APHW Conference Jointly with 1st AOGS Annual Meeting Volume II* (2004), pp. 668–676.
14. H. Vathananukij, Geoinformatic public domain model on Upper Ping River Basin, *Proceedings of Interdisciplinary Workshop on Multi-scale Governance of Forests, Village and Water in the Upper Ping River Basin, Northern Thailand* (2005), pp. 89–94.
15. H. Vathananukij, Geoinformatics and public domain system model on thailand mega prototype, *Proceedings of the fourth international Symposium on New Technologies for Urban Safety of Mega Cities in Asia* (2005), pp. 579–584.
16. P. Tongdeenok, B. Samakkee, N. Tangtham and C. Kankhajane, *Estimating evapotranspiration using remote sensing*, Thesis, *Graduate School, Kasetsart University* (2006), 138p.
17. R. Srinivasan, J. G. Arnold, W. Rosenthal and R. S. Muttiah, *Hydrologic Modeling of Texas Gulf Basin Using GIS* (USDA-ARS Temple, Texas, 1996), p. 535.
18. *The Use of GIS for Practical Hydrology* (United Nation Educational Scientific and Cultural Organization, 1994), p. 243.
19. *Textbook for Fifteenth IHP Course in 2005/International Hydrology Programme* (United Nation Educational Scientific and Cultural Organization, 2005), p. 238.

Advances in Geosciences
Vol. 11: Hydrological Science (2007)
Eds. Namsik Park *et al.*

STORAGE REALLOCATION FOR A MULTIPURPOSE RESERVOIR

JAEEUNG YI
Division of Environment
Construction and Transportation, Ajou University, Suwon-Si
Geonggi-Do, 443-749, Korea
jeyi@ajou.ac.kr

Reallocation procedure of multipurpose reservoir storage capacity between flood control and conservation is presented as an alternative for securing more water resources. Storage reallocation is an adaptive management mechanism for converting existing reservoirs to more beneficial uses without requiring physical alteration. This study is intended to develop a reservoir storage reallocation methodology that allows increased water supply storage while minimizing adverse impacts on flood control. The methodology consists of a flood control reservoir simulation for various design flows, flow routing from reservoir to a potential damage site, analyzing river carrying capacity, and reservoir yields estimation for reallocated storages. For the flood control model, a simulation technique called rigid ROM (Reservoir Operation Method) and HEC-5 are used. The approach is illustrated by applying it to a two-reservoir system on the Geum River basin in Korea. In particular, with and without the new project conditions are considered to analyze the trade-offs between the competing objectives.

1. Introduction

According to the water plan (2000–2020),[1] shortages in water supplies are expected to begin from the year 2011 in Korea. While a typical method to alleviate water shortage is the development of new water resources such as constructing reservoirs, this has become much more difficult now than in the past, due to economic, environmental, institutional, and political restrictions.

Storage reallocation of multipurpose reservoirs can be considered as an alternative to secure additional water resources from existing reservoirs currently in operation. Reservoir storage is usually allocated in accordance with various purposes established prior to construction. Normally it is not altered unless particular changes such as raising a dam are required, or project purposes have changed. Reservoir storage reallocation is an attractive adaptive management mechanism for converting existing reservoirs to more beneficial uses without requiring physical alteration.

Since the most multipurpose projects in Korea have been constructed between the 1960s and 1980s, substantial changes such as hydrologic conditions, water and energy demands, construction of additional projects in the same basin, and priority changes in water uses have occurred. An alternative for reservoir storage reallocation has not been given serious consideration, since the construction of new projects has been relatively easy in the past. However, as the situation has changed drastically since the 1990s, reservoir storage reallocations will be proposed more frequently. Reservoir storage reallocation is studied herein to consider these changes and to secure more water resources.

Wurbs and Cabezas[2] presented a hydrologic and economic evaluation procedure for analyzing proposed reallocations of reservoir storage between flood control, municipal, and industrial water supply. Patrick[3] developed a hydrologic and statistical approach to evaluate storage reallocation potentialities in general and the trade-offs and interactions among reservoir purposes. The US Army Corps of Engineers[4] identified situations in which flood pool volumes could be seasonally adjusted to derive greater economic benefits from existing reservoirs. Ford[5] developed a PC-based reservoir-operation simulator to provide information on trade-offs between competing demands. Johnson *et al.*[6] reviewed studies on the reallocation of reservoir storage capacity for reservoirs in the Corps of Engineers and the State of Texas.

The MODSIM model and the USGS MODFLOW are applied to the Upper Snake basin to quantify how much water reallocation is needed to accomplish flow augmentation objectives.[7] McMahon and Farmer[8] drew distinctions between the principles intended to ensure that federal water projects are formulated to meet all of the society's needs in the long term, and those policies and procedures devised decades ago to expeditiously ameliorate the immediate problems of the time.

This study aims at providing a proper reallocation of multipurpose reservoir storage to satisfy both conservation and flood control purposes by considering the past, present, and future conditions in a basin. The case of constructing new projects in the basin with existing downstream projects is given special consideration in this study, since it may allow reallocation of storage in an existing reservoir from flood control to conservation purposes. Reallocations between flood control and conservation storage, however, requires trade-offs between purposes. Since the importance of flood control has not been diminished in Korea, the goal is to present a reservoir storage reallocation methodology that allows increased water supply storage while minimizing adverse impacts on flood control.

2. Methods and Application

2.1. *The methodology*

The approach used in this study is as follows: First, alternative synthetic design flows with various selected return periods for an exiting reservoir are obtained with and without a new upstream reservoir. Second, these synthetic design flows are applied as inputs to perform flood control reservoir simulations for a number of alternative hypothetical reallocation plans by changing pool elevations between water supply and flood control. Third, the reservoir releases resulting from flood simulation are routed to a downstream potential damage site and flood peak discharges are obtained at that point with respect to each storage reallocation alternative of the reservoir. Fourth, a reservoir storage reallocation level is determined in accordance with a river carrying capacity which does not overtop levees at a potential damage site. Fifth, reservoir yields before and after storage reallocation are estimated and compared. Finally, the reservoir reallocation level is determined considering both flood control and conservation of the reservoir.

In order to evaluate and determine the reservoir storage reallocation plans, economic, legal, environmental, institutional and social analyses, as well as hydrologic analysis, are essential. Only the hydrologic analysis, however, is considered here to avoid additional levels of uncertainty in the problem. The two-reservoir system in the Geum River basin in Korea provides a case study for evaluating the potential for storage reallocations.

2.2. *Geum River basin study area*

The Geum River basin, the third largest basin in South Korea, is located in the middle western part of South Korea, which occupies some 9853 km^2. Its headwaters lie in rural areas, and it flows through a highly urbanized section of the city of Daejun and Gongju on its way to the Yellow Sea. The two major dams, Daechung and Yongdam, regulate the Geum River mainly of flood control and conservation purposes. Daechung Dam, constructed in 1981, lies 150 km upstream from the estuary impounding Lake Daechung

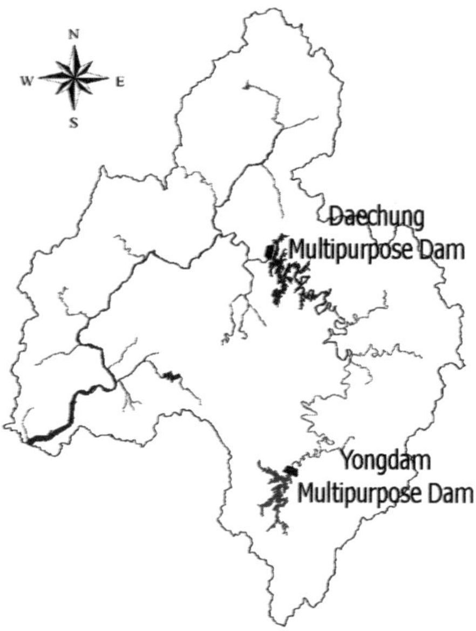

Fig. 1. Geum River basin.

Table 1. Details of the Yongdam and Daechung dams.

	Catchments area	
	Yongdam ($930\,km^2$)	Daechung ($4134\,km^2$)
Reservoir flood control		
Storage capacity ($10^6\,m^3$)	137	250
Active storage capacity ($10^6\,m^3$)	672	790
Project flood level (EL.m)	265.5	80.0
Normal pool level (EL.m)	263.5	76.5
Low pool level	228.5	60.0
Project effects		
Flood control ($10^6\,m^3$)	137	250
Water supply ($10^6\,m^3$)	650.4	1649.0

with a total storage capacity of $14.9 \times 10^9\,m^3$. Yongdam Dam, constructed in 2001, lies 190 km upstream from Daechung Dam, creating Lake Yongdam with a capacity of $8.15 \times 10^9\,m^3$. A schematic of the Geum River basin is given in Fig. 1, with details of the Daechung and Yongdam dams provided in Table 1. The normal pool level for the Daechung reservoir is set as EL.76.5 m throughout the year. The Daechung project can supply $1649 \times 10^6\,m^3$ of urban, industrial, and irrigation water as well as generating 238 GWh yearly. Also, a volume of $250\times10^6\,m^3$ is allocated for flood control during the rainy season from June to September. The Yongdam project can supply $650.4 \times 10^6\,m^3$ for urban, irrigation, and instream water demands annually, and up to $15.67\,m^3/s$ of water is diverted from the Yongdam reservoir out of the Geum River basin. A volume of $137 \times 10^6\,m^3$ is allocated for flood control purposes.

The reasons for the Daechung reservoir reallocation in the Geum River basin are as follows:

- Hydrologic conditions in the Geum River basin have changed due to the construction of the upstream Yongdam multipurpose project in 2001.
- Since more than two decades have elapsed since the construction of the Daechung multipurpose project in 1980, there have been substantial changes in the basin, such as the use of the floodplain, new levees, etc.
- Since the administrative capital of Korea is planned to move to Chungcheong province by 2014 which is located in the Geum River basin, according to President Roh's election pledge, water demands in the Geum River basin will be increased considerably.

2.3. *Design flows*

Hydrologic design for flow regulation projects is concerned with mitigating the adverse effects of floods. Design flows for the Daechung reservoir obtained from KOWACO,[9] with and without the upstream Yongdam reservoir, are used in this study. "With the Yongdam reservoir" means that the flood control capability of the Yongdam reservoir is considered in computing design flows for Daechung reservoir. Likewise, "without the Yongdam reservoir" means that the flood control capability of the Yongdam reservoir is not considered in computing design flows for Daechung reservoir.

2.4. *Reservoir operation for flood control*

The flood peak discharge at a potential damage site resulting from a flood of return period T, will be a function of the upstream reservoir flood storage capacity and the reservoir operation policy

$$Q_{pt} = f(S_f), \tag{1}$$

where S_f is the upstream reservoir flood storage capacity. Since the total storage of a reservoir is fixed, the flood peak discharge at a potential damage site is also a function of upstream active storage

$$Q_{pt} = f(S_a), \tag{2}$$

where S_a is the upstream reservoir active storage capacity.

For a given operating policy for flood flow releases, the function $f(\cdot)$ can be defined by routing a series of floods through the upstream reservoir with various flood storage capacities, according to various reallocation plans and operating policies, to a downstream potential damage site.

3. Applications

3.1. *Reservoir operation for flood season using rigid ROM*

In the flood season, multipurpose reservoir operation is usually based on emptying the flood control pools as rapidly as possible without contributing to a specified river carrying capacity at the downstream potential damage site. Simulation techniques such as auto ROM (Reservoir Operation Method), rigid ROM, and technical ROM are widely used for flood control reservoir operations in Korea. In auto ROM operation, a reservoir stores all flood inflows until the pool level reaches a specific target level, and then releases all inflows after reaching that level. Rigid ROM, using forecasted inflow hydrograph, releases at a constant rate until flood inflows reach the peak, and releases a constant magnitude after that. Also, using forecasted inflow hydrograph, technical ROM determines releases so as to utilize flood control storage from the current storage level to the flood level.

The flood control operation of the Daechung reservoir is simulated based on rigid ROM since KOWACO, the leading water management company in Korea, uses it in real situations.

The Daechung reservoir is simulated to operate for the existing and alternative hypothetical normal pool levels. Reallocation by raising and lowering the pool level by 0.5 m and 1.5 m is applied to examine the effects of alternative storage reallocation plans. That is, the study focuses on the following normal pool levels: EL.75 m, 76 m, 76.5 m, 77 m, and 78 m. The change in flood control and conservation storage levels according to alternative hypothetical normal pool levels is presented in Table 2. Reservoir flood control simulation is performed for each reallocation plan and for 50-, 100-, 200-, 500-year return period design floods and the PMF for the Daechung reservoir, considering the upstream Yongdam reservoir.

Table 2. Flood control and conservation storages according to alternative normal pool levels (MCM).

	Normal pool level (m)				
Storage	75	76	76.5	77	78
Conservation storage	695 (87.8%)	759 (95.8%)	792 (100.0%)	825 (104.2%)	894 (112.9%)
Flood control storage	345 (139.1%)	281 (113.3%)	248 (100.0%)	215 (86.7%)	146 (58.9%)

Table 3. Flood peak discharges at Gongju (m^3/s).

	Normal pool level (m)				
Flood frequency	75	76	76.5	77	78
50-year frequencies	6717	7121	7352	7578	8094
100-year frequencies	8078	8526	8772	9024	9561
200-year frequencies	9272	9749	10,023	10,296	10,895
500-year frequencies	10,880	11,401	11,694	11,973	12,632
PMF	22,419	22,783	22,975	23,180	23,631

3.2. *Flood peak discharges at the control point*

Gongju, with a population of 150,000, is located 50 km downstream of the Daechung reservoir. The control point selected within the city is a critical point in the Geum River basin. The river carrying capacity at this point is estimated to be 11,700 m^3/s, which corresponds to the levee height of EL.20.55 m.[10] Flood peak discharges at Gongju are determined from the flood routing method for various design flows and reallocation levels for the Daechung reservoir. The results are summarized in Table 3. The flood peak discharges at Gongju rise as the normal pool levels of the Daechung reservoir increase, or as the design flows of the Daechung reservoir are increased.

The relationships between the flood peak discharges at Gongju and reallocation alternatives are derived using a linear regression technique. The coefficients of determination for 50, 100, 200, 500 years, and the PMP design flows are 0.9984, 0.9987, 0.9986, 0.9987, and 0.9994, respectively. The derived relationships are presented in Fig. 2 and summarized in Table 4.

When Daechung Dam was constructed in 1981, it was designed to protect downstream wear from 100-year frequency flood flows. The flood control plan for the lower Geum River basin has been mainly a combination of levee and Daechung Dam until Yongdam Dam was constructed in 2001. With Yongdam Dam construction, it also contributed and improved

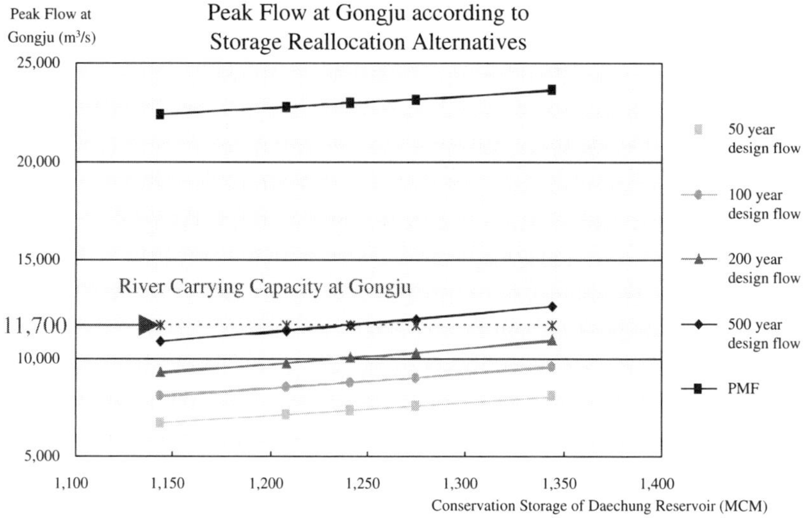

Fig. 2. Relationships between river carrying capacity and reallocation alternatives.

Table 4. Relationships between river carrying capacity and reallocation alternatives.

Frequency	Relationships between river carrying capacity and reallocation alternatives	R2
50	$Q_{p,50} = 6.9S_a - 1180.0$	0.9984
100	$Q_{p,100} = 7.4S_a - 427.0$	0.9987
200	$Q_{p,200} = 8.1S_a - 48.2$	0.9986
500	$Q_{p,500} = 8.7S_a - 853.0$	0.9987
PMP	$Q_{p,\mathrm{PMP}} = 6.1S_a - 15,480$	0.9994

flood control in the Geum River basin. From a flood control standpoint, flood peak discharges at Gongju from the 200-year design flow should not exceed the river carrying capacity at this point. Therefore, the appropriate conservation storage for the Daechung reservoir is estimated as $1350 \times 10^6\,\mathrm{m}^3$ from Fig. 2. This conservation storage corresponds to the normal pool level of 78 m.

3.3. *Estimating reservoir yield after storage reallocation*

Since reservoir storage reallocation cannot be determined from the flood control point of view alone, a change in reservoir yield due to reallocation is also analyzed. Since reservoir storage is usually maintained below the normal pool level except during flood season, the yield of an existing reservoir is less than the sum of the active reservoir storage and total inflows during the most severe drought period. Accordingly, reservoir simulation analysis considering reservoir inflows, water demands, evaporation, precipitation on the reservoir, diversion from the reservoir or downstream, etc., is widely used to assess the water yield of a reservoir.

The water supply yield for the Daechung reservoir needed to satisfy water demands for 2021 is estimated as $1649 \times 10^6\,\text{m}^3$ annually.[1]

Reservoir yield is estimated using the HEC-5 simulation model.[11] The simulation is performed based on 20 years of historic monthly streamflow data from 1981 to 2000. The HEC-5 model for the Geum River basin consists of two reservoirs, five main tributaries, eleven control points, one hypothetical control point for diversions from stream and reservoirs, a multiregional water supply system, and power plants.

Reservoir yield with 95% reliability is estimated by changing the normal pool level at the Daechung reservoir. The Daechung reservoir water yield for the original normal pool level, 76.5 m, is estimated as $1428.2 \times 10^6\,\text{m}^3$. Raising the normal pool level by 0.5 m and 1.5 m increases the water yield to $1446.6 \times 10^6\,\text{m}^3$ and $1465.0 \times 10^6\,\text{m}^3$, respectively. Lowering the normal pool level by 0.5 m and 1.5 m decreases the water yield to $1418.0 \times 10^6\,\text{m}^3$ and $1391.4 \times 10^6\,\text{m}^3$, respectively. Accordingly, if the normal pool level of the Daechung reservoir is increased to 78 m as suggested from a flood control standpoint, $36.8 \times 10^6\,\text{m}^3$ of the reservoir yield, which corresponds to 2.6% of the water yield for the original normal pool level, will be increased. The results for water yield are presented in Table 5.

3.4. *Results*

Based on the flood peak discharges at Gongju from the 200-year design flow, the appropriate conservation storage for the Daechung reservoir is

Table 5. Water yield of the Daechung reservoir.

	Normal pool level (m)				
Water yield ($10^6\,\text{m}^3$/y)	75	76	76.5	77	78
Rate against existing normal pool level (%)	1391.4 97.4	1418.0 99.3	1428.2 100.0	1446.6 101.3	1465.0 102.6

estimated as $1350 \times 10^6\,m^3$, and it corresponds to the normal pool level of 78 m. Consequently, if the Daechung reservoir is rearranged for 78 m normal pool level and operated jointly with the upstream Yongdam reservoir, it can satisfy both the 200-year design flow and flood peak discharges at Gongju. Daechung reservoir yield, when it is operated for 78 m normal pool level, is estimated using HEC-5. Raising the normal pool level by 0.5 m and 1.5 m increases the water yield by 1.3% and 2.6%, respectively. Lowering the normal pool level by 0.5 m and 1.5 m decreases the water yield by 0.7% and 2.6%, respectively.

4. Summary and Conclusions

Reservoir storage reallocation is proposed as an alternative for securing more water resources. To decide on the suitable reservoir reallocation level, alternative synthetic design flows with various return periods for the Daechung reservoir are obtained with and without the new upstream reservoir. These design flows are applied as inputs to perform flood control reservoir simulations for a number of alternative hypothetical reallocation plans by changing pool elevations between water supply and flood control. The reservoir releases resulting from the flood simulation are routed to a downstream potential damage site. Flood peak discharges at that point are obtained with respect to each storage reallocation alternative. A reservoir storage reallocation level is determined in accordance with a river carrying capacity which does not overtop levees at a potential damage site. Reservoir yields before and after storage reallocation are obtained and compared. Finally, the reservoir reallocation level considering both flood control and conservation of the reservoir is decided.

The following conclusions are obtained from this study. A new normal pool level of 78 m is recommended for the Daechung reservoir that is 1.5 m higher than the original normal pool level. And this level can still control the 200-year frequency design flood with a 24 h duration when the Daechung and Yongdam reservoirs are operated jointly. The reservoir yield with 95% reliability is increased by 3% from $1428 \times 10^6\,m^3$ with an existing normal pool level of 76.5 m to $1465 \times 10^6\,m^3$ with the new reallocated normal pool level of 78 m.

Since flood control reservoir operation may conflict with water conservation reservoir operation, reservoir storage reallocation may worsen the conflicts. In Korea, a spillway design flood flow of existing large reservoirs are changing from the previous 200-year design flow to the probable maximum flow (PMF), because flood protection is even more emphasized due to recent severe flood damages from typhoons in 2002 and 2003. That means, more accurate analyses are required to increase the normal pool level in the real situation.

References

1. Korean Ministry of Construction and Transportation, *Water Plan* (2001–2020) (2000).
2. R. A. Wurbs and L. M. Cabezas, Analysis of reservoir storage reallocations, *J. Hydrol.* **92** (1987) 77–95.
3. E. C. Patrick, A hydrologic and statistical evaluation of storage reallocation for multipurpose reservoir system operation, PhD thesis, Department of Civil Engineering, University of Texas A&M, College Station, Texas, USA (1988).
4. U.S. Army Corps of Engineers, Opportunities for reservoir storage reallocation, IWR Policy Study 88-PS-2, Water Resources Center, Institute for Water Resources, Ft. Bevoir, VA. 1988.
5. D. T. Ford Reservoir storage reallocation analysis with PC, *ASCE J. Water Res. Plan. Management* **116**(3) (1990) 402–416.
6. W. K. Johnson, R. A. Wurbs and J. E. Beegle, Opportunities for reservoir-storage reallocation, *ASCE J. Water Res. Plan. Management* **116**(4) (1990) 550–566.
7. R. L. Larson and J. Spinazoala, Conjunctive management analyses for endangered species flow augmentation alternatives in the Snake River, in *Proc. Watershed Management and Operations Management 2000*, 2000, Reston, VA, pp. 1–7.
8. G. F. McMahon and M. C. Farmer, Reallocation of federal multipurpose reservoirs: Principles, policy, and practice, *J. Water Res. Plan. Management* **130**(3) (2004) 187–197.
9. Korea Water Resources Corporation, Report on water supply capability of existing dam (Geum River basin) (1998).
10. Korea Water Resources Corporation, *Multipurpose Reservoir Operation Handbook* (2004).
11. U.S. Army Corps of Engineers, HEC-5, Simulation of flood control and conservation systems — User's manual version 8.0, Department of the Army Corps of Engineers, Institute for Water Resources Hydrologic Engineering Center, Davis, CA. 1998.

Advances in Geosciences
Vol. 11: Hydrological Science (2007)
Eds. Namsik Park *et al.*

SEASONAL TREND OF WATER BUDGET AT RAIN-FED PADDY FIELD IN CENTRAL-THAILAND

DAISUKE KOMORI*,††, MASATOSHI AOKI†, WONSIK KIM‡, SHIN MIYAZAKI§, JAEIL CHO*, SAMAKKEE BOONYAWAT¶, PANYA POLSON‖, SOMNIMIT PUKNGAM¶, PIYAPONG TONGDEENOK¶, SHINJIRO KANAE*, TAIKAN OKI* and KATUMI MUSIAKE**

**Institute of Industrial Science, University of Tokyo, 4-6-1 Komaba Meguro-ku, Tokyo 153-8505, Japan*

†*Tokyo University of Agriculture and Technology, 3-5-8 Saiwai-cho Fuchu-city, Tokyo 183-8509, Japan*

‡*National Institute for Agro-Environmental Sciences, 3-1-3 Kannondai Tukuba-city, Ibaraki, 305-8604, Japan*

§*National Institute for Environmental Studies, 16-2 Onogawa Tukuba-city, Ibaraki, 305-8506, Japan*

¶*Kasetsart University, 50 Phahonyothin Road Chatuchak, Bangkok 10900, Thailand*

‖*Hydrology and Water Management Center 2, Tha Thong Muang, Phitsanulok, 65000, Thailand*

***Fukushima University, 1 Kanayagawa Fukushima-city, Fukushima 960-1296, Japan*

††*d-komori@rainbow.iis.u-tokyo.ac.jp*

For understanding the interaction between vegetation and atmosphere over paddy field, a rain-fed paddy field, micrometeorological observation site, Sukhothai has been started since March 1997. This observation site, located in the Chaophraya river basin, is one of the sites in the framework of the GAME project. From the eight-year data analysis, it was found that the submerge process was different between 1999 and 2005. This indicates that water resource management is changing from purely rain-fed to carry water to farms such as artificial flood. From surface energy fluxes observations, it was found that the timing of decreasing submerge water is important for estimation of water budget on rain-fed paddy field. In the dry season, it was found that dew condensation is strongly related to evaporation with low surface temperature in the early morning and non-low amount of water vapor even in the dry season.

1. Introduction

As a part of the Global Energy and Water cycle EXperiment (GEWEX), the GEWEX Asian Monsoon Experiment (GAME) is being implemented to understand the role of the Asian monsoon in the global energy and water cycle and to improve the simulation and seasonal prediction of Asian monsoon patterns and regional water resources. For these objectives, we have been monitoring vapor flux, precipitation, evapotranspiration, radiative flux, and their seasonal, intraseasonal and interannual variation at the target area of the humid temperate area in Southeast Asia.

Paddy field is a typical agroecosystem in Asia including Japan, which accounted for a major share of world grain zone. Much scientific knowledge based on many researches is provided regarding these interactions over forest in the mid-latitude temperate zone. However, less knowledge is provided over crop in the low-latitude tropical zone. Therefore, it is pivotal investigating the interaction between vegetation and atmosphere over paddy field for understanding global environmental change including global warming.

For these objectives, a rain-fed paddy field, Sukhothai (17°04′N, 99°42′E) located at Chaophraya river basin, which is one of the observation sites in the framework of the GAME project, was selected for measuring heat/water vapor exchanges between biosphere and atmosphere in time and in space using a micrometeorological fluxes measurement method. Sensors were installed on a 10 m high tower to measure the evapotranspiration, sensible/latent heat and micrometeorological data since March 1997. Long-time continuous observation has been obtained since May 1999. In this paper, the eight-year data observed at rain-fed paddy field (1999–2006) are analyzed. We firstly analyze the water budget and surface energy fluxes at the experimental site, and then analyze the dew condensation in dry season.

2. Site Description and Instrumentation

The experimental site (PS site) is located at about 40 km from Sukhothai in the northwest part of Thailand. The PS site is a flat area of at least 100 km^2, and 194 m above sea level. The climate of the experimental area is divided into wet and dry seasons as a part of the Southeast Asia Monsoon region. Onsets of the wet and dry seasons are in May and November.

Rice planting is done throughout the months of May until July according to the submerge situation. Rice harvest is in early December. During the dry season after rice harvest, there is no cultivation at the PS site.

The energy balance Bowen ratio (EBBR) technique has been used for the measurement of sensible and latent heat fluxes at the PS site since

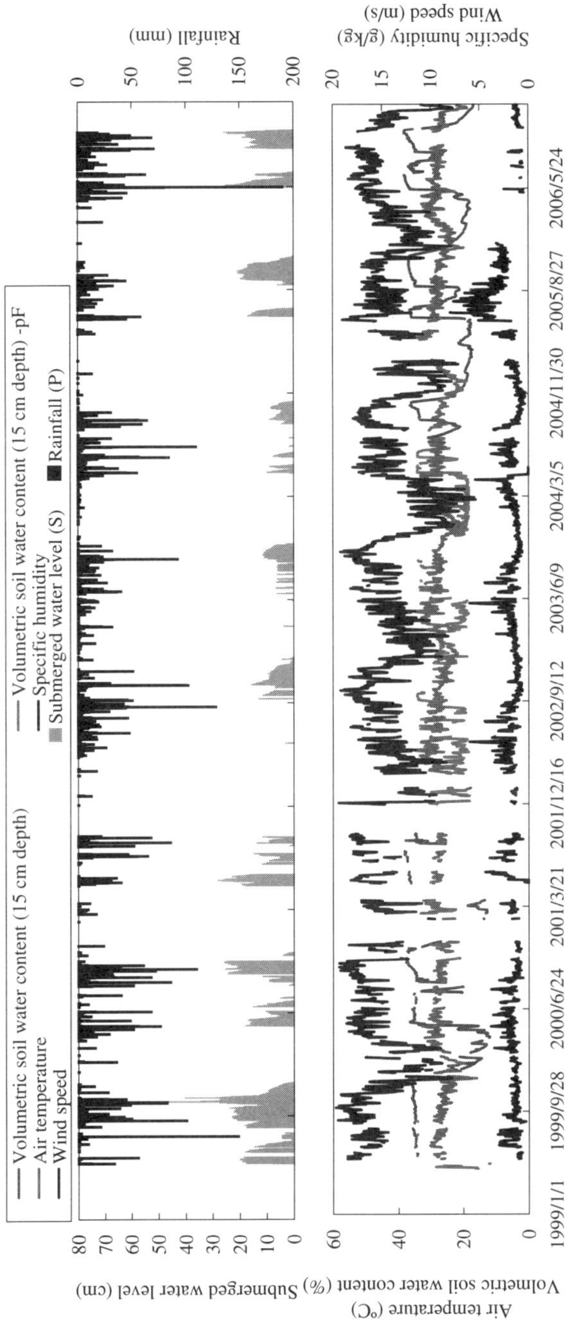

Fig. 1. The measured micrometeorological variables during 1999–2006.

March 1997. The eddy covariance (EC) technique has been used at the PS site since July 2004 for measurement of carbon dioxide (CO_2) flux. The fetch is at least 10 km long and horizontally flat enough for applications of two techniques. Fundamentally, micrometeorological instruments are installed on the southwest-facing side of the tower.

The measured meteorological variables during 1999–2006 are shown in Fig. 1. The achieved ratios of measured meteorological variables during 1999–2006 were 76% on a day basis. Although some solid blocks of measured meteorological variables were lost such as from the period from October to November in 2001, these reasonable values with respect to data acquisition[1] showed that typical data coverage ranges are between 65% and 75% for general field studies. In addition, volumetric soil water content data were also lost from December 2001 to July 2004 due to soil humidity sensors' gradual failure due to long usage. These missing data were supplemented with moisture characteristic curve at the PS site obtained from soil hydraulic model by van Genuchten.[2] By the analysis of moisture characteristic curve, soil type of the PS site fits the definition of loam.

3. Results and Discussion

3.1. *Consideration about water budget at PS site*

The following simple form of hydrologic equation was applied for consideration about water budget at the PS site:

$$P - E - (Q_{\text{out}} - Q_{\text{in}}) = \Delta S, \tag{1}$$

where P is precipitation, E is evapotranspiration, Q_{in} is inflow to the PS site such as water intake from rivers, Q_{out} is outflow from the PS site such as surface runoff and infiltration, and ΔS is difference of storage water (submerge water). The transitions of residual hydrologic equation ($P - E - \Delta S$) in 1999 and 2005 are displayed in Fig. 2. Evapotranspiration under submerged condition was regarded as the potential evapotranspiration calculated using the Penman equation.[3] The positive value of the residual hydrologic equation in Fig. 4 indicates that the outflow of water from the PS site as infiltration and runoff is greater than inflow of water as influent water.

In 1999, the residual ranged from −30 to 20 mm/day except during heavy rainfall and flood events such as in early November, and its average was −0.2 mm/day. In contrast, the residual in 2005 almost indicated a

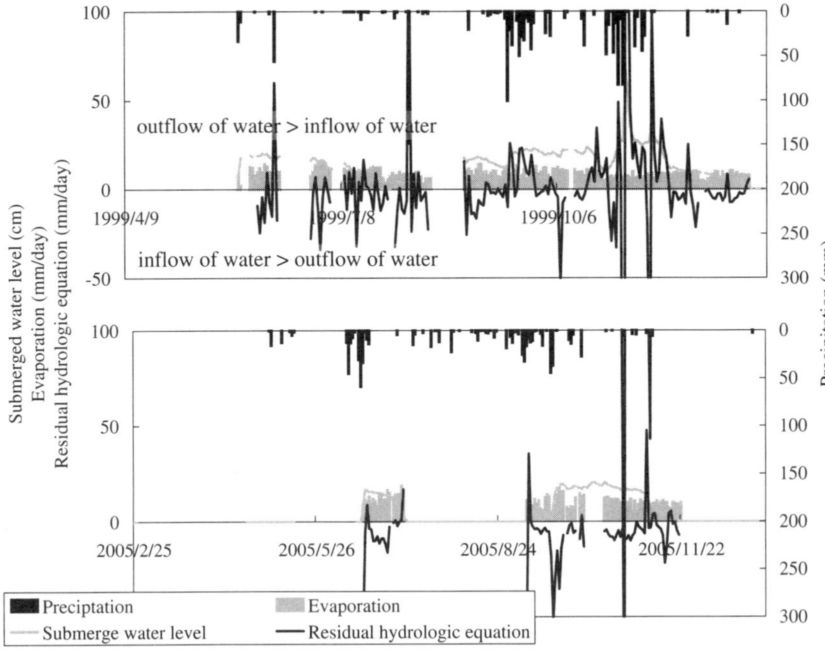

Fig. 2. The measured micrometeorological variables during 1999–2006.

negative value, and its average was −6.4 mm/day. This corresponds to about 65% of daily evaporation. From these data analysis using hydrologic equation, it was found that the submerge process was different in 1999 and 2005.

3.2. *Consideration about surface energy fluxes at PS site*

Figure 3 shows the transition of net radiation and the turbulent fluxes measured using the EC technique in 2005 and 2006. The amount of net radiation in wet season was larger than in dry season. This is because of the following reasons: (1) downward short-wave radiation in dry season is lower than that in wet season, because solar angle in dry season is lower than that in wet season; (2) the turbidity of air due to aerosols of burning agriculture waste during dry season. In wet season, latent heat flux reached $300\,W/m^2$ in the daytime. Sensible heat flux was around $50\,W/m^2$, which was equal to that of surface heat fluxes in soil and water. Latent heat flux in rice harvest season was in the same range of the sensible heat flux. In

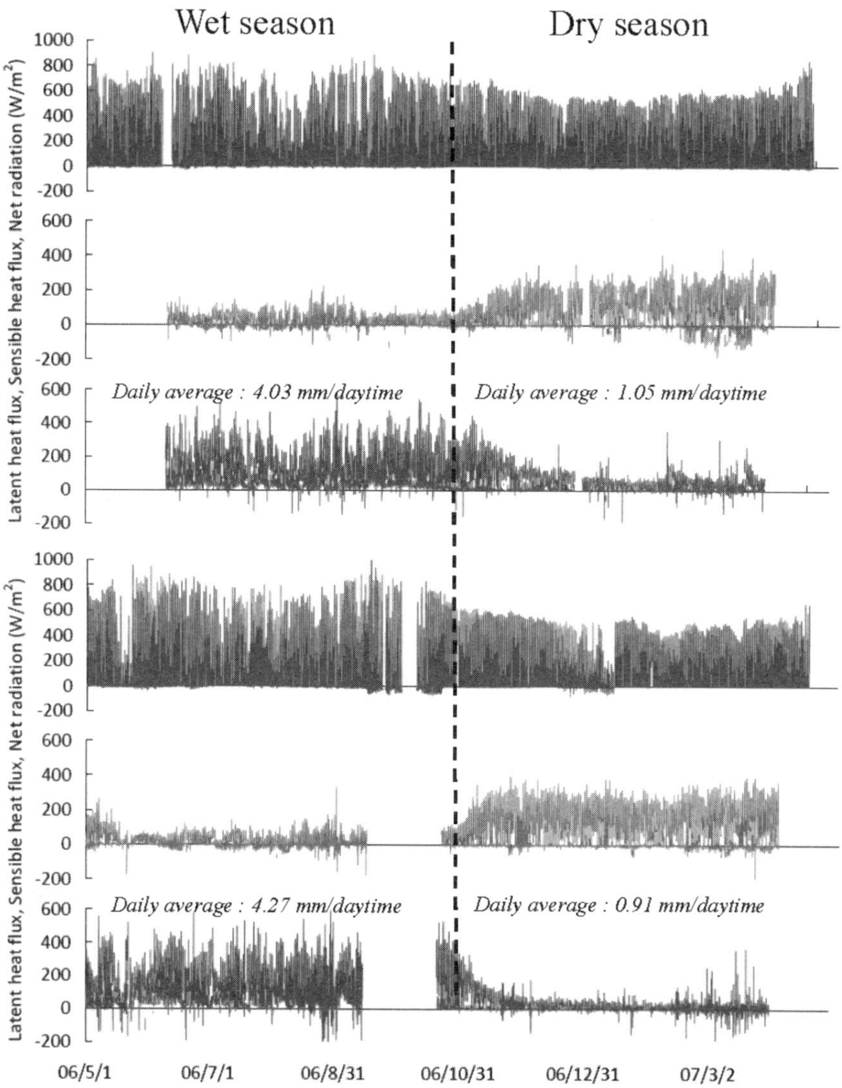

Fig. 3. The measured sensible and latent heat fluxes in 2005 (upper) and 2006 (lower).

dry season, the maximum of latent heat flux fell to 50–100 W/m^2, which was 2–3 times less than the sensible heat flux in dry season.

The paddy drying process in 2005 was delayed by 1 month than that in 2006. The comparison of energy balance on 1st December in each year

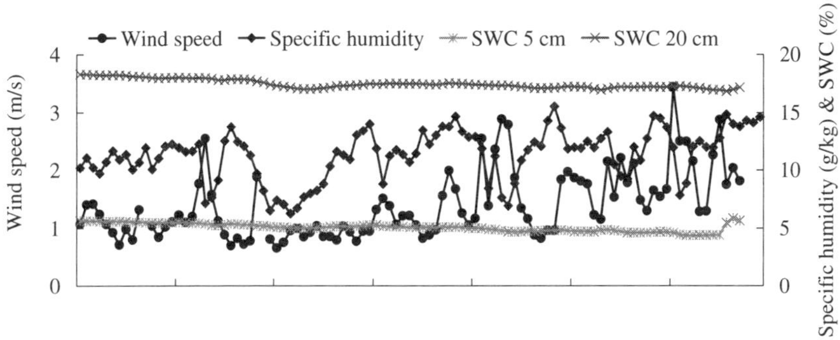

Fig. 4. The measured micrometeorological variables during dry season in 2007.

is as follows:

$$\text{Net radiation\_max} \fallingdotseq 620\,\text{W/m}^2 \quad \text{Bowen ratio\_max} \fallingdotseq 1.0 \quad (2005/12/1),$$
$$\text{Net radiation\_max} \fallingdotseq 590\,\text{W/m}^2 \quad \text{Bowen ratio\_max} \fallingdotseq 0.3 \quad (2006/12/1).$$

The daily difference of evapotranspiration on 1st December between 2005 and 2006 corresponds to $300\,\text{W/m}^2 \fallingdotseq 2.64\,\text{mm/daytime}$. The timing of decreasing submerge water is important for the estimation of water budget on rain-fed paddy field. The timing of increasing submerge water considering inflow water is also important.

3.3. *Dew condensation observation in dry season at the PS site*

From the surface energy fluxes observations (Fig. 3), it is clear that the daily average of evaporation in the dry season was about 0.91–1.05 mm/daytime in spite of no submerge water and low soil water contents. The measured meteorological variables from November to March 2007 are shown in Fig. 5. The decrease of soil humidity from January to March corresponded to only 0.4 mm of evaporation. In contrast, vapor pressure in the dry season was not relatively low which corresponded to about 60% of that in the rainy season. Therefore, low surface temperature in the early morning and non-low amount of water vapor even in the dry season caused oversaturated air conditions, which is dew condensation in the early morning. The 10 days' average of specific humidity is shown in Fig. 5. Specific humidity at the surface assumed 100% humidity for comparison between specific humidity

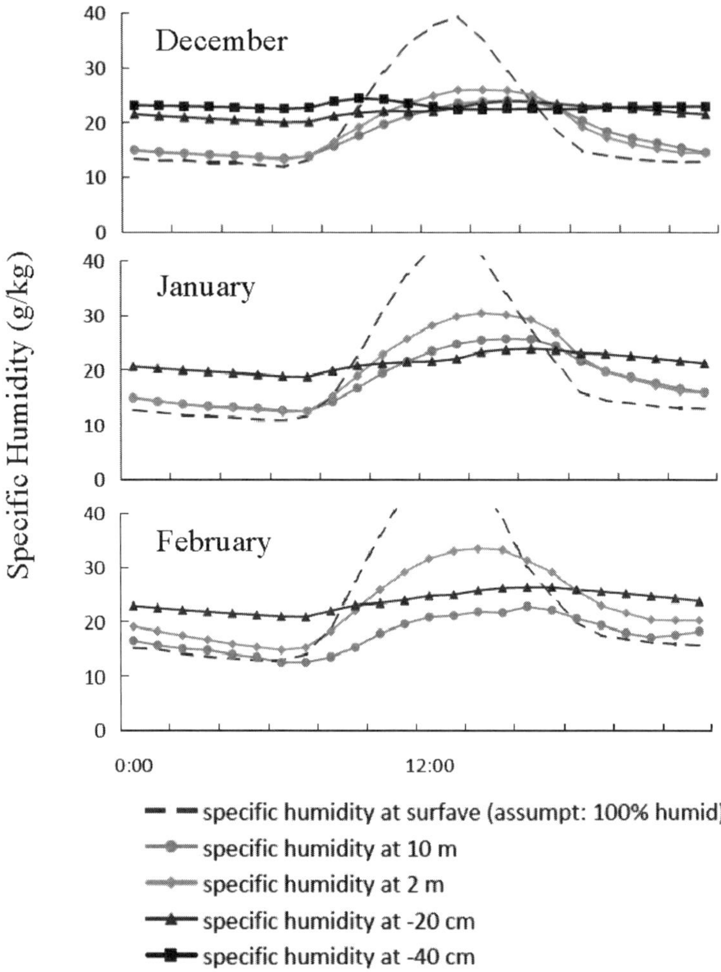

Fig. 5. The 10 days' average of specific humidity in the middle of December, January, and February 2007.

at another layer. Figure 7 indicated the dew condensation during night time, and the possibility of vapor transport from the soil layer to the surface.

As mentioned above, dew condensation was caused by oversaturated air conditions in the field in the early morning; therefore, it was considered that dew condensation was strongly related to evaporation in the dry season. For more understanding of the relation between dew condensation and evaporation in the dry season, weighting observation of dew condensation using reference surface was carried out from January to March in 2007.

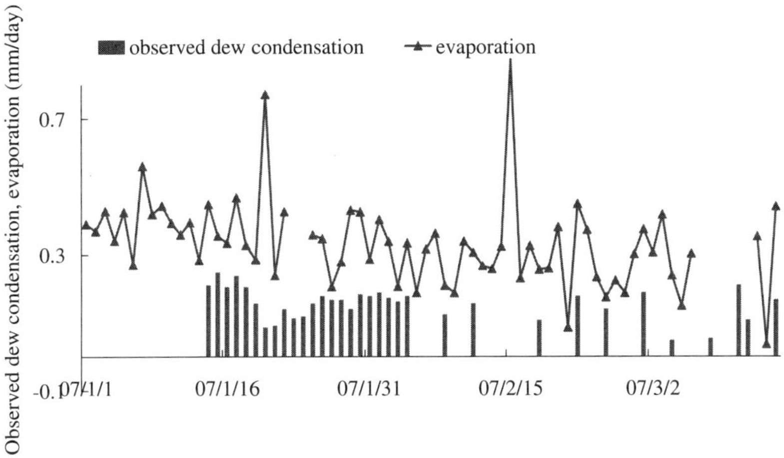

Fig. 6. The measured dew condensation in dry season 2007.

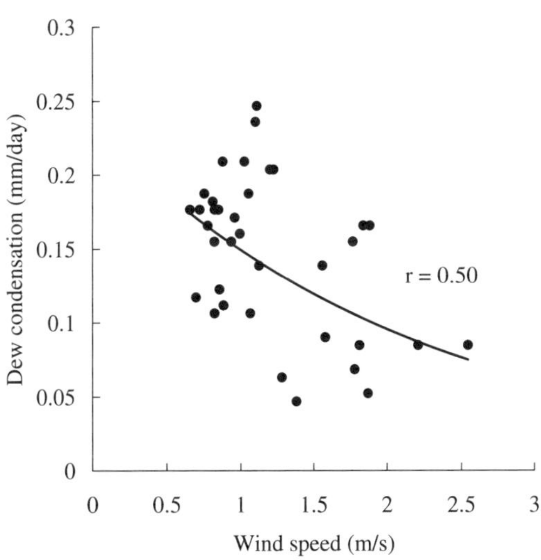

Fig. 7. The relation between dew condensation and wind speed.

The observed dew condensation and evaporation by the EC technique during the dry season are shown in Fig. 6. The observed dew condensation accounted for more than 50% of evaporation in the early dry season, and then the ratio of dew condensation in the evaporation decreased gradually from February with an increase of the lowest surface temperature. The

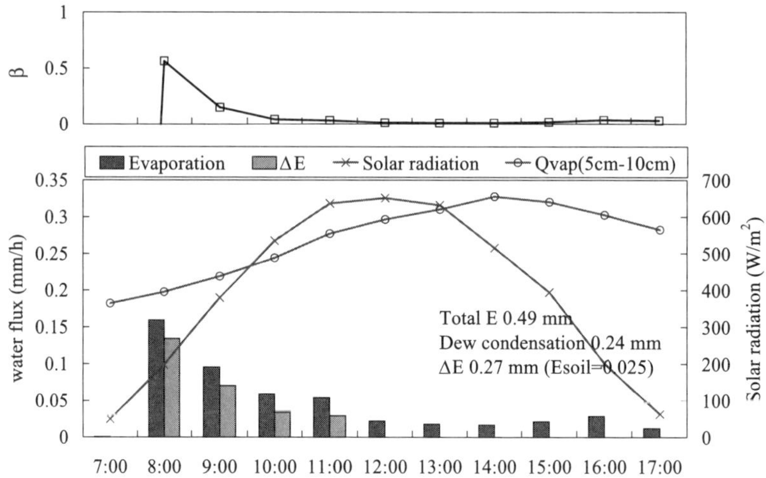

Fig. 8. The transition of water flux, solar radiation, and vapor flux in soil (Q_{vap}) on 17th January.

relation between dew condensation and wind speed is shown in Fig. 8. The negative correlation was recognized between dew condensation and wind speed. This indicates that the resistance of soil is stronger than the resistance of air on the condition of the dry-up surface.

Figure 8 shows the transition of water flux, solar radiation, and vapor flux in soil (Q_{vap}) on 17th January. The peak of evaporation is at 8:00 o'clock, irrespective of the peak of solar radiation. In contrast, the transition of Q_{vap}, which is ascribed to evaporation from soil, is virtually united in that of solar radiation counting time-lag of soil temperature. Moreover, β[a] which is the evaporation efficiency indicates dry-up process at the surface after dew condensation in the early morning. Given the average evaporation from 13:00 to 16:00 as evaporation from soil (Esoil), the difference between evaporation (ΔE) and Esoil indicates that dew condensation is strongly related to evaporation in the dry season.

[a] β was calculated by the following equation: $\beta \frac{E}{ga[hq_{\text{SAT}}(T_S - q)]}$ ρ is air density, $q_{\text{sat}}(Ts)$ is saturation specific humidity for the surface temperature T_s, q is the specific humidity at the surface and h is the relative humidity at the surface, ga is the exchange velocity between the surface and atmosphere (1.33×10^2 ms^{-1}) derived from submerge condition in May 2007.

4. Conclusion and Remarks

The rain-fed paddy field micrometeorological observation of Sukhothai has been started since March 1997. The achieved ratios of slow response data during 1999–2006 were 76% on a day-by-day basis. They are reasonable values with respect to data acquisition, because typical data coverage ranges between 65% and 75% for general field studies.

From the eight-year data analysis, it was found that the submerge process was different between 1999 and 2005. This indicates that water resource management is changing from purely rain-fed to carry water to farms such as artificial flood. In addition, rice planting is also changing from transplantation to direct-seeding recently according to the change in the water resource management.

From surface energy fluxes observations, the timing of the decrease in submerge water is important for the estimation of water budget on rain-fed paddy fields. The timing of the increase in submerge water considering inflow water is also important.

In the dry season, dew condensation is strongly related to evaporation with low surface temperature in the early morning and non-low amount of water vapor even in the dry season. Although we need to more carefully quantify the amount of observed dew condensation regarding reference surface usage, the observed dew condensation accounted for more than 50% of evaporation in the early dry season, and then the ratio of dew condensation in the evaporation decreased gradually from February with an increase in the lowest surface temperature.

References

1. E. Falge, D. D. Baldocchi, R. Olson, P. Anthoni, M. Aubinet, C. Bernhofer, G. Burba, R. Ceulemans, R. Clement, H. Dolman, A. Granier, P. Gross, T. Grunwald, D. Hollinger, N. Jensen, G. Katul, P. Keronen, A. Kowalski, C. T. Lai, B. E. Law, T. Meyers, J. Moncrieff, E. Moors, J. W. Munger, K. Pilegaard, U. Rannik, C. Rebmann, A. Suyker, J. Tenhunen, K. Tu, S. Verma, T. Vesala, K. Wilson and S. Wofsy, *Agri. Forest Meteorol.* **107** (2001) 43.
2. M. Th. Van Genuchten, *Soil Sci. Am. J.* **44** (1980) 892.
3. H. C. Penman, *Proc. Roy. Soc. Lond. A* **193** (1948) 120.

Advances in Geosciences
Vol. 11: Hydrological Science (2007)
Eds. Namsik Park *et al.*

STRATEGIC PLANNING AND MANAGEMENT OF WATER RESOURCES: A PRACTICAL APPROACH TO INTEGRATED WATER RESOURCES MANAGEMENT

TI LE-HUU
Water Resources Section, United Nations Economic and Social Commission for Asia and the Pacific (UNESCAP)
United Nations Building, Rajdamnern Nok Avenue
Bangkok 10200, Thailand
ti.unescap@un.org

Experiences in the application of the strategic planning and management of water resources to the formulation of integrated water resources management (IWRM) plans in 17 countries in Asia and the Pacific are described in this paper. The paper also discusses the context within which efforts were made to develop a new set of guidelines on strategic planning and management of water resources, on the basis of challenges in the planning, implementation, and management of water resources in the region. The paper also summarizes the findings in the regional efforts to improve tools and methodologies for IWRM and recommendations for further action.

1. Introduction

Integrated water resources management (IWRM) is a concept that has been conceived over the years for more effective development process in each country and for regional cooperation at the international level. At the international level, master planning was conceived at the Mar del Plata United Nations Water Conference in 1977 as a road map for water resources development and an essential mechanism for IWRM. Subsequently, the United Nations Conference on Environment and Development held in Rio de Janeiro in 1992 (also known as the Rio Summit), which endorsed the recommendations of the Dublin Principles on the necessity of integrated water management and on active participations of all stakeholders, from the highest levels of government to the smallest communities. The Rio Summit, with most of the Dublin Principles incorporated in Chap. 18 of the Agenda 21, led to a series of national and international efforts on the development of water resources development strategies.

However, as pointed out by M. Falkemark (1997) in her analytical summary for the 20th anniversary of the Conference that "Efforts done to integrate water into macroeconomic analysis have been rather unsatisfactory: attention has gone into allocation principles, neglected sources, return flows and quality problems." This fact led to the call by the UN General Assembly's Special Session in June 1997 for a dialogue among governments aiming "at building a consensus on the necessary actions, and... consider initiating *a strategic approach* for the implementation of all aspects of the sustainable use of freshwater for social and economic purposes..." The development of a set of guidelines on "Strategic Planning and Management of Water Resources in Asia and the Pacific" (D. Webster and Ti Le-Huu, 2002) to facilitate the application of a strategic approach was carried out in that context. The Guidelines can be accessed at websites: www.unescap.org/esd/water/ or www.spmwater-asiapacific.net. This set of guidelines on strategic planning and management (SPM) was applied to 17 developing countries in the region. The experiences of these applications together with the principles and concepts adopted in the development of SPM in the past three decades in the world and in the preparation of the above-mentioned Guidelines are main subjects of this paper.

2. Perception of IWRM Processes, Key Challenges, and Evolution

In order to reflect the regional experiences of Asia and the Pacific on the integration of water resources management into the socio-economic development processes, a regional survey was conducted by UNESCAP in 1998, the report of related findings, which can be accessed at the website www.spmwater-asiapacific.net pointed out the complexity of such processes, although there were important achievements toward making IWRM more effective as shown in Fig. 1.

2.1. *Key challenges*

As Asia and the Pacific is the region with the highest economic growth of the world, water resources is increasingly recognized as one of the key challenges to sustaining the process of socio-economic development of the region. For water resources managers, IWRM requires them to answer the

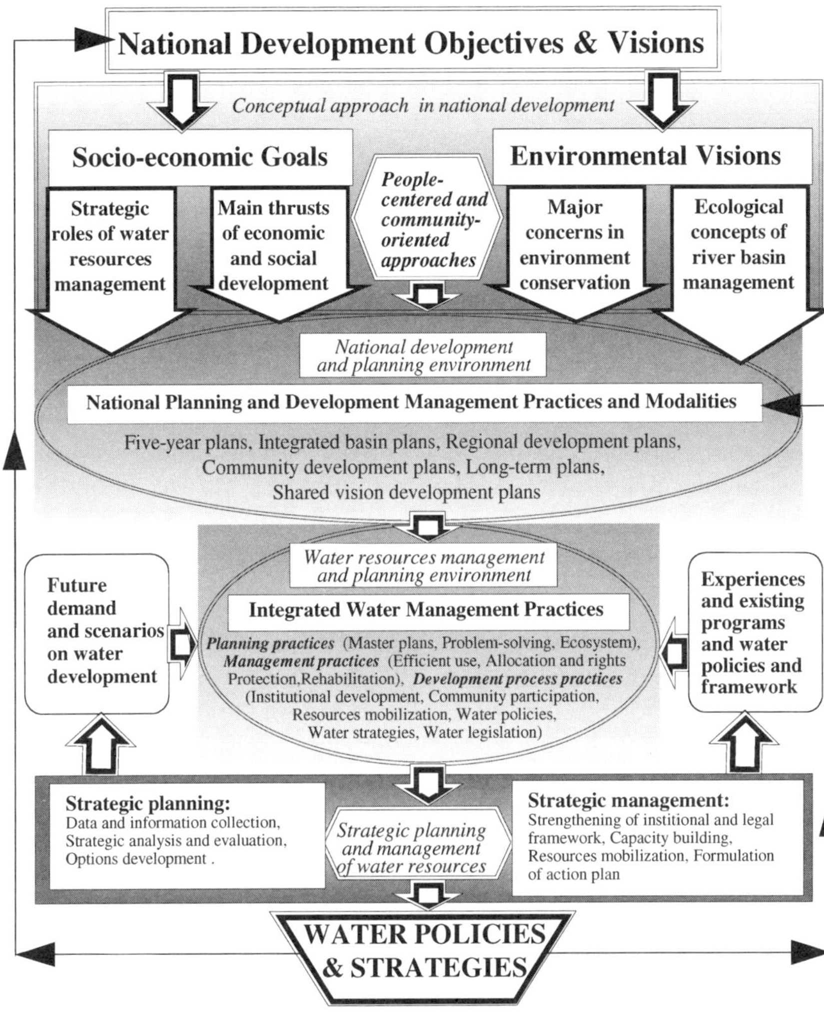

Fig. 1. Perception of the IWRM process at the national level.

following urgent questions among others:

- *How to effectively support, promote, and coordinate the multitude of development in such an active investment environment* to ensure a smooth and sustainable development process; *and*

- *How to identify suitable and efficient water and related projects to* maintain such high economic growth rates, amid the keen competition for not only the limited water resources, but also financial resources.

Apart from the above two urgent questions, key challenges that policy and decision-makers must address in the application of IWRM would include the following:

- Resource mobilization versus social equity.
- Coordination and stakeholders' participation.
- Continuity and consistency of development.

As shown in Fig. 1, an effective IWRM plan must be linked to the overall development process, guided by the national development objectives and visions, which is built on the three pillars: economic prosperity, social equity, and environmentally sound development. For such an effective process, IWRM must be supported by commitment of the top decision-making level of each country. The other challenges in IWRM would therefore include the following:

- IWRM requires tools that support leadership and maintain commitment of leaders.
- IWRM requires tools that would allow the use of the strength of existing planning practices and introduction of a new management mindset.
- IWRM requires tools to create a conducive environment for incorporating IWRM principles into the existing legal framework of water resources planning and development.

2.2. *Evolution of water resources management toward IWRM*

Over the decades, focus of IWRM planning process has changed significantly from planning activities to planning practices, approaches, and underlying philosophy. Apart from basin potentials and the needs of the riparian communities; the evolved planning approaches focus also on economic development opportunities, the common goals of basin development and priority areas of cooperation: toward a shared vision process. As water is a cross-sectoral issue, the adoption of a shared vision process is a natural and ultimate evolution of multi-sectoral and multi-stakeholders cooperation and coordination of efforts.

In parallel to the above evolution of the process, development of tools for water resources management has also made many important achievements, which can be summarized as follows:

- *Project-based approach*: project-focused tools to support decision-making (local optimization).
- *Natural resource-based approach*: Water balance models; basin-system simulation; system optimization models (master plans, basin plans).
- *Socio-economic approach*: Multi-criteria models; utility trade-off analysis; generalized optimization models (regional plans and strategies).
- *IWRM approach*: strategic planning and management models (integration of models including consensus building, conflict resolution, and prevention).

2.3. *IWRM and basin development planning*

As recommended by the Dublin principle, the basin management approach is increasingly adopted by many countries in the region. The experiences showed that the basin approach faces many difficulties in its application, due to the fact that there is a large variety of problems in river basin management and there are a large number of approaches and tools to be adopted in river basin management. One of the key issues that one must answer is "How to prioritize water resources management activities within limited resources and diverse interests and different perception of priorities." In these efforts, the issues of authority and most importantly accountability are becoming more and more evident. In the end, the process of river basin management or IWRM must be built on the concept of "shared vision," shared responsibility and win-win cooperation. It is therefore increasingly recognized that an effective implementation of IWRM requires adoption of "Strategic Planning and Management." In this process, it is obvious that for a successful application of SPM it is necessary that key stakeholders must change their "Mindsets" of water resources management. As SPM is a relatively new approach, capacity building is a necessity and instrumental to success of application of SPM.

3. Trends of SPM in Water Resources

SPM, which had started in military operations, has long been applied and developed in business development of the private sector. Its application has been relatively new, with the most significant efforts made in the management of the River Rhine, following the chemical accident in Basel, Switzerland in 1986 with tragic environmental consequences, which resulted in an important initiative for the River Rhine Action Program: Salmon 2000. With the success of this initial strategic management program, in 1995, the International Commission for the Protection of the Rhine (ICPR), ICPR established an *Action Plan on Flood Defense* (1998–2020): to protect people and goods against flooding while integrating ecological improvements of the Rhine and its floodplains. As an illustration of important key elements for an SPM process, the Rhine Action Plan on Flood Defense introduced a new perception for flood management; adopted more stringent principles for the integration of flood control and management into the national development process; established quantitative targets for the action plan; and identified resources required for the implementation. The new perception involves also spatial planning and land use, not only water resources professionals. The principles provided guidelines for the following two categories of the integration:

- *Integration of related sectoral measures*: (i) water management, (ii) spatial planning and urban development, (iii) nature protection, and (iv) agriculture and forestry.
- *Integration of preventive measures*: (i) water is part of the whole, (ii) store water, (iii) let the river expand, (iv) beware of the danger, and (v) integrated and concerted action.

The quantitative targets of the Action Plan were classified into four categories as important indicators for flood management:

- *Reduce damage risks*: no increase until 2000, reduction up to 10% by 2005 and up to 25% by 2020.
- *Reduce flood stages*: reduction of extreme flood stages downstream of the impounded part of the river up to 30 cm by 2005 and up to 70 cm by 2020.
- *Increase of awareness of floods*: risk maps for 50% of the floodplains and areas at flood risk by 2000 and 100% by 2005.
- *Improve flood forecasting system*: prolong forecasting period by 50% by 2000 and by 100% by 2005.

It is important to note the strong commitment of the related governments, from the resources required for a successful implementation of the Action Plan, which had been adopted by them. The commitment illustrated the importance of the shared vision and the win-win cooperation approach for the long term.

Another important development on SPM at the international level was the enactment of the Government Performance Results Act of the United States of America. This act provided the overall framework for the adoption and implementation of SPM in all public sectors at the Federal level, which has now been widely adopted also at the state level. Important achievements have been made in the implementation of the act for the water-related agencies, which can be accessed through the Internet.

4. IWRM: Strategic Planning and Management and Regional Efforts

4.1. *Development of the guidelines*

Following the decision of the United Nations General Assembly, adopted at its 19th Special Session in 1997, which called for the initiation of "a strategic approach for the implementation of all aspects of the sustainable use of freshwater for social and economic purposes," the promotion and development of strategic planning and management methodologies of water resources has become a major focus of regional cooperation in the program of work of UNESCAP. In that context, with funding from the United Nations Development Account, a project entitled "Capacity building in strategic planning and management of natural resources in Asia and the Pacific" was launched in 2000 aiming at assisting planners and decision-makers in developing strategies to address the trade-offs and complimentarities between policies on economic growth, social equity, and environmental objectives in three phases.

Under phase I of the project, which took place in 2000 and 2001, efforts were made to review achievements in water resources management in the region, including those of ESCAP to form the basis for preparing a set of guidelines on strategic planning and management of water resources. Important achievements made by the ESCAP secretariat identified in the review included the results of studies on "Integration of Water Resources Management into Economic and Social Development Plans in Asia and the Pacific" and "A Synthesis of Experience from the FAO-ESCAP Pilot Project on the Formulation of National Water Visions to Action" (see website www.spmwater-asiapacific.net). These results together with international experiences and inputs from invited water resources experts in the region were compiled to form the first draft of the Guidelines on Strategic Planning and Management of Water Resources, which was discussed at the Regional Workshop to Finalize the Guidelines on Strategic Planning and Management of Natural Resources Development, held in Bangkok in December 2001.

The draft guidelines was then posted on the website of ESCAP and used for training, during phase II of the project, at the five workshops for South-East Asia (July 2002), Central Asia and the Pacific (August 2002), South Asia (September 2002), and North-East Asia (October 2002). Additional inputs were received from participants of the five subregional workshops and the revised version, which included a new chapter on "Monitoring outcomes of strategic planning in water management: Using indicators effectively," was then posted on the ESCAP website in July 2003.

The guidelines were subsequently used in some 17 case studies in phase III in 2003 and 2004 for the development of strategic plans in the following countries: Kazakhstan, Kyrgyz Republic, Tajikistan, Turkmenistan, and Uzbekistan of Central Asia; China and Mongolia of North-East Asia; Fiji and Papua New Guinea of the Pacific; Pakistan and Sri Lanka of South Asia; and Lao PDR, Malaysia, Myanmar, Philippines, Thailand, and Vietnam of South-East Asia. This set of the guidelines is aimed primarily to assist decision makers, planners, and practicing water resources experts in their efforts to enhance the effectiveness in the formulation and implementation of strategic plans for water related organizations within the context of IWRM. The publication is also intended to serve as a reference for planners and practicing professionals working in other sectors when dealing with water resources development. On the basis of the joint efforts of UNESCAP and FAO described above, a joint website has been established by UNESCAP and FAO to continue the regional efforts through the hope of strengthening of the professional network of SPM practitioners at www.spmwater-asiapacific.net.

4.2. *Key features of the guidelines*

The key concepts adopted in the preparation of the guidelines were drawn from the experiences of the two main authors, Prof. Douglas Webster of the Asia Pacific Research Center of Stanford University, United States of America and the author of this paper, when they were involved in the development of a basin development plan for the Mekong River Commission in the early 1990s. These concepts included on the one hand, recent developments in strategic planning and management principles mentioned in Sec. 2 of this paper, and on the other hand, the principles of cross-sectoral management of water resources. It is on the basis of the latter principles, that the guidelines include two models of strategic planning and management. The titles adopted for these two models had evolved through the process of subregional application and were finally adopted as (1) the strategic functional planning model and (2) the strategic collaborative planning model. While the strategic functional planning model is basically the commonly accepted SPM, the strategic collaborative planning model was built around the key concept of water as a cross-sectoral issue, which requires collaborative efforts and the win-win approach.

The relationship between the two models is illustrated in Fig. 2, which underlines the importance of river basin management as well as coordination at the national level for socio-economic development. On the other hand, the proposed strategic collaborative planning model is to be guided by the Vision that the key agency responsible for collaborative efforts within its own legal mandate and subsequent course of action to develop and implement the SPM process as illustrated in Fig. 3.

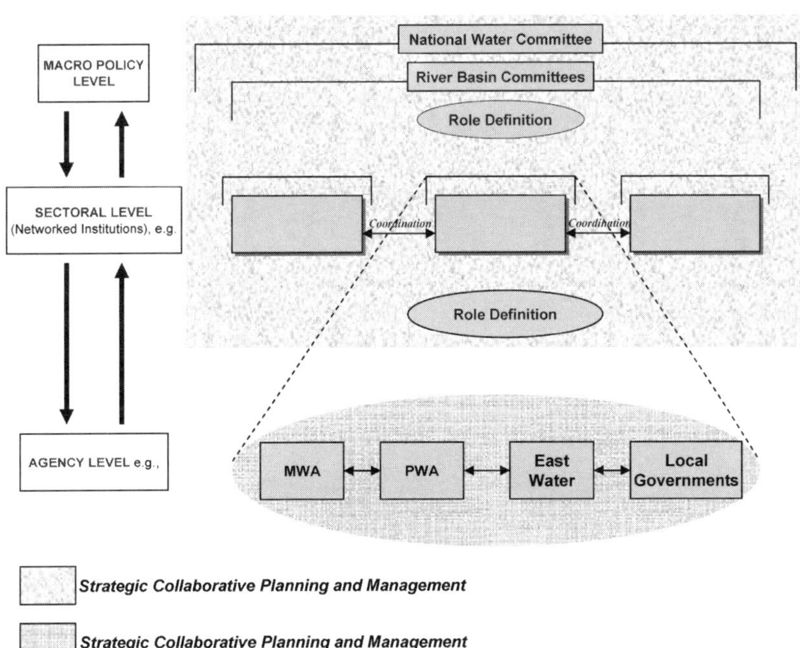

Fig. 2. Nested strategic planning: Water resources.

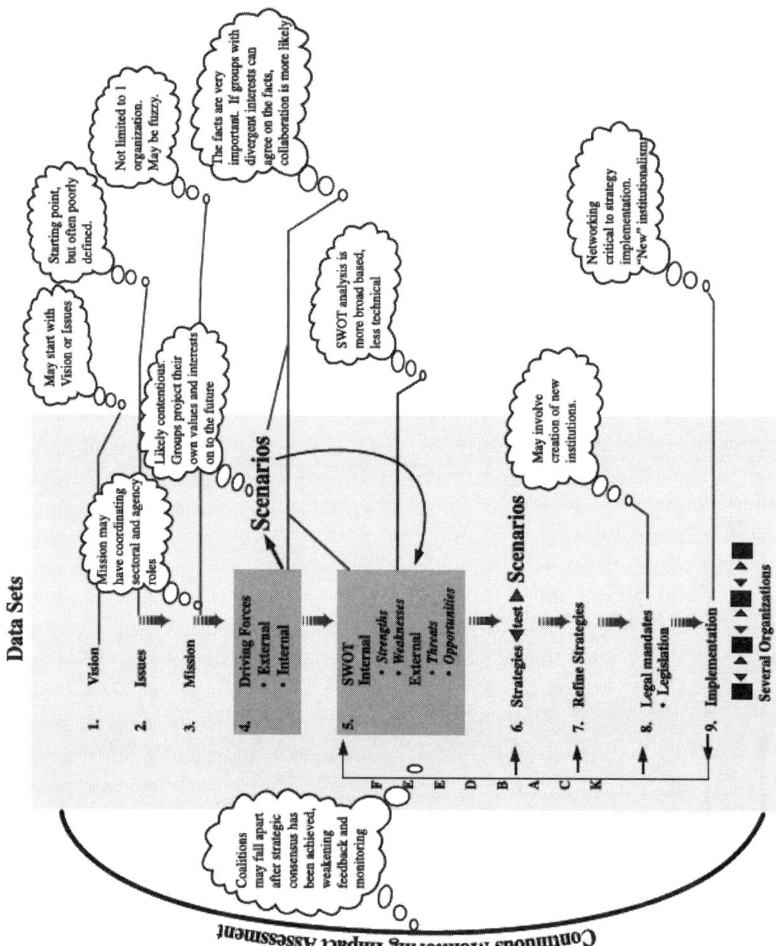

Fig. 3. Strategic collaborative planning model.

4.3. *Experiences of SPM application*

On the basis of the application to 17 case studies in the five subregions of Asia and the Pacific, 10 good practices have been compiled and published (see website www.unescap.org/esd/water or www.spmwater-asiapacific.net). These experiences pointed to the fact that the correct introduction of SPM could result in important change in the mindsets of key stakeholders and attract interest and then commitment of key leaders for the process of IWRM to start. Important lessons from these applications for further developments would include the following:

- IWRM is a complex process, which must be established at the national level in order to ensure effective support to economic growth and sustainable development.
- IWRM is a process which must be adapted according to the socio-economic conditions of each country.
- For an effective and sustainable development process, IWRM must shift from economic efficiency to ecological efficiency.
- Integrated river basin management is a practical approach toward IWRM, which must be conceived as an integral part of the development of an overall framework for IWRM at the national level in order to ensure consistency and continuity of efforts.
- Regional cooperation on the promotion of IWRM at the national level was recognized as a necessity to ensure a systematic and effective adoption of IWRM by countries in the region. Regional workshop at the senior most level of water resources professionals was considered to be instrumental to the establishment of a firm foundation for regional cooperation on IWRM at the national level.

5. Concluding Remarks

The above experiences of the application of SPM to IWRM pointed to the need to further strengthening of leadership in the region on IWRM. From these experiences, future activities could include the following activities, which could be developed under the framework of AOGS:

- Good practices at the national level be compiled and analyzed for possible replication. Analysis of good practices would include the reasons for selection, details of good practices, and the detailed process of key activities with benchmarking.

- For countries with many difficulties in the implementation of IWRM at the national level, it was recommended that various optional courses of action be developed for examination during the workshop. In addition, supporting tools were also recommended to be adopted along with optional courses of action.
- Monitoring of investment and results was considered to be an important aspect of IWRM implementation, on which future workshops could focus to promote different approaches to the promotion of IWRM at the national level. These approaches could include experiences of either a top-down process of IWRM at the national level to sectoral or areal level or a bottom-up process from sectoral level to the national level.
- The IWRM tools developed by leading organizations are recommended to be further simplified for less developed countries for easier applications.